MAPS OF THE ANCIENT SEA KINGS

Evidence of Advanced Civilization in the Ice Age

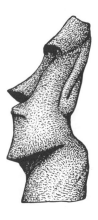

MAPS OF THE ANCIENT SEA KINGS

ISBN 0-932813-42-9

Printed in the United States of America

Published by
Adventures Unlimited Press
One Adventure Place
Kempton, Illinois 60946 USA

A Word of Appreciation

Discoveries are often made by persons who, having fastened onto suggestions made by others, follow them through. This is the case with this book, which is the result of seven years of intensive research undertaken as the result of a suggestion made by someone else.

That person is Captain Arlington H. Mallery. He first suggested that the Piri Re'is Map, brought to light in 1929 but drawn in 1513 and based upon much older maps, showed a part of Antarctica. It was he who made the original suggestion that the first map of this coast must have been drawn before the present immense Antarctic ice cap had covered the coasts of Queen Maud Land. His sensational suggestion was the inspiration for our research.

It is therefore with deep appreciation that I dedicate this book to Captain Arlington H. Mallery.

IP·ORBIS·VNIVERSALIS·IVXTA·PTO
RIC·VESPVCII·ALIORQVE·LVSTRAT

MAPS OF THE ANCIENT SEA KING
ZATION IN THE ICE AGE

, EVIDENCE OF ADVANCED CIVILI

BY CHARLES H. HAPGOOD, F.R.G.S.

The author wishes to express his gratitude to the following for permission to reprint material in this volume:

Abelard-Schuman, Limited, New York. *Greek Science in Antiquity* by Marshall Clagett. Copyright © 1955.

Material from *Admiral of the Ocean Sea* by Samuel Eliot Morison. Copyright © 1942 by Samuel Eliot Morison. Reprinted by permission of Atlantic–Little, Brown and Company, Boston.

E. J. Brill, Ltd., Leiden, The Netherlands. *Hallucinations Scientifiques (Les Portolans)* by Prince Youssouf Kamal. Copyright © 1937.

E. J. Brill, Ltd., Leiden, The Netherlands, and the Editorial Board. *Encyclopaedia of Islam*, edited by M. Th. Houtsma, et al. Copyright © 1936.

Cambridge University Press, New York. *Science and Civilization in China, Vol. III* by J. Needham. Copyright © 1959.

Harvard University Press, Cambridge, Massachusetts. *Travel and Discovery in the Renaissance, 1420–1620* by Boies Penrose. Copyright © 1955.

The National Library, Ankara, Turkey. *The Oldest Map of America* by Dr. A. Afet Inan, translated by Dr. Leman Yolac. Copyright © 1954.

The Royal Danish Geographical Society, Copenhagen, Denmark. *Ptolemy's Maps of Northern Europe: A Reconstruction of the Prototypes* by Gudmund Schütt. H. Hagerup, Copenhagen, 1917.

The Map Collectors Circle, London, England, for use of the Buache map of Antarctica.

Twayne Publishers, Inc., New York, for permission to trace the map of the mouths of the St. Lawrence River in *Explorations in America Before Columbus* by Hjalmar R. Holand. Copyright © 1956.

Maps adapted by Caru Studios, Inc., N.Y., N.Y., from drawings by Charles H. Hapgood.

MAPS OF THE ANCIENT SEA KINGS

Evidence of Advanced Civilization in the Ice Age

by Charles H. Hapgood

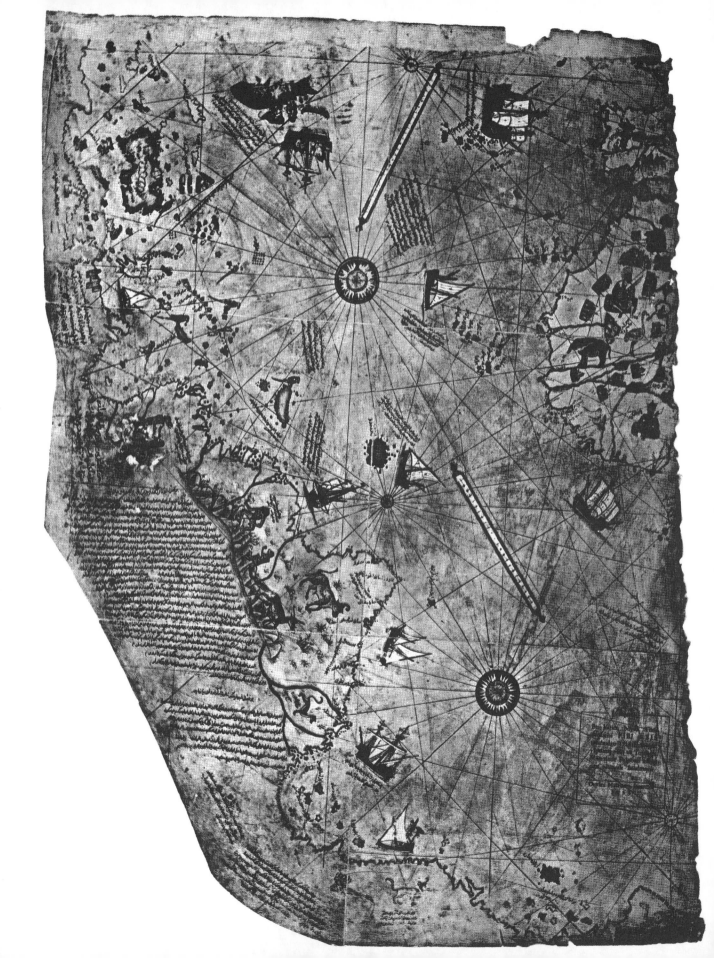

Preface

This book contains the story of the discovery of the first hard evidence that advanced peoples preceded all the peoples now known to history. In one field, ancient sea charts, it appears that accurate information has been passed down from people to people. It appears that the charts must have originated with a people unknown; that they were passed on, perhaps by the Minoans (the Sea Kings of ancient Crete) and the Phoenicians, who were for a thousand years and more the greatest sailors of the ancient world. We have evidence that they were collected and studied in the great library of Alexandria and that compilations of them were made by the geographers who worked there.

Before the catastrophe of the destruction of the great library many of the maps must have been transferred to other centers, chiefly, perhaps, to Constantinople, which remained a center of learning through the Middle Ages. We can only speculate that the maps may have been preserved there until the Fourth Crusade (1204 A.D.) when the Venetians captured the city. Some of the maps appear in the west in the century following this "wrong way" crusade (for the Venetian fleet was supposed to sail for the Holy Land!). Others do not appear until the early 16th Century.

Most of these maps were of the Mediterranean and the Black Sea. But maps of other areas survived. These included maps of the Americas and maps of the Arctic and Antarctic seas. It becomes clear that the ancient voyagers traveled from pole to pole. Unbelievable as it may appear, the evidence nevertheless indicates that some ancient people explored the coasts of Antarctica when its coasts were free of ice. It is clear, too, that they had an instrument of navigation for accurately finding the longitudes of places that was far superior to anything possessed by the peoples of ancient, medieval, or modern times until the second half of the 18th Century.

This evidence of a lost technology will support and give credence to many other evidences that have been brought forward in the last century or more to support the hypothesis of a lost civilization in remote times. Scholars have been able to dismiss most of that evidence as mere myth, but here we have evidence that cannot be dismissed. This evidence requires that all the other evidence that has been brought forward in the past should be reexamined with an open mind.

To the inevitable question, are these remarkable maps genuine, I can only reply that they have all been known for a long time, with one exception. The Piri Re'is Map of 1513 was only rediscovered in 1929, but its authenticity, as will be seen, is sufficiently established. To the further question, why didn't somebody else discover all this before, I can only reply that new discoveries usually seem self-evident, by hindsight.

C. H. H.

Contents

Foreword

The geographer and geologist William Morris Davis once discussed "The Value of Outrageous Geological Hypotheses." * His point was that such hypotheses arouse interest, invite attack, and thus serve useful fermentative purposes in the advancement of geology. Mr. Hapgood will agree, I am sure, that this book records a mighty proliferation of outrageous cartographical and historical hypotheses, as luxuriant as an equatorial vine. His hypotheses will "outrage" the conservative instincts of historically minded cartographers and cartographically minded historians. But while those in whom conservatism predominates will react to this book like bulls to red rags, those of radical, iconoclastic bent will react like bees to honeysuckle, and the liberals in between will experience a feeling of stimulating bafflement.

A map dating from 1513, and by the Turkish Admiral, Piri Re'is, is the seed from which the vine has grown. Only the western half of the map has been preserved. It shows the Atlantic coasts from France and the Caribbean on the north to what Hapgood (following Captain A. H. Mallery) holds to be Antarctica on the south; and, of course, the proposition that any part of Antarctica could have been mapped before 1513 is startling. But yet more startling are the further propositions that have arisen from the intensive studies that Mr. Hapgood and his students have made of this and other late medieval and early modern maps. These studies, which took seven years, have convinced him that the maps were derived from prototypes drawn in pre-Hellenic times (perhaps even as early as the last Ice Age!), that these older maps were based upon a sophisticated understanding of the spherical trigonometry of map projections, and—what seems even more incredible—upon a detailed and accurate knowledge of the latitudes and longitudes of coastal features throughout a large part of the world.

In my opinion, Mr. Hapgood's ingenuity in developing his basic concept regarding the accuracy of the maps is fascinating and accounts for the book's most valuable contribution. Whether or not one accepts his "identifications" and his "solutions," he has posed hypotheses that cry aloud for further testing. Besides this, his suggestions as to what might explain the disappearance of civilizations

---

* *Science*, vol. 63, 1926, pp. 463–468.

sufficiently advanced in science and navigation to have produced the hypothetical lost prototypes of the maps that he has studied raise interesting philosophical and ethical questions. Had "Sportin' Life" in *Porgy and Bess* read this book, he would have been inspired to sing: "it ain't nessa . . . it ain't nessa . . . it ain't necessarily *not* so."

John K. Wright,
Lyme, New Hampshire
June 7, 1965

John K. Wright, who did his undergraduate work at Harvard and also received his Ph.D. in history from that university, was with the American Geographical Society in New York for thirty-six years. He was director of the society for the last eleven years of his association with it. His latest work, Human Nature in Geography, *has just been published by the Harvard University Press.*

MAPS OF THE
ANCIENT SEA KINGS

Evidence of Advanced Civilization in the Ice Age

CHAP
TER I
THE
TREASURE
HUNT
BEGINS

In 1929, in the old Imperial Palace in Constantinople, a **map** was found that caused great excitement. It was painted on parchment, and dated in the month of Muharrem in the Moslem year 919, which is 1513 in the Christian calendar. It was signed with the name of Piri Ibn Haji Memmed, an admiral of the Turkish navy known to us as Piri Re'is.[1]

The map aroused attention because, from the date, it appeared to be one of the earliest maps of America. In 1929 the Turks were passing through a phase of intense nationalism under the leadership of Kemal, and they were delighted to find an early map of America drawn by a Turkish geographer. Furthermore, examination showed that this map differed significantly from all the other maps of America drawn in the 16th Century because it showed South America and Africa in correct relative longitudes. This was most remarkable, for the navigators of the 16th Century had no means of finding longitude except by guesswork.

Another detail of the map excited special attention. In one of the legends inscribed on the map by Piri Re'is, he stated that he had based the western part of it on a map that had been drawn by Columbus. This was indeed an exciting statement because for several centuries geographers had been trying without success to find a "lost map of Columbus" supposed to have been drawn by him in the West Indies. Turkish and German scholars made studies of the map. Articles were written in the learned journals, and even in the popular press.[2]

One of the popular articles, published in the *Illustrated London News* (1),[3]

---

[1] From his title, Re'is, "admiral." Pronounced "Peeree Ry-iss." See Note 1.

[2] See the Bibliography, Nos. 1, 2, 5, 6, 23, 27, 28, 36, 40, 61, 78, 83, 104, 105, 106, 109, 115, 117, 154, 181, 187, 208, 215.

[3] Figures referring to specific sources listed in the Bibliography are inserted in parentheses throughout the text. The first number indicates the correspondingly numbered work in the Bibliography, and a number following a colon indicates the page in the work.

caught the eye of the American Secretary of State Henry Stimson. Stimson thought it would be worthwhile to try to discover the actual source Piri Re'is had used, a map which had supposedly been drawn by Columbus and which might still be lying about somewhere in Turkey. Accordingly, he ordered the American Ambassador in Turkey to request that an investigation be made.[4] The Turkish Government complied, but no source maps were found.

Piri Re'is made other interesting statements about his source maps. He used about twenty, he said, and he stated that some of them had been drawn in the time of Alexander the Great, and some of them had been based on mathematics.[5] The scholars who studied the map in the 1930's could credit neither statement. It appears now, however, that both statements were essentially correct.

After a time, the map lost its public interest, and it was not accepted by scholars as a map by Columbus. No more was heard of it until, by a series of curious chances, it aroused attention in Washington, D.C., in 1956. A Turkish naval officer had brought a copy of the map to the U.S. Navy Hydrographic Office as a gift (although, unknown to him, facsimiles already existed in the Library of Congress and other leading libraries in the United States). The map had been referred to a cartographer on the staff, M. I. Walters.

Walters happened to refer the map to a friend of his, a student of old maps, and a breaker of new ground in borderland regions of archaeology, Captain Arlington H. Mallery. Mallery, after a distinguished career as an engineer, navigator, archaeologist, and author (130), had devoted some years to the study of old maps, especially old Viking maps of North America and Greenland. He took the map home, and returned it with some very surprising comments. He made the statement that, in his opinion, the southernmost part of the map represented bays and islands of the Antarctic coast of Queen Maud Land now concealed under the Antarctic ice cap. That would imply, he thought, that somebody had mapped this coast before the ice had appeared.

This statement was too radical to be taken seriously by most professional geographers, though Walters himself felt that Mallery might be right. Mallery called in others to examine his findings. These included the Reverend Daniel L. Linehan, S.J., director of the Weston Observatory of Boston College, who had been to Antarctica, and the Reverend Francis Heyden, S.J., director of the Georgetown University Observatory. These trained scientists felt confidence in Mallery. Father Linehan and Walters took part with Mallery in a radio panel discussion, sponsored by Georgetown University, on August 26, 1956. Verbatim copies of this broadcast were distributed and brought to my attention. I was impressed by the confidence placed in Mallery by men like Walters, Linehan, and Heyden, and, when I met Mallery himself, I was convinced of his sincerity and honesty. I had a

---

[4] See correspondence, Note 2.
[5] For a translation of all the legends on the map, see Note 3.

strong hunch that, despite the improbabilities of his general theories, and the lack, then, of positive proof, Mallery could well be right. I decided to investigate the map as thoroughly as I could. I therefore initiated an investigation at Keene State College.

This investigation was undertaken in connection with my classes at the college, and the students from the beginning took a very important part in it.[6] It has been my habit to try to interest them in problems on the frontiers of knowledge, for I believe that unsolved problems provide a better stimulation for their intelligence and imagination than do already-solved problems taken from textbooks. I have also long felt that the amateur has a much more important role in science than is usually recognized. I teach the history of science, and have become aware of the extent to which most radical discoveries (sometimes called "breakthroughs") have been opposed by the experts in the affected fields. It is a fact, obviously, that every scientist is an amateur to start with. Copernicus, Newton, Darwin were all amateurs when they made their principal discoveries. Through the course of long years of work they became specialists in the fields which they *created*. However, the specialist who starts out by learning what everybody else has done before him is not likely to initiate anything very new. An expert is a man who knows everything, or nearly everything, and usually thinks he knows everything important, in his field. If he doesn't think he knows everything, at least he knows that other people know less, and thinks that amateurs know nothing. And so he has an unwise contempt for amateurs, despite the fact that it is to amateurs that innumerable important discoveries in all fields of science have been due.[7] For these reasons I did not hesitate to present the problem of the Piri Re'is Map to my students.

---

[6] See Acknowledgments.

[7] The late James H. Campbell, who worked in his youth with Thomas A. Edison, said that once, when a difficult problem was being discussed, Edison said it was too difficult for any specialist. It would be necessary, he said, to wait for some amateur to solve it.

CHAP
TER II

THE
SECRETS
OF THE
PIRI RE'IS
MAP

When our investigation started my students and I were amateurs together. My only advantage over them was that I had had more experience in scientific investigations; their advantage over me was that they knew even less and therefore had no biases to overcome.

At the very beginning I had an idea—a bias, if you like—that might have doomed our voyage of discovery before it began. If this map was a copy of some very ancient map that had somehow survived in Constantinople to fall into the hands of the Turks, as I believed, then there ought to be very little in common between this map and the maps that circulated in Europe in the Middle Ages. I could not see how this map could be *both* an ancient map (recopied) and a medieval one. Therefore, when one of my students said this map resembled the navigation charts of the Middle Ages, at first I was not much interested. Fortunately for me, I kept my opinions to myself, and encouraged the students to begin the investigation along that line.

We soon accumulated considerable information about medieval maps. We were not concerned with the land maps, which were exceedingly crude. (See Figures 1 and 2.) We were interested only in the sea charts used by medieval sailors from about the 14th Century on.[1] These "portolan" [2] maps were of the Mediterranean and Black Seas, and they were good. An example is the Dulcert Portolano of 1339. (Fig. 3.) If the reader will compare the pattern of lines on this chart with that on the Piri Re'is Map (Frontispiece) he will see that they are similar. The only difference is that, while the Dulcert Portolano covers only the Mediterranean and the Black Seas, the Piri Re'is Map deals with the shores of the entire Atlantic Ocean. The lines differ from those on modern maps. The lines do not resemble

---

[1] Maps in this book, except where it is otherwise indicated, are taken from the Vatican Atlas (139) or that of Nordenskiöld (146).

[2] The term "portolan" or "portolano" apparently derived from the purpose of the sea charts, which was to guide navigators from port to port.

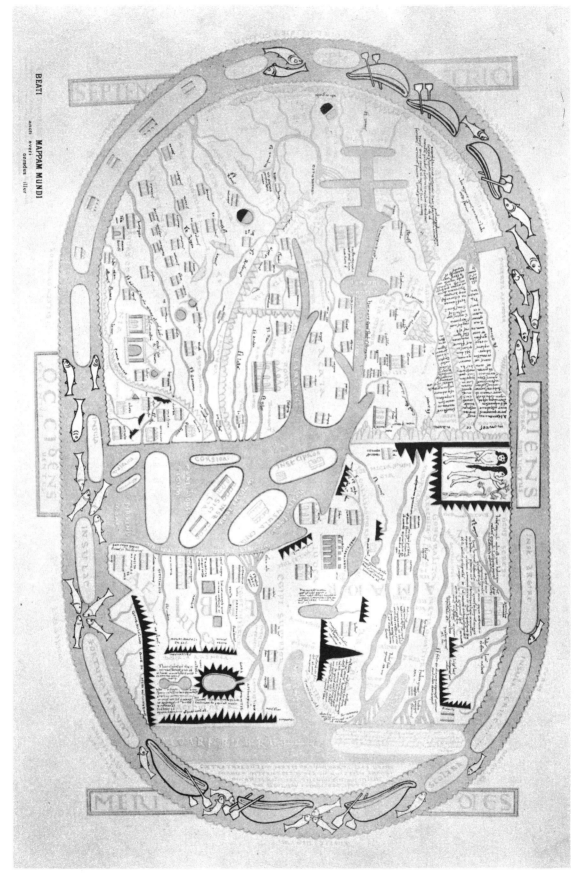

Figure 1. The S. Osma Beatus Medieval World Map.

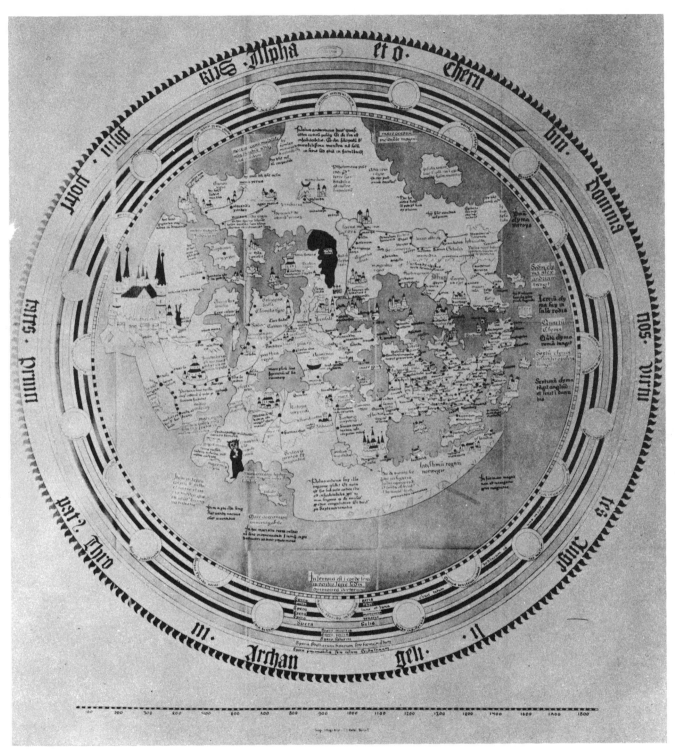

Figure 2. The Andreas Walsperger Map of 1448.

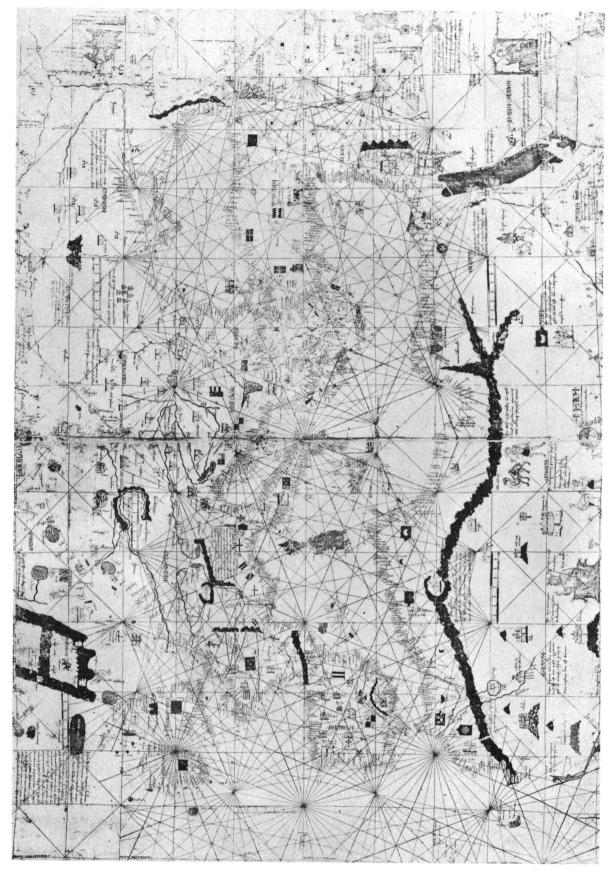

Figure 3. The Dulcert Portolano of 1339.

the modern map's lines of latitude and longitude that are spaced at equal intervals and cross to form "grids" of different kinds. Instead, some of the lines, at least, on these old maps seem to radiate from centers on the map, like spokes from a wheel. These centers seem to reproduce the pattern of the mariner's compass, and some of them are decorated like compasses. The radiating "spokes" are spaced exactly like the points of the compass, there being sixteen lines in some cases, and thirty-two in others.

Since the mariner's compass first came into use in Europe about the time that these charts were introduced, most scholars have concluded that the charts' design must have been intended to help medieval sailors sail by the compass. There is no doubt that medieval navigators did use the charts to help them find compass courses, for the method is described in a treatise written at the time (89, 179, 200). However, as we continued to study these medieval charts, a number of mysteries turned up.

We found, for example, that one of the leading scholars in the field did not believe that the charts originated in the Middle Ages. A. E. Nordenskiöld, who compiled a great Atlas of these charts (146) and also wrote an essay on their history (147), presented several reasons for concluding that they must have come from ancient times. In the first place, he pointed out that the Dulcert Portolano and all the others like it were a great deal too accurate to have been drawn by medieval sailors. Then there was the curious fact that the successive charts showed no signs of development. Those from the beginning of the 14th Century are as good as those from the 16th. It seemed as though somebody early in the 14th Century had found an amazingly good chart which nobody was to be able to improve upon for two hundred years. Furthermore, Nordenskiöld saw evidence that only *one* such model chart had been found and that all the portolanos drawn in the following centuries were only copies—at one or more removes—from the original. He called this unknown original the "normal portolano" and showed that the portolanos, as a body, had rather slavishly been copied from this original. He said:

> The measurements at all events show: (1) that, as regards the outline of the Mediterranean and the Black Sea, all the portolanos are almost unaltered copies of the same original; (2) that the same scale of distance was used on all the portolanos (147:24).

After discussing this uniform scale that appears on all the portolanos, and the fact that it appears to be unrelated to the units of measurement used in the Mediterranean, except the Catalan (which he had reason to believe was based on the units used by the Carthaginians), Nordenskiöld further remarks:

> . . . It is therefore possible that the measure used in the portolanos had its ultimate origin in the time when the Phoenicians or Carthaginians ruled over the

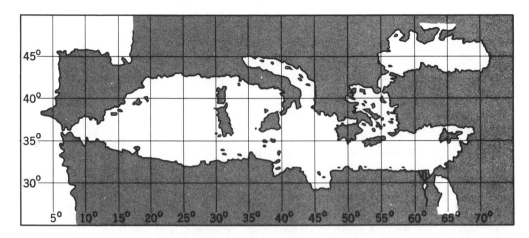

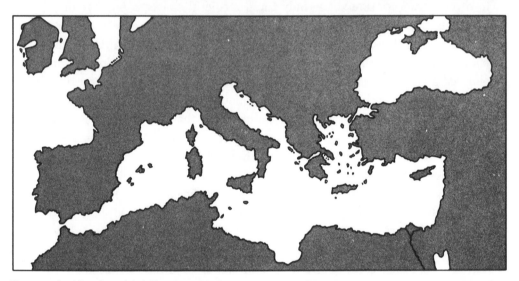

Figure 4. Nordenskiold's comparison of Ptolemy's Map of the Mediterranean (top) with the Dulcert Portolano.

navigation of the western Mediterranean, or at least from the time of Marinus of Tyre . . . (147:24).[3]

Nordenskiöld inclined, then, to assign an ancient origin to the portolanos. But this is not all. He was quite familiar with the maps of Claudius Ptolemy which had survived from antiquity and had been reintroduced in Europe in the 15th Century. After comparing the two, he found that the portolanos were much better than Ptolemy's maps. He compared Ptolemy's map of the Mediterranean and the Black Seas with the Dulcert Portolano (Fig. 4) and found that the superiority of the portolano was evident.

---

[3] Marinus of Tyre lived in the 2nd Century A.D. and was the predecessor of the geographer Claudius Ptolemy.

Let us stop to consider, for a moment, what this means. Ptolemy is the most famous geographer of the ancient world. He worked in Alexandria in the 2nd Century A.D., in the greatest library of the ancient world. He had at his command all the accumulated geographical information of that world. He was acquainted with mathematics. He shows, in his great work, the *Geographia* (168), a modern scientific mentality. Can we lightly assume that medieval sailors of the *fourteenth century*, without any of this knowledge, and without modern instruments except a rudimentary compass—and without mathematics—could produce a more scientific product?

Nordenskiöld felt that there had been in antiquity a geographic tradition superior to the one represented by Ptolemy. He thought that the "normal portolano" must have been in use *then* by sailors and navigators, and he answered the objection that there was no mention of such maps by the various classical writers by pointing out that in the Middle Ages, when the portolan charts were in use, they were never referred to by the Schoolmen, the academic scholars of that age. Both in ancient and in medieval times the academic mapmaker and the practical navigator were apparently poles apart. (See Figs. 5, 6, 7, 8.) Nordenskiöld was forced to leave the problem unsolved. Neither the medieval navigators nor the known Greek geographers could have drawn them. The evidence pointed to their origin in a culture with a higher level of technology than was attained in medieval or ancient times.[4]

All the explanations of the origins of the portolan charts were opposed by Prince Youssouf Kamal, a modern Arab geographer, in rather violent language:

Our incurable ignorance . . . as to the origin of the portolans or navigation charts known by this name, will lead us only from twilight into darkness. Everything that has been written on the history or the origin of these charts, and everything that will be said or written hereafter can be nothing but suppositions, arguments, hallucinations. . . . (107:2)[5]

Prince Kamal also argued against the view that the lines on the charts were intended to facilitate navigation by the compass:

As for the lines that we see intersecting each other, to form lozenges, or triangles, or squares: these same lines, I wish to say, dating from ancient Greek times, and going back to Timosthenes, or even earlier, were probably never drawn . . . to give . . . distances to the navigator. . . .

The makers of portolans preserved this method, that they borrowed from the

---

[4] The Arabs, famous for their scientific achievements in the Early Middle Ages, apparently could not have drawn them either. Their maps are less accurate than those of Ptolemy. (See Fig. 5.)

[5] My translation from the French.

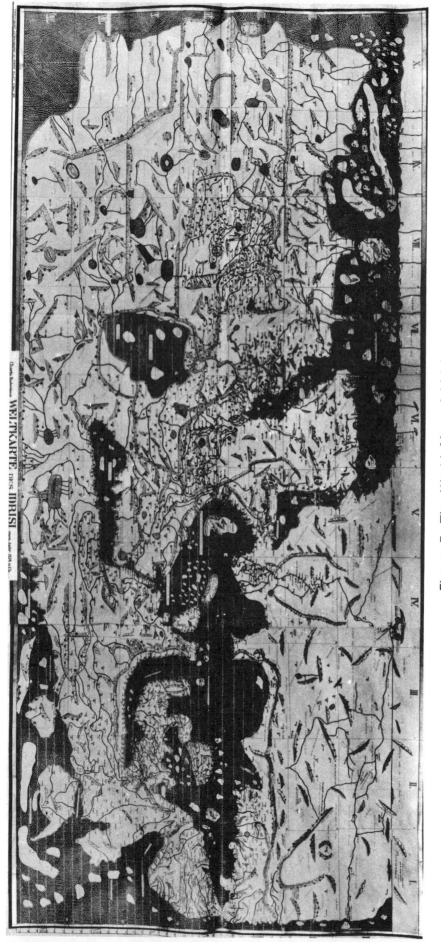

Figure 5. The World Map of Idrisi.

Figure 6. The World Map of Ptolemy.

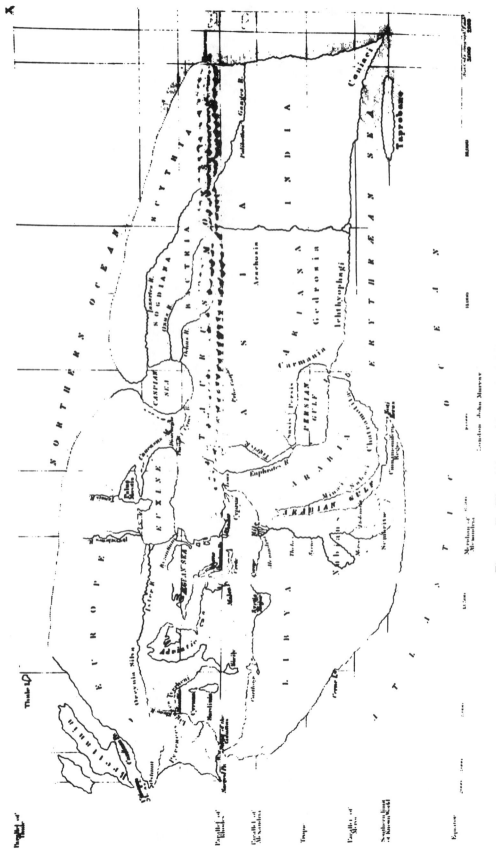

Figure 7. The World Map of Eratosthenes.

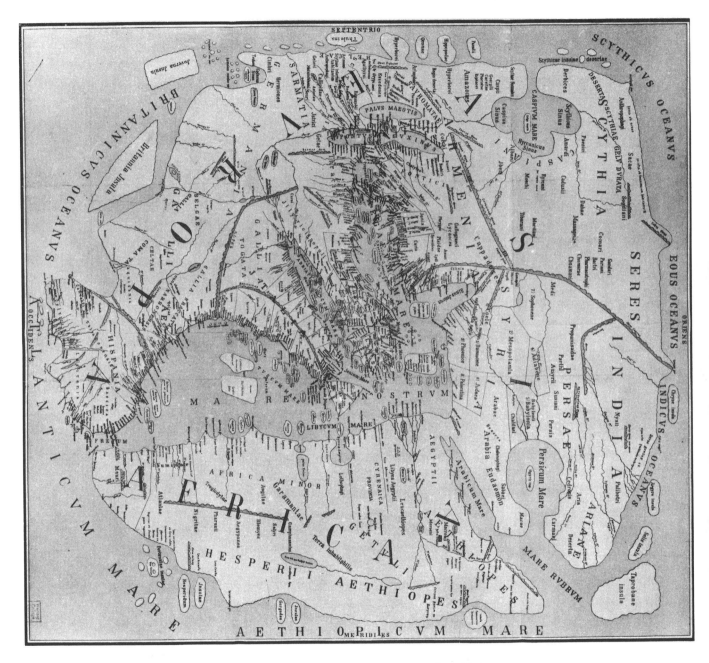

Figure 8. The World Map of Pomponius Mela.

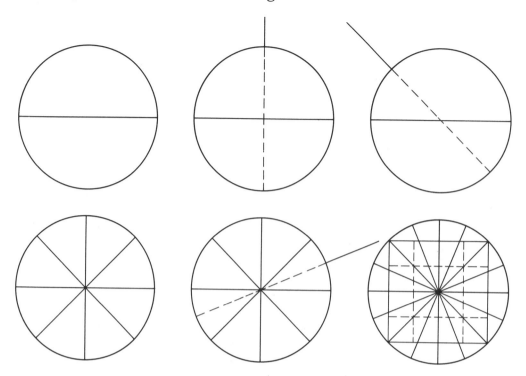

Figure 9. The Eight-Wind System in the Portolan Charts.

Rectangular grids could be constructed with the circular or polar type of projection used in the Piri Re'is and other portolan charts. In this diagram we see how Livengood, Estes, and Woitkowski solved the problem of the construction of the main grid of the Piri Re'is Map (see Fig. 12). (Redrawn by Isroe)

ancient Greeks or others, more probably and rather to facilitate the task of drawing a map, rather than to guide the navigator with such divisions. . . . (107:15–16)

In other words, the portolan design was an excellent design to guide a mapmaker either in constructing an original map or in copying one, because of the design's geometrical character.

Early in our investigation, three of my students, Leo Estes, Robert Woitkowski, and Loren Livengood, decided to take this question—the purpose of the lines on the portolan charts—as their special project. They journeyed to Hanover, New Hampshire, to inspect the medieval charts in the Dartmouth College Library. On their return, one of them, Loren Livengood, said he thought he knew how the charts had been constructed.

The problem was to find out, from the lines actually found on the charts, whether it might be possible to construct a grid of lines of latitude and longitude such as are found on modern maps. In other words, the problem was to see if this portolan system could be *converted* to the modern one.

Livengood's approach was simple. Without actually realizing the importance of his choice, he put himself in the position of a mapmaker rather than of a navigator. That is, he saw the problem not as one of finding a harbor, but of actually constructing a map. He had never heard of Prince Kamal, but he was adopting the Prince's view of the purpose of the lines. The probable procedure of the mapmaker, Livengood speculated, was first to pick a convenient center for his map and then determine a radius long enough to cover the area to be mapped. With this center and radius the mapmaker would draw a circle.

Then he would bisect his circle, again and again, until he had sixteen lines from the center to the periphery at equal angles of 22½°.[6]

The third step would be to connect points on the perimeter to make a square, with four different squares possible.

The fourth step would be to choose one of the squares, and draw lines connecting the opposite points, thus making a map grid of lines at right angles to each other. (Fig. 9)

Now, although the scholars agreed that the portolan charts had no lines of latitude and longitude, it stood to reason that if one of the vertical lines (such as the line through the center) was drawn on True North, then it would be a meridian of longitude, and any line at right angles to it would be a parallel of latitude. Assuming that a projection similar to the famous Mercator projection, in which all meridians and parallels are straight lines crossing at right angles, underlay these maps (see Fig. 10), then all parallel vertical lines would be meridians of longitude, and all horizontal lines would be parallels of latitude.[7]

Applying this idea to the Piri Re'is Map, we could see that the mapmaker had selected a center, which he had placed somewhere far to the east of the torn edge of our fragment of the world map,[8] and had then drawn a circle around it. He had bisected the circle four times, drawing sixteen lines from the center to the perimeter, at angles of 22½°, and he had also drawn in all the four possible squares, perhaps with the intention of using different squares for drawing grids for different parts of the map, where it might be necessary to have different Norths.[9]

---

[6] These angles could also be bisected, if desired, resulting in thirty-two points on the periphery, at angles of 11¼°.

[7] See Note 4 and Note 5.

[8] The complete map included Africa and Asia. It was, according to Piri Re'is, a map "of the seven seas" (see Note 3). In addition to the loss of the eastern part, there was also originally a northern section, which was detached and lost. I am indebted to Dr. Alexander Vietor, of Yale, for this observation.

[9] Since the earth is round, and the portolan design was apparently based on a flat projection (that is, apparently on plane geometry) which could not take account of the spherical surface, the parallel meridians would deviate further and further from True North the farther they were removed from the center of the map. The portolan design could compensate for this, however, as we shall see in the next chapter, by using different Norths.

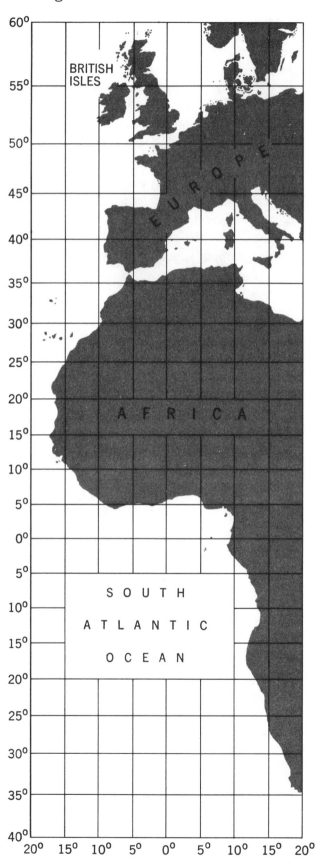

Figure 10. A Map of the eastern shores of the Atlantic on the Mercator Projection.

Compare the meridian of 20° West Longitude with the "Prime Meridian" of the Piri Re'is Map (Fig. 18).

It was Estes who originally pointed out to us that the portolan design had the potentiality of having several different Norths on the same map.

Now the next question was: Which was the right square for us? That is, which (if any) of the squares that could be made out of the design of the Piri Re'is Map was correctly oriented to North, South, East, and West?

Estes found the solution. Comparing the Piri Re'is Map with a modern map (Figs. 10, 11, 12) he found a meridian on the modern map that seemed to coincide very nearly with a line on the Piri Re'is Map—a line running north and south close to the African coast, in about 20° W longitude, leaving the Cape Verde Islands to the west, the Canaries to the east, and the Azores to the west.

Estes suggested that this line might be our prime meridian, a line drawn on True North. All lines parallel to this (assuming, of course, that the underlying projection resembled in some degree the Mercator projection) would also be meridians of longitude; all lines at right angles would be parallels of latitude. The meridians and parallels thus identified, provisionally, on the Piri Re'is Map, formed a rectangular grid, as shown in Fig. 12.

The only difference between this large rectangular grid actually found on the Piri Re'is Map and the grids of modern maps was that the latter all carry registers of degrees of latitude and longitude, with parallels and meridians at equal intervals, usually five or ten degrees apart. We could convert the Piri Re'is grid into a modern grid if we could find the precise latitudes and longitudes of its parallels and meridians. This, we found, meant finding the exact latitude and longitude of each of the five projection centers in the Atlantic Ocean, through which the lines of Piri Re'is' grid ran.

At the beginning of our inquiry I had noticed that these five projection centers had been placed at equal intervals on the perimeter of a circle, though the circle itself had been erased (Fig. 11). I had also noticed that converging lines were extended from these points to the center, beyond the eastern edge of the map. This, it seemed to me at the time, was a geometrical construction that should be soluble by trigonometry. I did not then know that, in the opinion of all the experts, there was no trigonometrical foundation to the portolan charts.

Not knowing that there was not supposed to be any mathematical basis for the portolanos, we now made the search for it our main business. I realized from the start that to accomplish this we would have to discover first the precise location of the center of the map, and then the precise length of the radius of the circle drawn by the mapmaker. I was fortunate in having a mathematician friend, Richard W. Strachan, at the Massachusetts Institute of Technology. He told me that, if we could obtain this information for him, he might be able, by trigonometry, to find the precise positions of the five projection points in the Atlantic Ocean on the Piri Re'is Map, in terms of modern latitude and longitude. This would enable us to draw a modern grid on the map, and thus check every detail of it accurately. Only in this way, of course, could we verify the claim of Mallery regarding the Antarctic sector of the map.

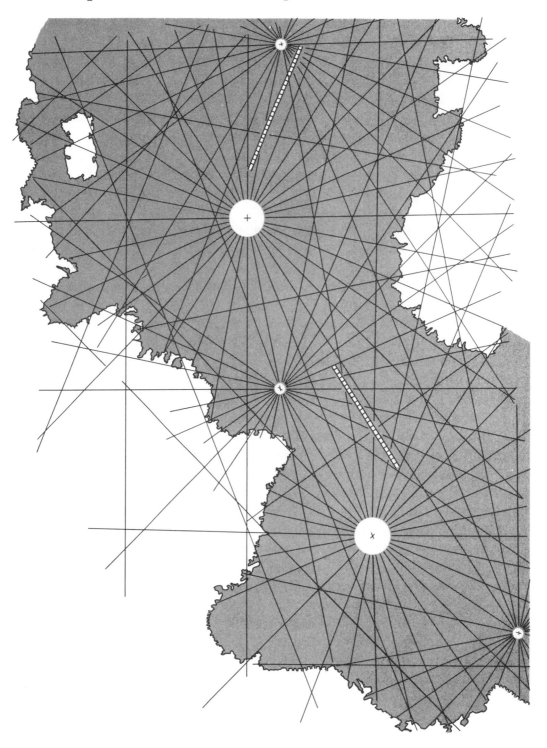

Figure 11. The Piri Re'is Map: the lines of the Portolan Design traced from the facsimile.

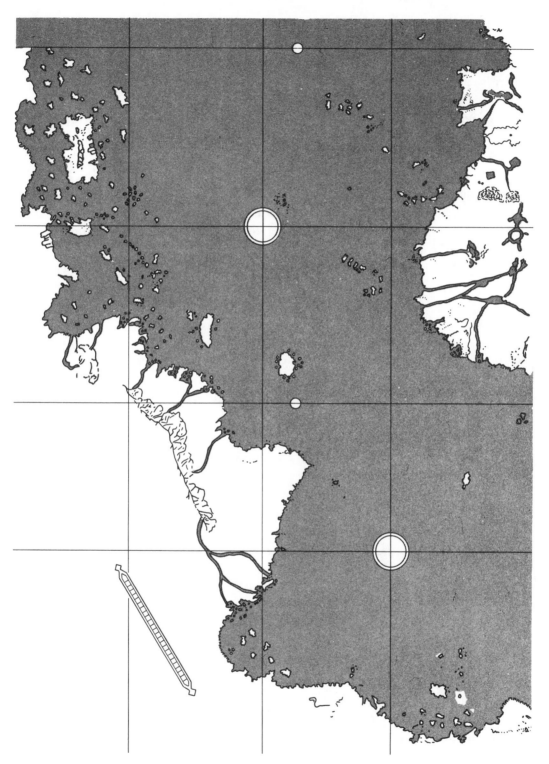

Figure 12. The Piri Re'is Map: the Main Grid of the Portolan Design traced from the facsimile.

The search for the center of the map lasted about three years. We thought from the beginning that the lines extending from the five projection points probably met in Egypt. We used various methods to project the lines to the point where they would meet. Our first guess for the center of the map was the city of Alexandria. This appealed to me because Alexandria was long the center of the science and learning of the ancient world. It seemed likely that, if they were drawing a world map, the Alexandrian geographers might naturally make their own city its center.

However, this guess proved to be wrong. A contradiction appeared. The big wind rose in the North Atlantic looked as if it were meant to lie on the Tropic of Cancer. One of the lines from this center evidently was directed toward the center of the map. But we noticed that this line was at right angles to our prime meridian. This meant, of course, that it was a parallel of latitude. Now, the Tropic of Cancer is at $23\frac{1}{2}°$ North Latitude, and therefore the parallel from the wind rose would reach a center in Egypt at $23\frac{1}{2}°$ North. But Alexandria is not at that latitude at all. It lies in 31° North. Therefore Alexandria could not be the center of our circle.

We looked at the map of ancient Egypt to find, if we could, a suitable city on the Tropic of Cancer that might serve as a center for the map. (We were still attached to the idea that the center of our map should be some important place, such as a city. Later, we were emancipated from this erroneous notion.)

Looking along the Tropic of Cancer, we found the ancient city of Syene, lying just north of the Tropic, near the present city of Assuan, where the great dam is being built. Now we recalled the scientific feat of Eratosthenes, the Greek astronomer and geographer of the 3rd Century B.C., who measured the circumference of the earth by taking account of the angle of the sun at noon as simultaneously observed at Alexandria and at Syene.

We were happy to change our working theory and adopt Syene as the center of the map. With the help of hindsight, we could now see how reasonable it was to place the center of the map on the Tropic, an astronomically determined line on the surface of the earth. The poles, the tropics, and the equator can be exactly determined by celestial observations, and they have been the bases of mapmaking in all times. Syene, too, was an important city, suitable for a center. A good "proof" of this center for the map was constructed by two students, Lee Spencer and Ruth Baraw. Only at the end of our inquiry did we find that Syene was not, after all, exactly the center.

The matter of the radius caused us much more trouble. At first, there appeared to be absolutely no way of discovering its precise length. However, some of my students started talking about the Papal Demarcation Line—the line drawn by Pope Alexander VI in 1493, and revised the next year, to divide the Portuguese from the Spanish possessions in the newly discovered regions (Fig. 13). On the Piri Re'is Map there was a line running north and south, passing through the

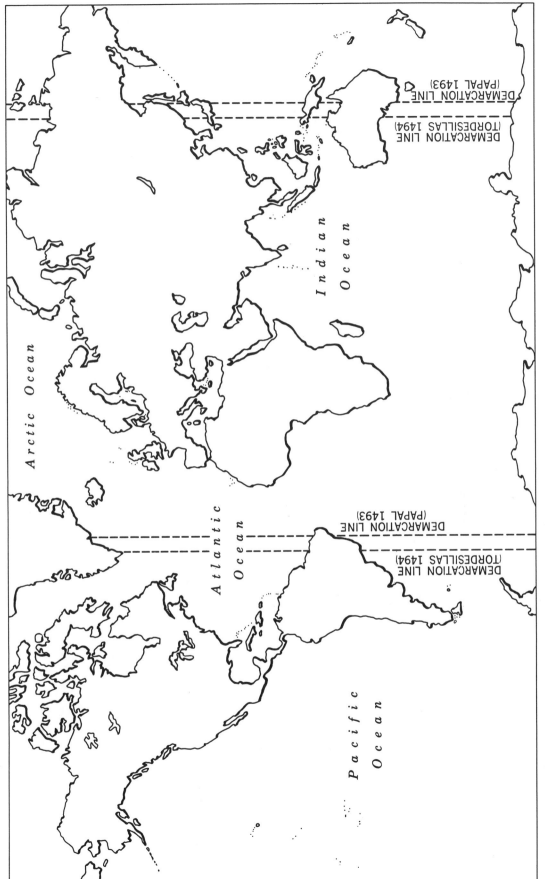

Figure 13. The Papal Demarcation Lines of 1493 and 1494.

northern wind rose and then through Brazil at a certain distance west of the Atlantic coast. This line appeared to be identical, or nearly identical, with the Second Demarcation Line (of 1494), which also passed through Brazil. Piri Re'is had mentioned the Demarcation Line on his map, and we reached the conclusion that this line, if it was the Demarcation Line, could give us the longitude of the northern wind rose and thus the length of the radius of the circle with its center at Syene.

The Papal Demarcation Line of 1494 is supposed to have been drawn north and south at a distance of 370 leagues west of the Cape Verde Islands. Modern scholars have calculated that it was at 46° 30′ West Longitude (140:369). We therefore assigned this longitude to the northern wind rose, and thus obtained our first approximate guess as to the length of the radius of the circle. According to this finding the radius was 79° in length (32½ plus 46½). This result was wrong by 9½°, as we later discovered, but it was close enough for a starter.

At this stage, our findings were too uncertain to justify an attempt to apply trigonometry to the problem. Instead, we tested our results directly on an accurate globe provided by Estes. We made our test by actually drawing a circle, with Syene as the center, and the indicated radius, and then laying out the lines from the center to the perimeter, 22½° apart, beginning with one to the equator. The result seemed pretty good, and we were sure we were on the right track.

It was lucky that we got so far before we discovered that our interpretation of the Demarcation Line on the map was wrong. This fact was finally brought home to us by two other students, John F. Malsbenden and George Batchelder (Fig. 14). They had been bending over the map during one of our long night sessions[10] when suddenly Malsbenden straightened up and exclaimed indignantly that all our work had been wasted, that the line we had picked out was not the right one. In an inscription on his map which we had overlooked Piri Re'is had himself indicated an entirely different line. It was the first line, the line of 1493, and it did not go through the wind rose at all. The mistake, however, had served its purpose. It was true enough that the line we had picked out on the Piri Re'is Map represented neither line; nevertheless it was close enough to the position of the Demarcation Line of 1494 to give us a first clue to the longitude.

Another error that turned out to be very profitable was the assumption we made, during a certain period of time, that perhaps our map was oriented not to True North, but to Magnetic North. Later, we were to find that many, if not most, of the portolanos were indeed oriented, very roughly, to Magnetic North. Some writers on the subject had argued, as already mentioned, that the lines on the portolan charts were intended only for help in finding compass directions, and were therefore necessarily drawn on Magnetic North.[11]

---

[10] Interest in the map was so keen that the students would come to my apartment in the evening, and sometimes argue until the small hours.

[11] See the Bibliography, Nos. 89, 116, 143, 179, 199, 200, 223.

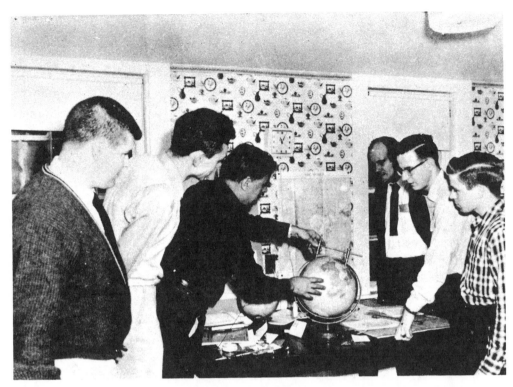

Figure 14. An argument over the Piri Re'is Map: left to right, Leo Estes, Frank Ryan, Charles Hapgood, Clayton Dow, John Malsbenden, George Batchelder.

In the interest of maximum precision, I wanted to find out how the question of Magnetic North might affect the longitude of the Second Demarcation Line, which now determined our radius. If the Demarcation Line lay at 46° 30′ West Longitude at the Cape Verde Islands, it would, with a magnetic orientation, lie somewhat farther west at the latitude of the northern wind rose, and this would affect the radius. We spent time trying to calculate how much farther west the line would be. This in turn involved research to discover the amount of the compass declination (the difference between True and Magnetic North) today in those parts of the Atlantic, and speculation as to what might have been the amount of the variation in the days of Piri Re'is or in ancient times. We found ourselves in a veritable Sargasso Sea of uncertainties and frustrations.

Fortunately, we were rescued from this dead end by still another wrong idea. I noticed that the circle drawn with Syene as a center, and with a radius to the intersection of the supposed Second Demarcation Line with the northern wind rose, appeared to pass through the present location of the Magnetic Pole. We then allowed ourselves to suppose (nothing being impossible) that somebody in ancient times had known the location of the Magnetic Pole and had deliberately selected a radius that would pass through it. Shaky as this assumption might have been, it was at least better than the Demarcation Line, since in ancient times nobody could have had an idea of a line that was only drawn in 1494 A.D. The Magnetic Pole is, however, very unsatisfactory as a working assumption because it does not stay in one place. It is always moving, and where it may have been in past times is anybody's guess.

In the middle of this I read Nordenskiöld's statement that the portolan charts were drawn on True North, and not on Magnetic North (146:17). In this Nordenskiöld was really mistaken, unless he meant that the charts had *originally* been drawn on True North and then had been *reoriented* in a magnetic direction. But his statement impressed us, and then I observed, looking again at the globe with our circle drawn on it, that the circle that passed through the Magnetic Pole also passed very close indeed to the True Pole. Now, you may be sure, we abandoned our magnetic theory in a hurry, and adopted the working assumption that perhaps someone in ancient times knew the true position of the Pole, and drew his radius from Syene on the Tropic of Cancer to the Pole. Again, hindsight came to our support. As in the case of the Tropic of Cancer, the Pole was astronomically determined: It was a precisely located point on the earth's surface.

It appeared to us that we had swum through a murky sea to a safe shore. We had now reached a point where it would be feasible to attempt a confirmation of the whole theory by trigonometry. We were proceeding now on the following asumptions: (1) The center of the projection was at Syene, on the Tropic of Cancer and at longitude 32½° East; (2) the radius of the circle was from the Tropic to the Pole, or 66½° in length, and (3) the horizontal line through the middle projection point on the map (Point III) was the true equator. By comparison with the African coast of the Gulf of Guinea, this line, indeed, appears to be very close to the position of the equator. Nevertheless, this was not merely an assumption but also guesswork. We could not *know*, either, that the ancient mapmaker had precise information as to the size of the earth, which would be necessary for correctly determining the positions of the poles and the equator. Such assumptions could be only working assumptions, to be used for purposes of experiment and discarded if they proved wrong. They were, however, the best assumptions we had been able to come up with so far, and assumptions we had to have to work with.

We could now give our mathematician, Strachan, the data he required for a mathematical analysis. He calculated the positions of all the five projection centers on the Piri Re'is Map to find their precise locations in latitude and longitude.[12] He used our assumed equator as his base line of latitude. I have tried to explain this in Fig. 15. Here I have drawn the first radius from the center of the projection to the point of intersection of the assumed equator with the perimeter of the circle. I then have laid out the other radii at angles of 22½° northward and south-

[12] For the final determinations of these positions see Figure 18.
For the calculations see Appendix.

ward. In this way, our assumption that this equator is precisely correct controls the latitudes to be found for the other four projection points. The assumed equator is the base line for latitude, just as Syene is the reference point for longitude.

Strachan initially computed the positions of the five projection points both by spherical and by plane trigonometry. At each successive step, with varying assumptions as to the radius of the projection and the position of its center, he did the same thing, but in every case the calculations by plane trigonometry made

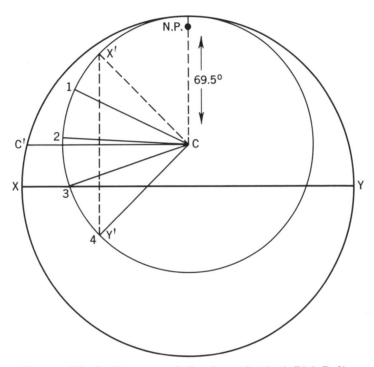

Figure 15. A diagram of the hypothetical Piri Re'is
projection, as based on the equator.

sense—that is, plane trigonometry made it possible to construct grids that fitted the geography reasonably well, while the calculations by spherical trigonometry led to impossible contradictions. It became quite clear that our projection had been constructed by plane trigonometry.[13]

Once we had precise latitudes and longitudes for the five centers on the Piri Re'is map, we could construct a modern type of grid. The total difference of latitude between Point I and Point V, divided by the millimeters that lay between them on our copy of the map (we used a tracing of our photograph of the map), gave us the length of the degree of latitude in millimeters. To check on any possible irregularities we measured the length of the degree of latitude separately

---

[13] See Note 6 for a comparison of the results in one case.

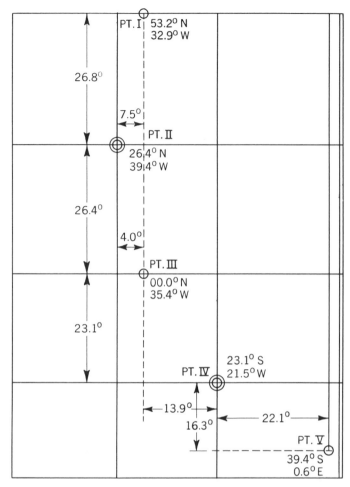

Figure 16. Mode of calculating the length of the degree for the Piri Re'is Map.

between each two of the five points. We followed the same procedure with the longitude, as illustrated in Fig. 16. The lengths of the degrees of latitude and longitude turned out to be practically the same; we thus appeared to have a square grid. In doing this we disregarded the scales actually drawn on the map, since there was no way of knowing when or by whom they had been drawn, or what units of distance they had represented.

The next step was to learn how to draw a grid, not at all an easy task. It was not a particularly complicated task, but it demanded a very high level of accuracy and an extreme degree of patience. Fortunately, one of my students, Frank Ryan, was qualified for the job. He had served in the Air Force, had been stationed at Westover Air Force Base in Massachusetts, and had been assigned to the Cartographic Section of the 8th Reconnaissance Technical Squadron, under a remarkable officer, Captain Lorenzo W. Burroughs. The function of the unit at that time was to prepare maps for the use of the United States Air Force's Strategic Air

Command, known as SAC. Later, it was attached to the 8th Air Force. Needless to say, the personnel of that unit were competent to serve the demanding requirements of the Air Force, as far as mapmaking was concerned, and Frank Ryan had been intensively trained in the necessary techniques. He had had the experience of being drafted into the Air Force: now he had the experience of being drafted again, to draw our grid.

Later Ryan introduced me to Captain Burroughs, and I visited Westover Air Force Base. The captain offered us his fullest cooperation in preparing a draft map with the solution of the projection, and virtually put his staff at our disposal. The co-operation between us lasted more than two years, and a number of officers and men gave us very valuable assistance.[14] Later both Captain Burroughs and his commanding officer, Colonel Harold Z. Ohlmeyer reviewed and endorsed our work (Note 23).

The procedure for drawing the grid was as follows: All the meridians were drawn parallel with the prime meridian, at intervals of five degrees, and all parallels were drawn parallel with the assumed equator, at intervals of five degrees. These lines did not turn out in all cases to be precisely parallel with the other lines of the big grid traced from the Piri Re'is Map, but this was understandable. The effect might have resulted from warping of the map, or from carelessness in copying the lines from the ancient source map Piri Re'is used. We had to allow for a margin of error here, for we could not be sure that no small errors had crept in when the equator or the prime meridian was recopied. Here, as in other respects, we simply had to do the best we could with what we had.[15]

When the grid was drawn, we were ready to test it. We identified all the places we could on the map and made a table comparing their latitudes and longitudes on the Piri Re'is Map with their positions on the modern map. The errors in individual positions were noted and averages of them made (Table 1). The Table is, of course, the test of our solution of the Piri Re'is projection.

But I must not get ahead of my story. We found that some of the positions on the Piri Re'is Map were very accurate, and some were far off. Gradually we became aware of the reasons for some of the inaccuracies in the map. We discovered that the map was a composite, made up by piecing together many maps of local areas (perhaps drawn at different times by different people), and that errors had been made in combining the original maps. There was nothing extraordinary about this. It would have been an enormous task, requiring large amounts of money, to survey and map all at once the vast area covered by the Piri Re'is Map. Undoubtedly local maps had been made first, and these were gradually combined, at different times, into larger and larger maps, until finally a world map was attempted. This long process of combining the local maps, so far as the sur-

---

[14] See Acknowledgments.
[15] See also Strachan's discussion, Note 8.

viving section of the Piri Re'is Map is concerned, had been finished in ancient times. This theory will, I believe, be established by what follows. What Piri Re'is apparently did was to combine this compilation with still other maps—which were probably themselves combinations—to make his world map.

The students were responsible for discovering many of the errors. Lee Spencer and Ruth Baraw examined the east coast of South America with great care and found that the compiler had actually omitted about 900 miles of that coastline. It was discovered that the Amazon River had been drawn twice on the map. We concluded that the compiler must have had two different source maps of the Amazon, drawn by different people at different times, and that he made the mistake of thinking they were two different rivers. We also found that besides the equator upon which we had based our projection (so far as latitude was concerned) there was evidence that somebody had calculated the position of the equator differently, so that there were really two equators. Ultimately we were able to explain this conflict. Other important errors included the omission of part of the northern coast of South America, and the duplication of a part of that coast, and of part of the coasts of the Caribbean Sea. A number of geographical localities thus appear twice on the map, but they do not appear on the same projection. For most of the Caribbean area the direction of North is nearly at right angles to the North of the main part of the map.

As we identified more and more places on our grid, and averaged their errors in position, we found all over the map some common errors that indicated something was wrong with the projection. We concluded that there must still be errors either in the location of the center of the map, in the length of the radius, or both. There was no way to discover these probable errors except by trying out all reasonable alternatives by a process of trial and error. This was time consuming and a tax on the patience of all of us. With every change in the assumed center of the map, or in the assumed radius, Strachan had to repeat the calculations, and once more determine the positions of the five projection points. Then the grid had to be redrawn and all the tables done over. As each grid in turn revealed some further unidentified error, new assumptions had to be adopted, to an accompaniment of sighs and groans. We had the satisfaction, however, of noting a gradual diminution of the errors that suggested that we were approaching our goal.

Among the various alternatives to Syene as the center of the map we tried out, at one stage, the ancient city of Berenice on the Red Sea. This was the great shipping port for Egypt in the Alexandrian Age, and it, too, lay on the Tropic of Cancer. Berenice seemed to be a very logical center for the map because of its maritime importance. We studied the history of Berenice, and everything seemed to point to this place as our final solution. But then, as in an Agatha Christie murder mystery, the favorite suspect was proved innocent. The tables showed the assumption to be wrong, for in this case the errors were even increased. We had to give up Berenice, with special regrets on my part because of the beauty of the name.

Now we went back to Syene, but with a difference. The tables showed that the remaining error in the location of the center of the map was small. Therefore we tried out centers near Syene, north, east, south and west, gradually diminishing the distances, until at last we used the point at the intersection of the meridian of Alexandria, at 30° East Longitude, with the Tropic. This finally turned out to be correct.

Immediately hindsight began to make disagreeable comments. Why hadn't we thought of this before? Why hadn't we tumbled to this truth in the beginning? It combined all the most reasonable elements: the use of the Tropic, based on astronomy, and the use of the meridian of Alexandria, the capital of ancient science. Later we were to find that all the Greek geographers based their maps on the meridian of Alexandria.

Remaining errors in the tables suggested something wrong with the radius. We knew, of course, that our assumption that the mapmaker had precise knowledge of the size of the earth was doubtful. It was much more likely that he had made some sort of mistake. We therefore tried various lengths. We shortened the radius a few degrees, on the assumption that the mapmaker might have underestimated the size of the earth, as Ptolemy had. This only increased the errors. Then we tried lengthening the radius. The entire process of trial and error was repeated with radii 7°, 5°, 2°, and 1° too long. Finally we got our best results with a radius extended three degrees. This meant that our radius was not 66.5°, the correct number of degrees from the Tropic to the Pole, but 69.5°. This error amounted to an error of 4½ per cent in overestimating the size of the earth.

A matter of great importance, which we did not realize at all at the time, was that we were, in fact, finding the length of the radius (and therefore the length of the degree) with reference mainly to longitude. I paid much more attention to the average errors of longitude than I did to the errors of latitude. I was especially interested in the longitudes along the African and South American coasts. Our radius was selected to reduce longitude errors to a minimum while not unduly increasing latitude errors. As it turned out, this emphasis on longitude was very fortunate, for it was to lead us to a later discovery of considerable importance.

With regard to the overestimating of the circumference of the earth, there was one geographer in ancient times who made an overestimate of about this amount. This was Eratosthenes. Does this mean that Eratosthenes himself may have been our mapmaker? Probably not. We have seen that the Piri Re'is Map was based on a source map originally drawn with plane trigonometry. Trigonometry may not have been known in Greece in the time of Eratosthenes. It has been supposed that it was invented by Hipparchus, who lived about a century later.

Hipparchus discovered the precession of the equinoxes, invented or at least described mathematical map projections, and is generally supposed to have developed both plane and spherical trigonometry (58:49; 175:86).[16] He accepted Eratosthenes' estimate of the size of the earth (184:415) though he criticized Eratosthenes for not using mathematics in drawing his maps.

We must interfere in this dispute between Hipparchus and Eratosthenes to raise an interesting point. Did Hipparchus criticize his predecessor for not using mathematically constructed projections on which to place his geographical data? If so, his criticism looks unreasonable. The construction of such projections requires trigonometry. If Hipparchus himself developed trigonometry, how could he have blamed Eratosthenes for not using it a century before? Hipparchus' own books have been lost, and we really have no way of knowing whether the later writers who attributed trigonometry to Hipparchus were correct. Perhaps all they meant, or all *he* meant or said in his works, was that he had *discovered* trigonometry. He might have discovered it in the ancient Chaldean books whose star data made it possible for him to discover the precession of the equinoxes.

But this is speculation, and I have a feeling that it is very much beside the point. If Hipparchus did in fact develop both plane and spherical trigonometry, the Piri Re'is Map, and the other maps to be considered in this book, are evidence suggesting that he only rediscovered what had been very well known thousands of years earlier. Many of these maps must have been composed long before Hipparchus. But it is not possible to see how they could have been drawn as accurately as they were unless trigonometry was used. (See Note 7.)

We have additional confirmation that the Piri Re'is projection was based on Eratosthenes' estimate of the size of the earth. The Greeks had a measure of length, which they called the stadium. Greek writers, therefore, give distances in stadia. Our problem has been that they never defined this measure of length. We have no definite idea, therefore, of what the stadium was in terms of feet or meters. Estimates have varied from about 350 feet to over 600. Further, we have no reason to even suppose that the stadium had a standard length. It may have differed in different Greek states and also from century to century.

A great authority on the history of science, the late Dr. George Sarton of Harvard, devoted much attention to trying to estimate the length of the stadium used by Eratosthenes himself at Alexandria in the 3rd Century B.C. He concluded that the "Eratosthenian stadium" amounted to 559 feet (184:105).[17]

The solution of the Piri Re'is projection has enabled us to check this.

---

[16] However, a knowledge of plane trigonometry has been attributed to Appolonius, an earlier Greek scientist, by Van der Waerden (216). The date of its origin appears, then, unknown.

[17] That is, there were about 9.45 Eratosthenian stadia to a mile of 5,280 feet, which figures out to 558.88 feet per stadium.

Presumably, it proves the amount of the overestimate of the earth's circumference to be 4½ per cent (or very nearly that). Eratosthenes gave the circumference of the earth as 252,000 stadia. We checked the length of his stadium by taking the true mean circumference of the earth (24,800 miles), increasing this by 4½ per cent, turning the product into feet, and dividing the result by 252,000. We got a stadium 547 feet long.

Now, if we compare our result with that of Sarton, we see that there is a difference of only 12 feet, or about 2 per cent. It would seem—again by hindsight—that we could have saved all our trouble by merely adopting Eratosthenes' circumference and Sarton's stadium. We could then have drawn a grid so nearly like the one we have that the naked eye could not have detected the difference.

The next stage, which came very late, was our realization that if Eratosthenes' estimate of the circumference of the earth was used for drawing Piri Re'is' source map, and if it was 4½ per cent off, then the positions we had found by trigonometry for the five projection points on the map were somewhat in error both in latitude and longitude. It was now necessary to redraw the whole grid to correct it for the error of Eratosthenes. We found that this resulted in reducing all the longitude errors until they nearly vanished.

This was a startling development. It could only mean that the Greek geographers of Alexandria, when they prepared their world map using the circumference of Eratosthenes, had in front of them source maps that had been drawn *without the Eratosthenian error*, that is, apparently without any discernible error at all. We shall see further evidence of this, evidence suggesting that the people who originated the maps possessed a more advanced science than that of the Greeks.

But now another perplexing problem appeared. The reduction of the longitude errors left latitude errors that averaged considerably larger. Since accurate longitude is much more difficult to find than accurate latitude, this was not reasonable. There had to be some further undetected error in our projection.

We started looking for this error, and we found one. That is, we found an error. It was not quite the right one; it did not solve our problem, but it helped us on the way. As already mentioned, we had found the positions of the five projection points by laying out a line first from the center of the projection to the intersection of the circle with the line on the Piri Re'is map running horizontally through the middle projection point, Point III, assuming this to be the equator. We had used this assumed equator as our base line for latitude. (See Fig. 15.)

When we laid out the projection in this way, we had not yet realized that the mapmaker was much more likely to have drawn his first radius from the center of the map directly to the pole and not to the equator. (See Fig. 17.) If he did this, since his length for the degree was wrong, then his equator must be off a number of degrees. This required new calculations, and still another grid.

At first, this new grid seemed to make matters worse, especially on the coast of Africa. The equator seemed to pass too near the Guinea coast by approximately

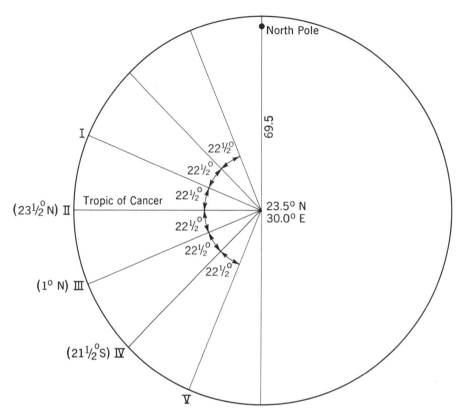

Figure 17. A diagram of the hypothetical Piri Re'is projection
as based on the North Pole.

five degrees. My heart sank when this result became apparent, but I am thankful that I persisted in redrawing the grid despite the apparent increase in the errors, for the result was a discovery of the very greatest importance.

At first I thought that the African coast (and that of Europe) had simply been wrongly placed too far south on the projection. But I soon saw that if the African coast appeared too far south on the corrected projection, the French coast was in more correct latitude than before. There was simply, I first concluded, an error in scale. Piri Re'is, or the ancient mapmaker, had used too large a scale for Europe and Africa. But why, in that case, though latitudes were thrown out, did longitudes remain correct?

I finally decided to construct an empirical scale for the whole coast from the Gulf of Guinea to Brest to see how accurate the latitudes were relative to one another. The result showed that the latitude errors along the coasts were minor. It was obvious that the original mapmakers had observed their latitudes extremely well. From this it became apparent that those who had originally drawn this map of these coasts had used a different length for the degree of latitude than for the degree of longitude. In other words, the geographers who designed the square portolan grid for which we had discovered the trigonometric solution, had apparently applied their projection to maps that had originally been drawn with another projection.

What kind of projection was it? Obviously it was one that took account of the fact that, northward and southward from the equator, the degree of longitude in fact diminished in length as the meridians drew closer toward the poles. It is possible to represent this by curving the meridians, and we see this done on many modern maps. It is also possible to represent this by keeping the meridians straight and spacing the parallels of latitude farther and farther apart as the distance from the equator increases. The essential point is to maintain the ratio between the lengths of the degrees of latitude and longitude at every point on the earth's surface.

Geographers will, of course, instantly recognize the projection I have described here. It is the Mercator projection, supposedly invented by Gerard Mercator and used by him in his Atlas of 1569 (Note 5). For a time we considered the possibility that this projection might have been invented in ancient times, forgotten, and then rediscovered in the 16th Century by Mercator (Note 15). Further investigation showed that the device of spreading the parallels was found on other maps, which will be discussed below.

I was very reluctant to accept without further proof the suggestion that the Mercator projection (in the full meaning of that term) had been known in ancient times. I considered the possibility that the difference in the length of the degree of latitude on the Piri Re'is Map might be *arbitrary*. That is, I thought it possible that the mapmaker, aware of the curvature of the earth, but unable to take account of it as is done in the Mercator projection by spherical trigonometry, had simply adopted a *mean* length for the degree of latitude, and applied this length over the whole map without changing the length progressively with each degree from the equator.

Strangely enough, shortly after this, I found that, according to Nordenskiöld, this is precisely what Ptolemy had done on his maps (see Note 9). In Nordenskiöld's comparison of the maps of the Mediterranean and Black Sea regions as drawn by Ptolemy and as shown on the Dulcert Portolano (Fig. 4), we see that he has drawn the lines of Ptolemy's projection in this way. This is, of course, another indication of the ancient origin of Piri Re'is' source map.

This is not quite the end of the story. We shall see, in subsequent consideration of the De Canerio Map of 1502, that the oblong grid, used by Ptolemy and found on the Piri Re'is Map, has its origin in an ancient use of spherical trigonometry.

These successive discoveries finally enabled us to draw a modern grid for most of the Piri Re'is map, as shown in Figure 18.

THE PIRI RE'IS MAP OF 1513

IN ALL THE WORLD THERE IS NO OTHER MAP LIKE THIS MAP—PIRI RE'IS

GRID A

GRID B

GRID C

GRID D

PRIME MERIDIAN

PIRI REIS EQUATOR

EQUATOR OF GRID B

PRIME MERIDIAN OF GRID B

EQUATOR OF ORIGINAL
TRIGONOMETRIC PROJECTION

Heavy lines represent Piri Reis
grid traced from photograph of
the original

Omission of coastline
(A) loss of 4.5° longitude
going West

Here the Prime Meridian of
Grid B, the equator of Grid
B, and a Meridian of the
main Grid intersect.

51.4° N
36.9° W

23.5° N
42.6° W

4.4° S
36.9° W

96° W

$11\frac{1}{4}$°

96° M.

The Grids

The longitudes of Grid A are de-

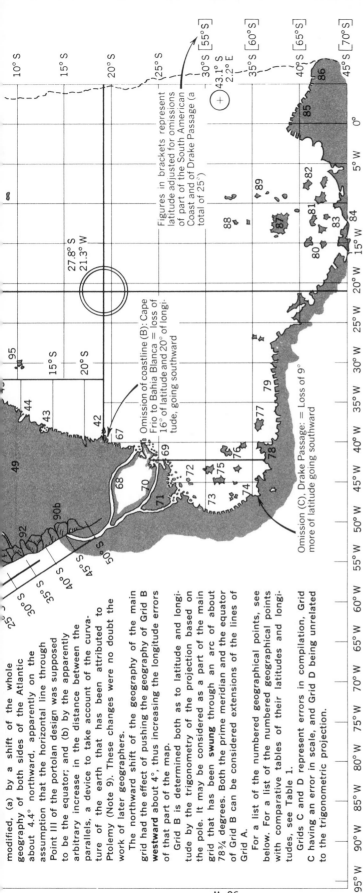

modified, (a) by a shift of the whole geography of both sides of the Atlantic about 4.4° northward, apparently on the assumption that the horizontal line through Point III of the portolan design was supposed to be the equator; and (b) by the apparently arbitrary increase in the distance between the parallels, a device to take account of the curvature of the earth that has been attributed to Ptolemy (Note 9). These changes were no doubt the work of later geographers.

The northward shift of the geography of the main grid had the effect of pushing the geography of Grid B **westward** about 4°, thus increasing the longitude errors of that part of the map.

Grid B is determined both as to latitude and longitude by the trigonometry of the projection based on the pole. It may be considered as a part of the main grid that has been **swung** through an arc of about 78¾ degrees. Both the prime meridian and the equator of Grid B can be considered extensions of the lines of Grid A.

For a list of the numbered geographical points, see below. For a list of the numbered geographical points with comparative tables of their latitudes and longitudes, see Table 1.

Grids C and D represent errors in compilation, Grid C having an error in scale, and Grid D being unrelated to the trigonometric projection.

Omission of coastline (B): Cape Frio to Bahia Blanca = loss of 16° of latitude and 20° of longitude, going southward

Omission (C). Drake Passage: = Loss of 9° more of latitude going southward

Figures in brackets represent latitude adjusted for omissions of part of the South American Coast and of Drake Passage (a total of 25°)

27.8° S
21.3° W

43.1° S
2.2° E

1. Annobon Islands
2. Cavally River
3. Cape Palmas
4. St. Paul River
5. Mano River
6. Freetown
7. Bijagos Islands
8. Gambia River
9. Dakar
10. Senegal River
11. Cape Blanc
12. Cape Juby
13. Sebu River
14. Gibraltar
15. Guadalquivir River
16. Cape St. Vincent
17. Tagus River
18. Cape Finisterre
19. Gironde River
20. Brest

21. Cape Verde Islands
22. The Canary Islands
23. Madeira Islands
24. The Azores
25. Cuba
  (a) Gulf of Guacanayabo
  (b) Guantanamo Bay
  (c) Bahia de Nipe
  (d) Bahia de la Gloria
  (e) Camaguey Mountains
  (f) Sierra Maestra Mountains
26. Andros Island
27. San Salvador (Watling)
28. Isle of Pines
29. Jamaica
30. Hispaniola
   (Santo Domingo, Haiti)
31. Puerto Rico
32. Rio Moroni
33. Corantijn River

34. Essequibo River
35. Orinoco River
36. Gulf of Venezuela
37. Pt. Gallinas
38. Magdalena River
39. Gulf of Uraba
40. Honduras (Cape Gracias a Dios)
41. Yucatan
42. Cape Frio
43. Salvador
44. San Francisco River
45. Recife (Pernambuco)
46. Cape Sao Rocque
47. Rio Parahyba
48. Bahia Sao Marcos
49. Serras de Gurupi, de Desordam, de Negro
50. The Amazon (No. 1) Para River
51. The Amazon (No. 2) Para River
52. The Amazon (No. 2) western mouth

53. Island of Marajo
54. Essequibo River
55. Mouths of the Orinoco
56. Peninsula of Paria
57. Martinique
58. Guadaloupe
59. Antigua
60. Leeward Islands
61. Virgin Islands
62. Gulf of Venezuela
63. Magdalena River
64. Atrato River
65. Honduras (Cape Gracias a Dios)
66. Yucatan
67. Bahia Blanca
68. Rio Colorado
69. Gulf of San Mathias
70. Rio Negro (Argentina)
71. Rio Chubua
72. Gulf of San Gorge

73. Bahia Grande
74. Cape San Diego (near the Horn)
75. Falkland Islands
76. The South Shetlands
77. South Georgia
78. The Palmer Peninsula
79. The Weddell Sea
80. Mt. Ropke, Queen Maud Land
81. The Regula Range
82. Muhlig-Hofmann Mountains
83. Penck Trough
84. Neumeyer Escarpment
85. Drygalski Mountains
86. Vorposten Peak
87. Boreas, Passat Nunataks
88. Tristan d'Acunha
89. Gough Island
77. South Georgia
95. Fernando da Naronha

Figure 18

CHAP
TER III
THE
PIRI RE'IS
MAP
IN
DETAIL

In undertaking a detailed examination of the Piri Re'is Map of 1513, I shall break down the map into sections representing originally separate source maps of smaller areas, which appear to have been combined in a general map by the Greek geographers of the School of Alexandria.[1]

With regard to each of the source maps, which I shall refer to as "component maps," since they are the parts of the whole, I will identify such geographical points as are evident in themselves, or are rendered plausible by their position on the trigonometric grid, and will find their errors of location.

Since in some cases the component maps were not correctly placed on the general map, we have two sorts of errors: those due to mistakes in compilation of the local maps into the general map and those due to mistakes in the original component maps. These can be distinguished because if a component map is misplaced, all the features of that map will be misplaced in the same direction and by the same amount. If the general error is discovered and corrected, then the remaining errors will be errors of the original local maps. We have discovered that in most cases the errors on the Piri Re'is Map are due to mistakes in the compilation of the world map, presumably in Alexandrian times, since it appears, as we shall see, that Piri Re'is could not have put them together at all. The component maps, coming from a far greater antiquity, were far more accurate. The Piri Re'is Map appears, therefore, to be evidence of a decline of science from remote antiquity to classical times.

1. The western coasts of Africa and Europe, from Cape Palmas to Brest, including the North Atlantic islands (Cape Verdes, Canaries, Azores, and Madeira) and some islands of the South Atlantic.

Longitudes, as well as latitudes, along the coasts are seen to be remarkably accurate (see Table 1). The accuracy extends also to the North Atlantic island groups as a whole, with an exception in the case of Madeira.

---

[1] I do not wish to exclude the possibility, however, that another reasonable explanation for the source of the compilation may some day be forthcoming.

The accuracy of longitude along the coast of Africa, where it is greatest, might be attributed simply to our assumptions as to the center and radius of the projection, but for two considerations. First, the assumption regarding the length of the radius (that is, the length of the degree) was not reached with reference to the coast of Africa, but with reference to the width of the Atlantic and the longitude of the coast of South America. It will be seen from our map (Fig. 18) and from Table 1 that both these coasts, separated by the width of the Atlantic, are in approximately correct relative longitude with reference to the center of the projection on the meridian of Alexandria. This seems to mean that the original mapmaker must have found correct relative longitude across Africa and across the Atlantic from the meridian of Alexandria to Brazil.

It is also important that most of the islands are in equally correct longitude. The picture that seems to emerge, therefore, is one of a scientific achievement far beyond the capacities of the navigators and mapmakers of the Renaissance, of any period of the Middle Ages, of the Arab geographers, or of the known geographers of ancient times. It appears to demonstrate the survival of a cartographic tradition that could hardly have come to us except through some such people as the Phoenicians or the Minoans, the great sea peoples who long preceded the Greeks but passed down to them their maritime lore.

The accuracy of placement of the islands suggests that they may have been found on the ancient source map used by Piri Re'is. The "discoveries" and mapping of these islands by the Arabs and Portuguese in the 15th Century may not, then, have been genuine discoveries. It is possible that the 15th Century sailors really found these islands as the result of accidental circumstances (being blown off course, etc.). On the other hand, nothing excludes the possibility that source maps used by Piri Re'is, dating from ancient times, were known in some form to people in Europe. Possibly some of the early voyages to some of these islands, particularly the Azores, were undertaken to confirm the accuracy of the old maps. It is hardly, if at all, possible that these 15th Century navigators could have found correct longitude for the islands. All they had to go by were rough guesses of courses run, based on the direction and force of the wind, and the estimated speed of their ships. Such estimates were apt to be thrown off by the action of ocean currents and by lateral drift when the ship was trying to make to windward.

A good description of the problem of finding position at sea is given by a 16th Century writer quoted by Admiral Morison in his *Admiral of the Ocean Sea:*

"O how God in His omnipotence can have placed this subtle and so important art of navigation in wits so dull and hands so clumsy as those of these pilots! And to see them inquire, one of the other, 'how many degrees hath your honor found?' One says 'sixteen,' another 'a scant twenty' and another 'thirteen and a half.' Presently they ask, 'How doth your honor find himself with respect to the land?' One says, 'I find myself forty leagues from land,' another 'I say 150,' another says 'I find myself this morning 92 leagues away.' And be it three or three hundred nobody agrees with anybody else, or with the truth." (140:321–322)

In the days of Piri Re'is no instruments existed by which the navigator at sea could find his longitude. Such an instrument did not appear for another 250 years, when the chronometer was developed in the reign of George III. It does not seem possible to explain the accuracy of longitude on the Piri Re'is Map in terms of navigational science in the time of Piri Re'is.

The case for latitude is somewhat different. Latitude could be determined in the 15th and 16th Centuries by astronomical observations. However, observations taken by trained people with proper equipment were one thing, and observations taken by explorers were quite another. Morison says that Columbus made serious mistakes in finding latitude. Speaking of the First Voyage he says: ". . . We have only three latitudes (all wrong) and no longitude for the entire voyage" (140:157). He describes one of Columbus' attempts to find his latitude as follows:

On the night of Nov. 2 (1492) two days before the full moon, he endeavored to establish his position by taking the altitude of the North Star with his wooden quadrant. After applying the slight correction he decided that Puerto Gibara, actually in Lat. 21° 06′ N, was in 42° N, the Latitude of Cape Cod (140:258).

For a long time after the four voyages of Columbus we find the latitudes of Cuba and Haiti wrong on the maps of the time. Almost all mapmakers put the islands above rather than below the Tropic of Cancer.[2]

To return to the problem of longitude, Morison remarks that the only method of finding longitude known in the 16th Century was by the timing of eclipses, but that nobody was successful in applying it. He says:

The only known method of ascertaining longitude in Columbus' day was by timing an eclipse. Regiomontanus's *Ephemerides* and Zacuto's *Almanach Perpetuum* gave the predicted hours of total eclipse at Nuremberg and Salamanca respectively, and if you compared those with the observed hour of the eclipse, wherever you were, and multiplied by 15 to convert time into arc (1 hour of time = 15° of Longitude) there was your longitude west of the Almanach maker's meridian. Sounds simple enough, but Columbus, with two opportunities (1494 and 1503) muffed both, as did almost everyone else for a century. (140:185–186)

Morison describes in an interesting manner the failure of an attempt to find the longitude of Mexico City in 1541 (twenty-eight years after Piri Re'is drew his map):

At Mexico City in 1541 a mighty effort was made by the intelligentsia to determine the longitude of the place by timing two eclipses of the moon. The im-

---

[2] Among such maps I may list the Juan de la Cosa Map of 1500 (Fig. 19), the Cantino Map of 1502, the so-called Bartholomew Columbus Map of 1503 (Fig. 22), the Waldseemuller Maps of 1507 and 1516, the Ruysch Map of 1508 (Fig. 24), the Robert Thorne Map of 1527 (Fig. 20) and the Miguel Vilanovano Map of 1535. There are many others.

posing result was 8h 2m 32s (= 120° 38′ west of Toledo) but the correct difference of longitude between the two places is 95° 12′, so the Mexican savants made an error of 25½°, or 1450 miles! Even in the 18th Century Pere Labat, the earliest writer (to my knowledge) who gives the position of Hispaniola correctly, adds this caveat: "I only report the longitude to warn the reader that nothing is more uncertain, and that no method used up to the present to find longitude has produced anything fixed and certain" (140:186).

With this backwardness of the 16th Century science of navigation, I cannot see how the accuracy of the Piri Re'is Map can be explained, either as to latitude or longitude.[3] Figures 19–24 illustrate the poor qualities of the maps that were drawn at this time.

With regard to latitude there are several complications in the Piri Re'is Map. Its history—that is, the history of the source map used by him for the Atlantic coasts—must have been a long one, for several different stages of mapmaking are reflected in it.

We thought at first that the horizontal line running through Point III represented the equator of the projection. This would involve a design as in Fig. 16, with the line from the center to the point of intersection of this horizontal line with the perimeter of the circle as the determining line; in reference to this line all the other radii were laid out at angles of 22½° to the north and south. Our results indicate that something like this was done; that is, that the source map we are discussing (embracing Africa, Europe, and some of the islands) was at some time during the history of the map placed on the projection in this way. It might have been done visually, or empirically, so to speak, simply by placing the African coast of Guinea at the correct distance north of this central line, which was taken for the equator. This was an error, with reference to the mathematical projection, the equator of which in fact lies nearly five degrees north of this line, as shown in Fig. 17.

We concluded that an error was made here because some of the Piri Re'is Map (to be discussed below) in fact is in line with the equator of the trigonometric projection. It seems probable that we are dealing here with the work of different people who redrew the map at different times with different ideas. The large wind roses in the North and South Atlantic Oceans, apparently identifying the Tropics, may have been superimposed on the map by the geographers who made the error in placing their source map. Figure 17 shows that according to the trigonometric projection the northern projection point lies on the Tropic of Cancer,

[3] **W. H. Lewis, in his "The Splendid Century" (Doubleday, 1957, pp. 227–228), quotes an extract from the memoirs of the Abbé de Choisy (1644–1724) on the difficulty of finding longitude a century and a half after Piri Re'is: "Father Fontenay lectures on navigation, and shows that not only is longitude undiscovered, but why it is undiscoverable. . . ."**

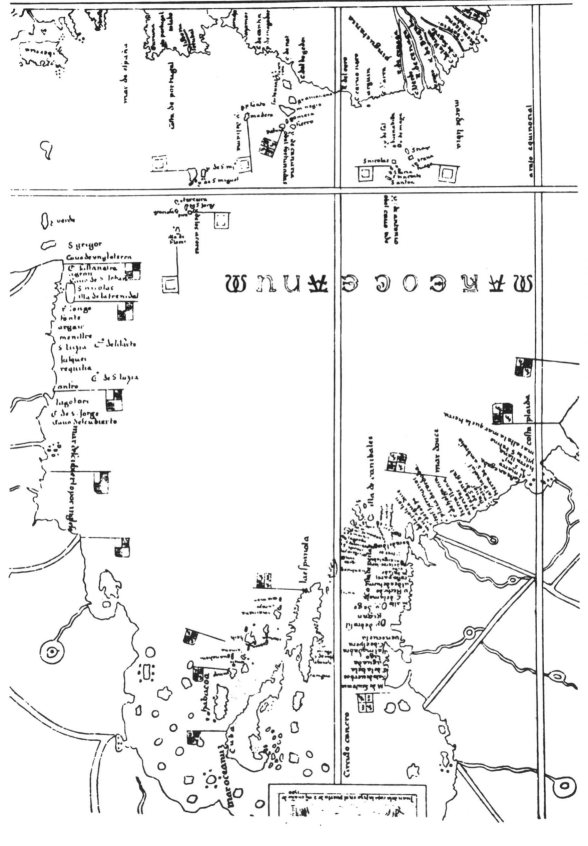

Figure 19. The Juan de la Cosa Map of 1500.

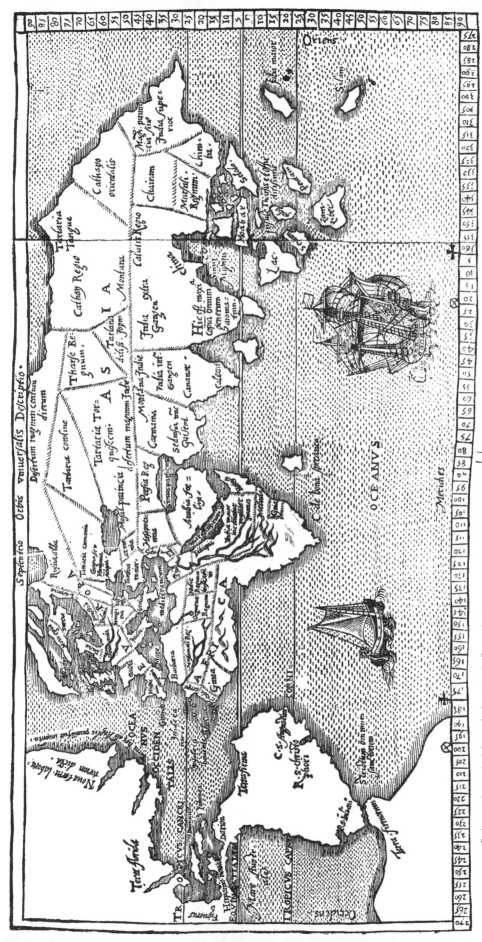

Figure 20. The Robert Thorne Map of 1527.

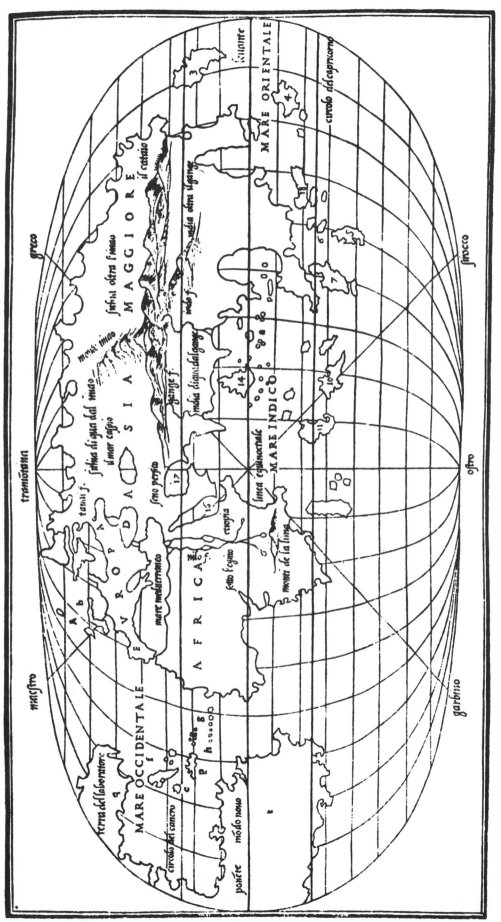

Figure 21. The Benedetto Bordone Map of 1528.

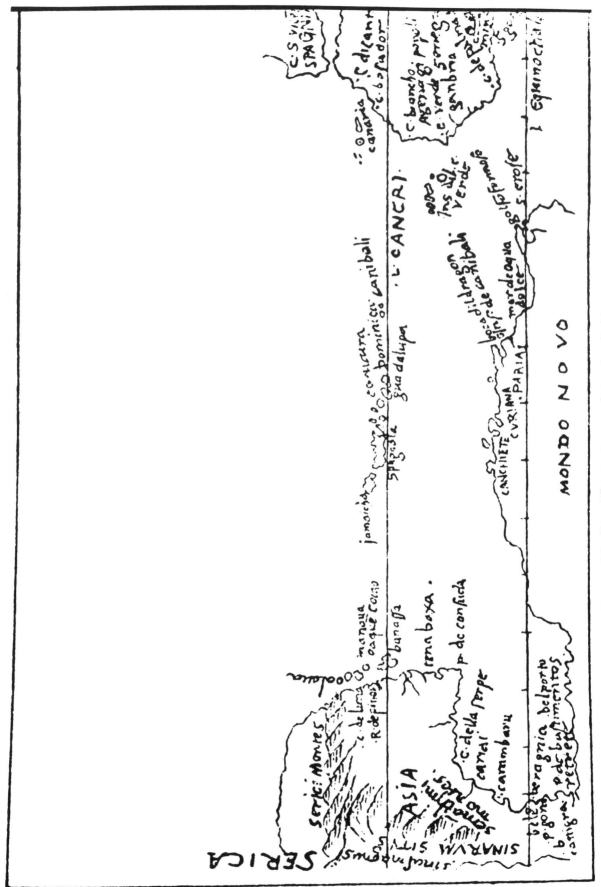

Figure 22. The "Bartholomew Columbus" Map.

Figure 23. The Ptolemaeus Basilae Map of 1540.

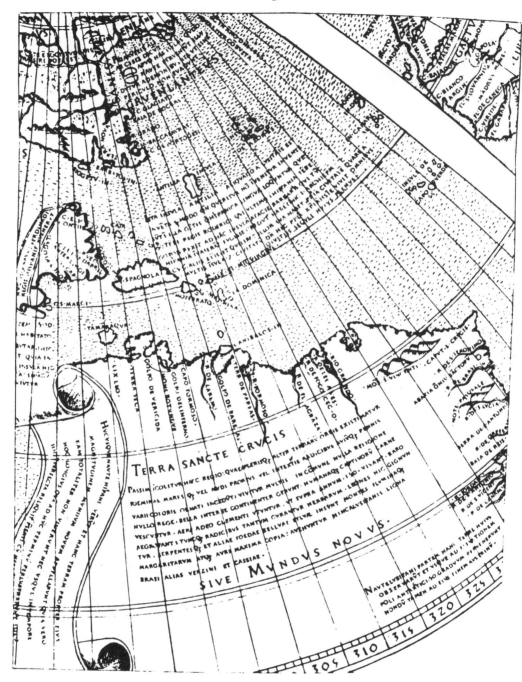

Figure 24. The Joannes Ruysch Map of America of 1508.

Note, Figures 19–24: This selection of maps drawn in the Age of Discovery illustrates the weaknesses of the cartographic science of the period. So far as relative distances, land shapes, and particularly longitude are concerned these maps are much inferior to the Piri Re'is Map. None of these maps suggests the use of trigonometry.

while the southern one does not coincide closely with the Tropic of Capricorn; the altered projection shows the northern projection point several degrees off the Tropic of Cancer while the southern projection point lies fairly close to the Tropic of Capricorn. The geometry of the projection makes it impossible for both projection points to coincide with the respective Tropics by either arrangement. It may be noted, however, that whoever converted the latitude from the trigonometric projection to make the equator coincide with the central projection point on the perimeter of the circle did not alter the longitudes of the coasts, which are close to correct by either system. We may conclude that the whole source map embracing Europe and Africa was simply shoved due North about five degrees. This would have produced some minor errors in longitude, but of too small a scale to be noticeable to us. Piri Re'is himself may have made the change.

Some of the islands included on this component map may have been added by Piri Re'is on the basis of accounts of recent explorations. Those islands which appear to be seriously misplaced in latitude or longitude may have been such additions. On the other hand, it may be the case with some (as perhaps with Madeira and Fernando da Naronha) that their errors in latitude may be due to having been ignored when the component map was shifted northward.

All things considered, I am inclined to believe that Piri Re'is himself shifted the component map northward to make it agree with his view that the line through the central projection point (Point III) must represent the equator. If he did this, it indicates a good knowledge on his part of the latitudes of the Guinea coast. As we shall see, this knowledge was available as the result of the Portuguese exploration of that coast in the 15th Century. This exploration resulted in some careful scientific observations of latitude such as were not available from accounts of the American explorations.

To sum up, then, this part of the Piri Re'is Map suggests that Piri Re'is had a source map of Africa, Europe, and the Atlantic islands, based on maps probably drawn originally on some sort of trigonometric projection adjusted to the curvature of the earth. By default of any alternative, we seem forced to ascribe the origin of this part of the map to a pre-Hellenic people—not to Renaissance or Medieval cartographers, and not to the Arabs, who were just as badly off as everybody else with respect to longitude, and not to the Greeks either. The trigonometry of the projection (or rather its information on the size of the earth) suggests the work of Alexandrian geographers, but the evident knowledge of longitude implies a people unknown to us, a nation of seafarers, with instruments for finding longitude undreamed of by the Greeks, and, so far as we know, not possessed by the Phoenicians, either.

2. A special projection in the Caribbean, including part of the coast of South America.

The Caribbean part of the Piri Re'is Map offered us the greatest difficulties. It seemed entirely out of line. The coast appeared to trend the wrong way. It looked at first like some of the very worst mapmaking imaginable. From our studies of the portolan projection, however, I was prepared to accept the possibility of there being more than one North on this map. Estes had pointed out that the portolan design permitted a change of North from one part of a map to another, if and when it became desirable to move from one square, or grid, to one of the others that the design made possible.

I was looking at the map one day when I suddenly found that by twisting my head to one side, I could make some sense of the Caribbean section. I saw that there was indeed another North in this area. I assumed to start with that it might be integrated with the mathematics of the world projection. It had already become evident to us that it was theoretically possible to take any one of the map's projection points, whose positions were now known, and repeat the portolan design by drawing a circle with this point as a center, and then constructing a grid within it exactly as with the world projection. This would be a satellite grid, and any North line could be chosen to suit the mapmaker's convenience.

To solve this problem it was necessary to locate a North line, that is, a prime meridian. By identifying on the map a number of geographical localities which lay at the same latitude on the modern map of the Caribbean, I drew a rough parallel. I then looked for—and found—a line on the Piri Re'is projection at right angles to this. The line I found came down from Projection Point I at the top of the map and bisected what looked like the Peninsula of Yucatán. The angle of this line to the meridians of the main part of the map was $78\frac{3}{4}°$; this meant that it lacked one compass point ($11\frac{1}{4}°$) of being at a right angle to the north of the rest of the map.

Gradually it became possible to extend the mathematical system of the whole projection to this part of the map. The common point was Projection Point I, which we had located at 51.4° North Latitude and 36.9° West Longitude. We assumed this point to be at the same latitude in both parts of the map. Since the length of the degree was (by assumption) the same, we could lay out parallels of latitude at five-degree intervals down to zero, which was, then, the equator of this special projection. Latitude was thus integrated mathematically with the world projection.

We found, after a number of tests, that the Ptolemaic spacing of parallels had also been applied in this component map.

The longitude problem presented much greater difficulty. Our first solutions were largely guesswork. Finally, the problem was solved by dropping a line, from the intersection of the prime meridian of our Caribbean section with the equator of that section, to the bottom of the map, where it intersected the register of longitude of the main grid extended westward. The longitude of the point of intersection at the bottom of the map became the longitude of our local prime meridian, and thus both the latitudes and longitudes of the Caribbean section were determined (See Fig. 18).

Now, if the reader will visualize the entire Caribbean grid as suspended from Point I, hanging down with no place to put its feet, and then *swung* through an arc of 78¾°, he should get the idea. Since the swing of the projection is so exact, and since, as the tables show (Table 1), the latitudes and longitudes of the identifiable places around the Caribbean are remarkably accurate, we are sure that the accuracy of this special projection is not a coincidence.

Perhaps the reader may wonder at the mapmaker's reason for resorting to this device. The only answer I am able to suggest—and it is only a guess—is that he may have had ancient maps (maps ancient *then*) of the Caribbean area, with ample notations of latitude and longitude, but drawn, like a modern map, on some sort of spherical projection. Perhaps because he was unfamiliar with spherical trigonometry, he may have been forced to treat the round surface of the earth as a series of flat planes. He therefore had to have different norths in areas that were too far removed from each other in longitude. He was clever enough to work out a scheme by which he could preserve the accuracy of latitude and longitude in the Caribbean. He had to find just the right angle for North that would achieve this purpose, and he did so. But it is probable that he did not achieve the full accuracy of his ancient sources.

Strong support for this hypothesis is provided by a comparison of the Piri Re'is Map with a modern map of the world drawn on a polar equidistant projection (see Figs. 25, 26, 27). This map was drawn for the use of the Air Force during World War II. It was centered at Cairo, Egypt, because an important U.S. air base was located there. Since Cairo is not far from the center of the Piri Re'is world projection, this modern map gives us a good idea of what the world would look like on a projection of this kind centered on Egypt. If we look at Cuba on this equidistant map, we notice that it appears to run at right angles to a latitude line drawn through Cairo. In other words, if we regard the map as representing a flat surface, then Cuba runs north and south, just as it seems to run with reference to the main projection of the Piri Re'is Map. Furthermore, in both cases we see Cuba much too far north.

How is this to be explained? What else can we conclude but that the mapmaker, confronted by a spherical projection he did not understand, had to translate his geographical data (latitudes and longitudes of places in the Caribbean) into terms of a flat surface? This contains the implication, of course, that spherical trigonometry must have been known ages before its supposed invention by Hipparchus in the second century B.C. It also raises another question: How did it

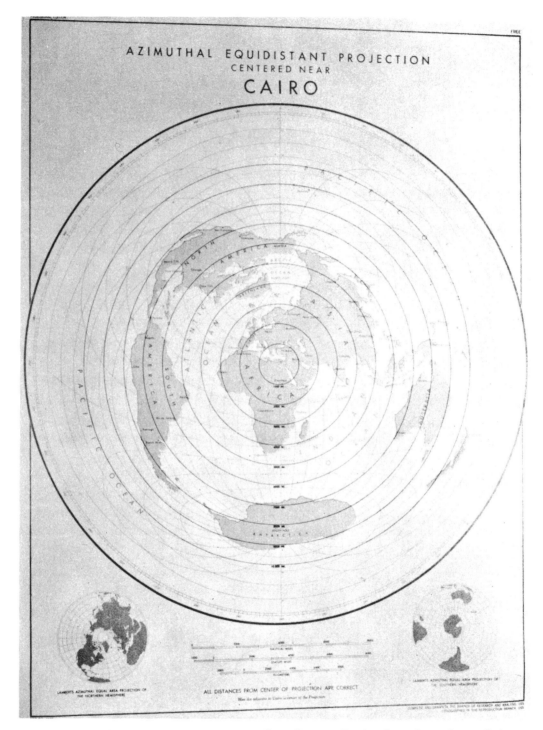

Figure 25. Map of the world on an Equidistant Projection, based on Cairo, Egypt. (United States Air Force)

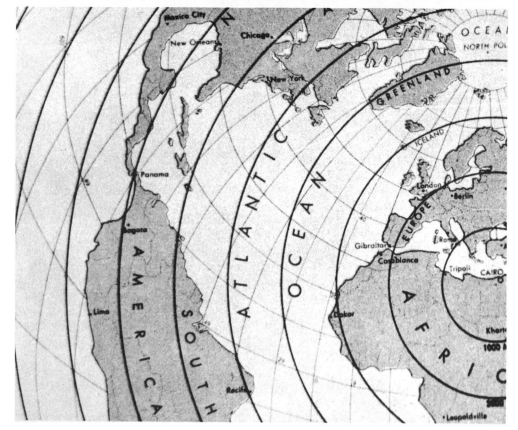

Figure 26. Map of the world on an Equidistant Projection. (Section for comparison with the Piri Re'is Map of 1513.)

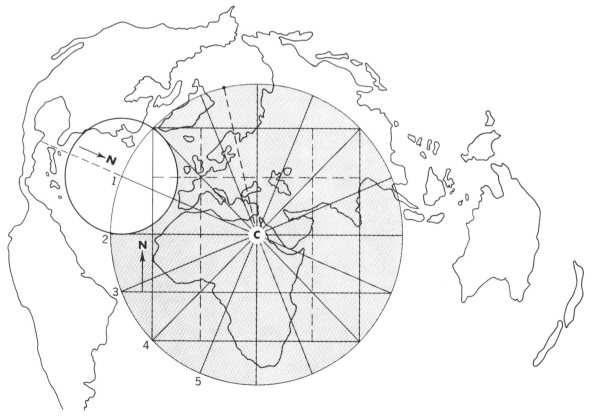

Figure 27. The Piri Re'is Projection imposed on the Equidistant Projection of the world.

happen that a world map, apparently drawn ages before Hipparchus, was centered on Egypt? Can we ascribe such advanced knowledge to the early Egyptians? If not, to whom? I do not apologize for raising such thorny questions. They are at present unanswerable. Perhaps they will be answered some day.

To sum up, then, our mapmaker was faced with the problem of indicating True North both for the Atlantic and for the Caribbean area, which extends much farther west. Since the portolan projection is a rectangular projection and the earth is round, it is evident that you cannot extend it through many degrees of longitude without getting to a place where the meridians will not point north at all. The geometrical scheme of the portolan projection, with several possible Norths, was the only way to solve this problem. But there had to be mathematical calculations. Only by trigonometry could the correct angle for the Caribbean prime meridian be found.

The peculiar projection for the Caribbean area permits some conclusions as to the probable history of the map as a whole. In the first place it is clear that Piri Re'is could not have constructed this part of his world map. Such a thing as two Norths on the same map was unheard of in the Renaissance. To Piri Re'is, the idea of changing the direction of north in the middle of the ocean would be lunacy, and all the mapmakers of the age would have looked at the matter the same way. But even if he had the idea, even if he knew some trigonometry (of which there is no evidence)[4] he still could not have drawn the map, because neither he nor, as far as is known, anyone else at that time had any information as to the longitudes of places in the Caribbean.

What applies to Piri Re'is applies also to Columbus. Columbus could not have drawn any part of the map included in the special grid because for him, as for Piri Re'is, there could be only one North on a map. It is possible, however, that this special grid may provide a solution to one of the problems of Columbus' first voyage.

Let us suppose that Columbus had a copy of this map of the Caribbean, as it appears on the Piri Re'is Map. (Piri Re'is himself believed this was the case.) Perhaps the map showed the Azores, or even some part of the European coast, so that by simple measurement Columbus was able to get an idea of the scale of the map and the distance across the ocean to the Caribbean islands.

We know he had some sort of map and that he had an idea of how soon he would find land. But we also know that he did not find land where he had expected

---

[4] Fortunately we possess, in Piri Re'is' extensive treatise on the geography of the Mediterranean, the "Kitabe Bahriye" (145a), a large number of maps personally drawn by him. Their characteristics are most interesting. Like Arab maps generally, they are good pictures. But they lack any sort of projection. They do not even carry scales of distance. They do not show the compass directions of the portolan charts. See Figs. 28–33.

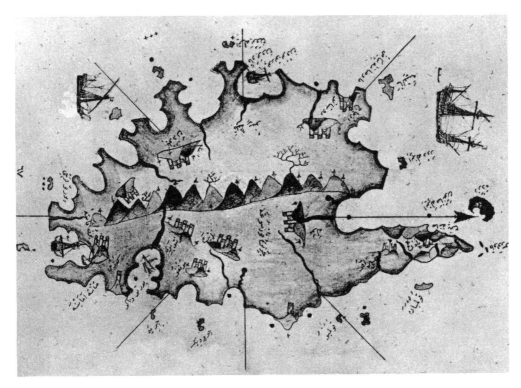

Figure 28. The Piri Re'is Map of Corsica (from the Bahriye).

Figure 29. A modern map of Corsica.

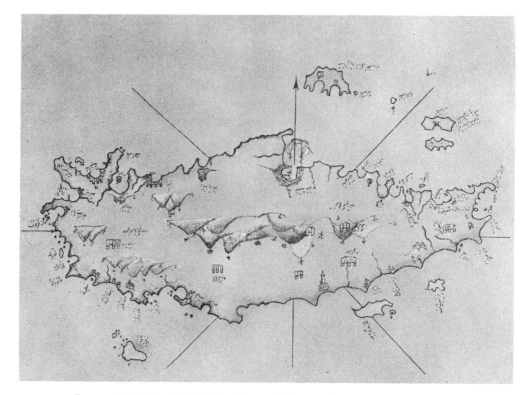

Figure 30. The Piri Re'is Map of Crete (from the Bahriye).

Figure 31. A modern map of Crete.

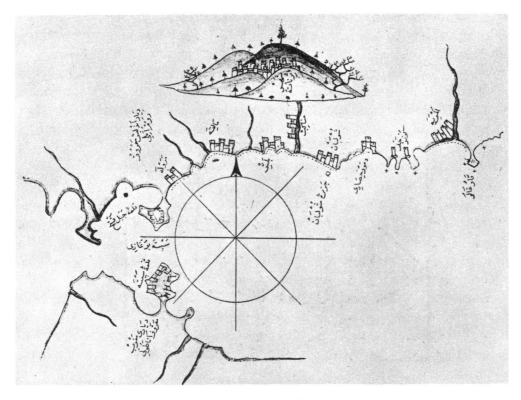

Figure 32. The Piri Re'is Map of the western Mediterranean and Gibraltar (from the Bahriye).

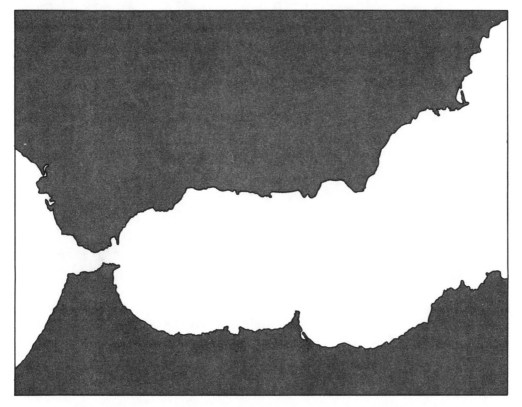

Figure 33. A modern map of the western Mediterranean.

to find it. Instead, he had to sail about one thousand miles farther and was faced with a threatened mutiny of his crew. Finally he made a landfall at the island of San Salvador (Watling Island) or some other island nearby.

Now, if you look at San Salvador on our map (Fig. 39) and note its longitude on the main grid of the map, you will see that it lies just west of the 60th meridian on that grid instead of at 74½° West Longitude where it actually should be. But if you swing the map around and find the longitude of the island on the special Caribbean grid, it turns out to be at 80.5° West. The trouble that Columbus ran into may now be understood. His error in not understanding the map he had may have led to a mistake of about 14° or about 840 miles in his estimate of the distance across the Atlantic, and thus nearly caused the failure of his expedition.

Let us consider the probabilities of Columbus' having carried with him from Spain a copy of this component map of the Caribbean. He need not have had with him the entire source map used by Piri Re'is, including South America. The evidence is that he did not suspect that a continent lay to the south of the Caribbean until he ran into the fresh water of the Orinoco out at sea.

We have seen that Piri Re'is, in all probability, had ancient maps at his disposal in Constantinople. It is quite possible that copies of some of these had reached the West long before his day. Greek scholars fleeing from the Turks

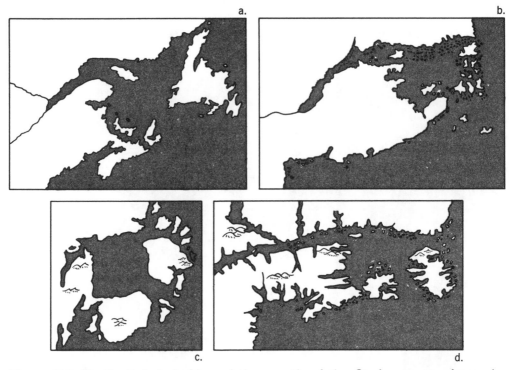

Figure 34. Martin Behaim's Map of the mouth of the St. Lawrence, drawn in 1492 before the return of Columbus from his first voyage, as compared with later maps, (a) modern map, (b) Sebastian Cabot, 1544, (c) Behaim Globe, 1492, (d) Lescarbot map of 1606. After Hjalmar R. Holand, in "Explorations in America Before Columbus," New York, Twayne, 1956.

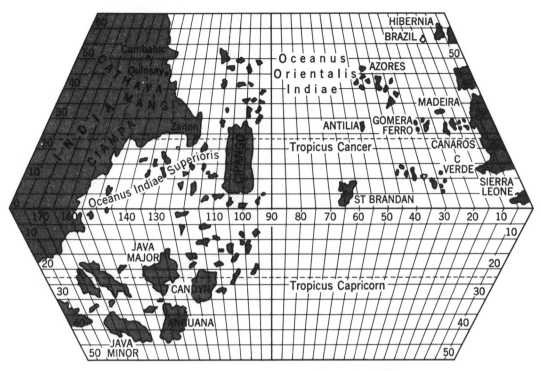

Figure 35. The Toscanelli Map of 1474.

brought thousands of Greek manuscripts to Italy before the fall of Constantinople in 1453. Much earlier still, in the year 1204, a Venetian fleet, supposedly intended to carry a crusade to the Holy Land, attacked and captured Constantinople. For about sixty years afterward Italian merchants had access to map collections in Constantinople.

We have reason to believe that good maps of the St. Lawrence River were available in Europe before Columbus sailed in 1492. In Fig. 34 we see a map of the river and the islands near its mouth that the mapmaker Martin Behaim placed on a globe he made and completed before Columbus returned from his first voyage. Columbus was not an ignorant mariner, as some people seem to imagine. He was quite at home in Latin, which indicated some education, and he was a cartographer by trade. It is known that he traveled widely in Europe, always on the lookout for maps. His voyage was not a sudden inspiration; it was a deeply settled objective, one followed with perseverance for many years, and it required, above all, maps. The historian Las Casas said that Columbus had a world map, which he showed to King Ferdinand and Queen Isabella, and which, apparently, convinced them that they should back Columbus.

Many have thought that this map may have been the map said to have been sent to Columbus by the Italian scholar Toscanelli (see Fig. 35). But a Soviet scientist has presented a strong argument against this, including evidence that the Toscanelli letter to Columbus, accompanying the map, was a forgery (209). In any case, the Toscanelli Map, whether Columbus had it or not, is a very poor map.

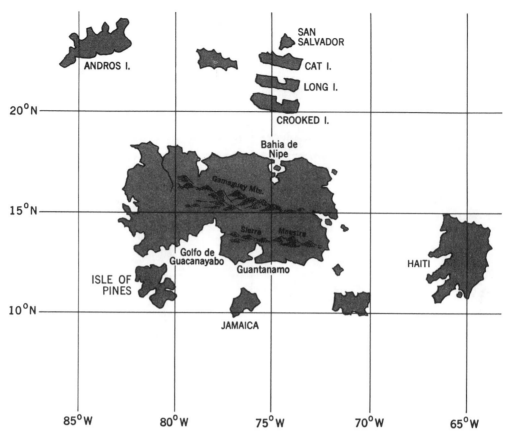

Figure 36. Cuba according to the Piri Re'is Map.

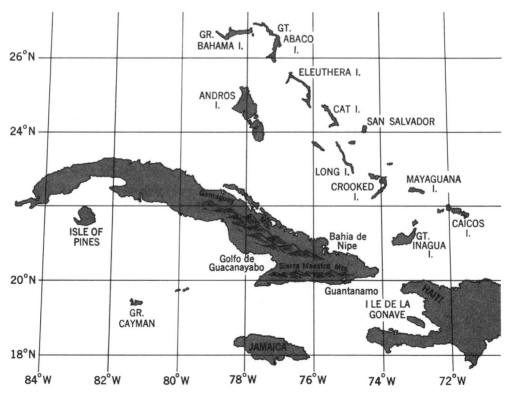

Figure 37. A modern map of Cuba and the adjacent islands.

Cuba on the Piri Re'is Map presents some very interesting problems.

In the first place, Cuba was wrongly labeled *Espaniola* (Hispaniola, the island now comprising Haiti and the Dominican Republic) by Piri Re'is. This error was accepted by Philip Kahle who studied the map in the 1930's (106). Nothing could better illustrate how ignorant Piri Re'is was of his own map. The mislabeling of Cuba also clearly shows that all he did was to get some information verbally from a sailor captured by his uncle, or from some other source, and then try to fit the information to a map already in his possession, a map he may have found in the Turkish Naval Archives, which possibly inherited it from the Byzantine Empire. In Figures 36, 37 I have compared the island I have identified on the Piri Re'is Map as Cuba with a modern map of that island.

This comparison shows that what we have in this island on the Piri Re'is Map is a map of Cuba, but a map only of its eastern half. We can identify a number of points around the coasts and in the interior. The western half is missing, but, as if to compensate for this, the island is shown at twice the scale of the rest of the map, so that it subtends about the correct amount of longitude for the whole island. Oddly enough, there is a complete western shoreline where the island is cut off, as if, when the map was drawn, all of western Cuba was still beneath sea level. We observe that some islands are shown in the west in the area now occupied by western Cuba.

There is good evidence that a map of a thus truncated Cuba was well known in Europe before the first voyage of Columbus. In Figure 38 I have compared the Cuba of the Piri Re'is Map with the island labeled "Cipango" on the Behaim Globe (completed before Columbus' return from his first voyage), on the Toscanelli Map, and on the Bordone Map of 1528.

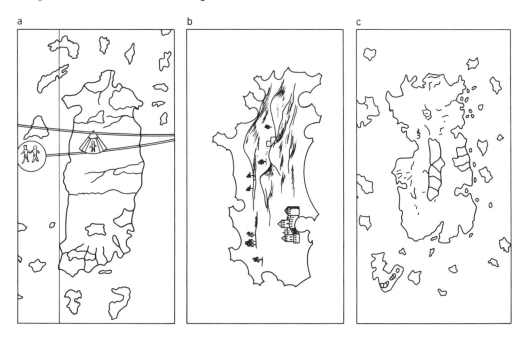

Figure 38. The Piri Re'is version of Cuba (c) as compared with versions of the island of "Cipango." (Behaim Globe, a, Bordone, b)

Figure 39. The Piri Re'is World Map of 1528 (surviving fragment).

It seems quite clear that Bordone's island, which of the three most closely resembles the Piri Re'is island, was not inspired by the current information on Cuba. Cuba on the maps made by the 16th Century explorers in no way resembles the island on the Piri Re'is Map. (See, for example, Fig. 39, the Piri Re'is Map of 1528. Here Piri Re'is represents Cuba in a form typical of the other maps of the day. He had evidently abandoned his ancient maps.)

In view of the possibility that an ancient map of the eastern half of Cuba may have been circulating in Europe before Columbus' first voyage, it becomes increasingly easy to accept the idea that Columbus may have found a good map,

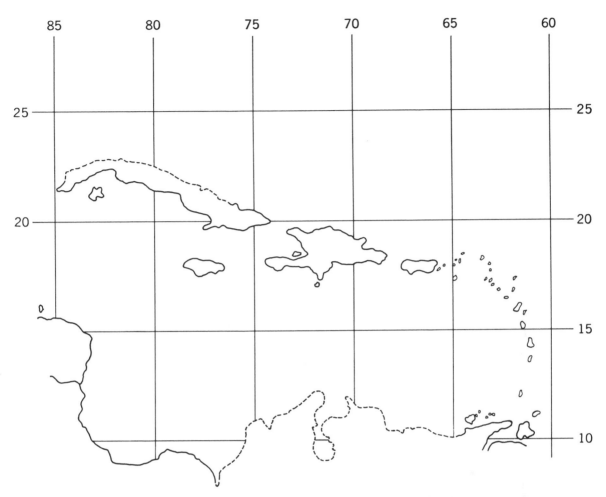

Figure 40. The coasts Columbus visited (solid lines).

at least the Caribbean section of the Piri Re'is Map, and that this may actually have led him to America. In view of these facts one of my students, Lee Spencer, revised the old verse:

> In Fourteen Hundred Ninety Two
> Columbus sailed the ocean blue.
> With maps in hand drawn long before
> He headed straight for Cuba's shore.
> Much fame he gained, so I am told,
> For he proved true the maps of old.

The Piri Re'is representation of Cuba suggests that the Caribbean section of his map was itself a compilation of originally separate local maps. One of these may be identified in the map of Hispaniola.

Here we have still another North. The arrow on our map indicates the direction of north for Hispaniola and some adjacent islands. It does not agree with the Norths either of the main grid or of the Caribbean grid; it is not, so far as we can see, integrated with the trigonometric projection. Columbus could not have placed it on the map (assuming he had it) because, if there was one thing Columbus could determine, it was north, and he would therefore have aligned Hispaniola with the rest of the Caribbean islands on the main grid of the map.

3. A map of the Atlantic coast of South America, from Cape Frio northward to the Amazon, with an error in scale.

On the Piri Re'is Map, South America consists of a compilation of various local maps differing in scale and in orientation. This particular component map is on too small a scale, as shown by the inset grid, Fig. 18, but is in correct longitude. It is possible that we can partially reconstruct the story of this map.

It was, in the first place, an accurate map of the coast. But it seems that the mapmaker may have been operating under the impression that Point IV of the world projection pattern lay on the Tropic of Capricorn, and he placed this component map so that its southern end lay on this assumed tropic. This left its northern end too far south, because of an error in the scale. The mapmaker, however, may have been unaware of this because of a failure to identify the river shown there as the Para River, one of the mouths of the Amazon. According to my interpretation, the map does show the course of the Amazon coming down to its Para River mouth, but it does not show the Island of Marajo. The map would suggest that it may come from a time when the Para River was the main or only mouth of the Amazon, and when the Island of Marajo may have been part of the mainland on the northern side of the river. If the mapmaker knew of the Island of Marajo as existing in his time he might not identify the river on his source map with the Amazon. We shall see evidence shortly that he did know of the existence of the Island.

The evidence for my interpretation of this part of the map is in the agreement of the inset grid on our tracing with the topography, as shown in Table 1.

4. A map of the Amazon and the Island of Marajo, correctly placed on the equator of the trigonometric projection; some other component maps.

One part of the Piri Re'is Map that seems to date without modification from the time when a trigonometric projection was used to compile a world map from various local maps is a map of the Amazon with a very good representation of the Island of Marajo. Here both mouths of the Amazon are shown. The upper one, the mouth of the Amazon proper, is shown about 10° north of the river just suggested as the Para River, on the inset grid of our map. It lies about 5° north of the line used as the equator for Africa and Europe. Interestingly, both the duplications of the Amazon suggest the actual course of the river, while all the representations of it in the later maps of the 16th Century bear no resemblance to its real course. Moreover, the excellent representation of the Island of Marajo is quite unique. Nothing like it can be found on any map of the 16th Century until after the official discovery of the island in 1543. Where could Piri Re'is have got his accurate conception of this island? If he had somehow obtained the information as to its shape, how could he have placed it correctly both in latitude and longitude, with reference to a mathematical projection of which he was almost certainly ignorant?

This Island of Marajo did quite a bit of drifting after Piri Re'is' day. It turned up on Mercator's 1569 Map of South America, but here we find it placed at the mouth of the Orinoco! (See Fig. 41.)

To the east of the South American coast the Piri Re'is Map shows a large island where no such island now exists. One might suppose that this island was imagined by Piri Re'is in the same way that many mythical islands were placed on other maps of the Renaissance. Piri Re'is actually does have such an island, which he names "Antillia" (No. 94 on our map). It looks artificial. But the island we have now to deal with does not have this artificial look. It has the appearance of a real island, with harbors, and islands off the coast. Some of the photographs show highlands around the coasts (indicated by deeper color) and a large central plain.

The fact that this island has more behind it than the imagination of one cartographer is suggested by another map, presented to the Paris Academy of Sciences in 1737, and associated with the name of the French geographer Philippe Buache, a member of the Academy. On this map we see an island, very roughly the shape and size of the island on the Piri Re'is Map, placed directly on the equator! Between this island and the coast of Africa we see another island, where none now exists. The map has indications that these islands were even then *former*, not *present*, islands. The coasts are hatched, suggesting approximations. Inside these coasts smaller islands are indicated, as if they were remnants left by the submergence of the larger islands. Indications are given that the Cape Verde and Canary Islands were once connected with the mainland of Africa, and other island groups are shown in the North Atlantic where none now exist (Fig. 42).

What is this map? Does it illustrate a legend of submerged islands in the Atlantic? If it does, then certain facts about the locations of these islands are relevant. One is that the big island on the Piri Re'is Map is located right over the Mid-Atlantic Ridge (formerly called the Dolphin Ridge) at the spot where two tiny islands, the Rocks of St. Peter and Paul, jut up above the sea, just north of the equator and about 700 miles east of the coast of Brazil. (See Fig. 43.) Another fact is that the island to the east of the corresponding island on the Buache Map is located just over the Sierra Leone Rise, a mountain range on the ocean bottom. Finally, as the reader can see, a cross-section of the equatorial Atlantic, from South

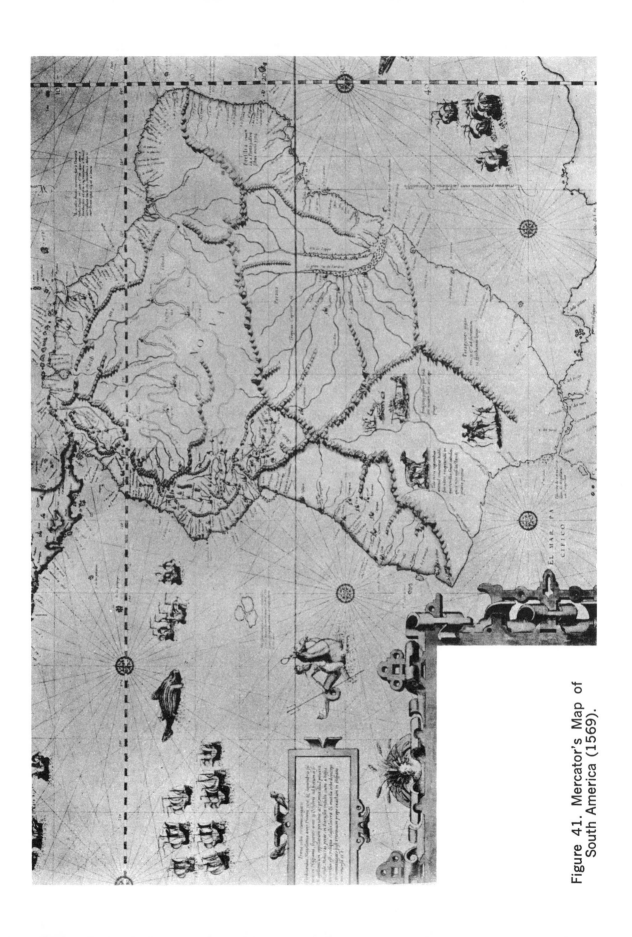

Figure 41. Mercator's Map of South America (1569).

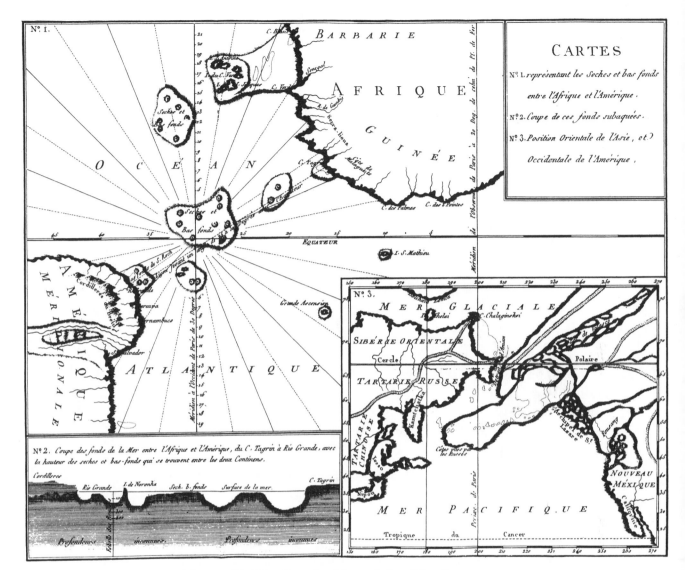

Figure 42. The Buache Map of the Atlantic.

America to Africa, shows both the Mid-Atlantic Ridge and the Sierra Leone Rise, although rather crudely.

Many will dismiss these facts as mere coincidence. Admittedly, there is no proof; but I feel strongly that something more than coincidence is involved. Figure 43 shows the present ocean bottom.

Another component map, which may be briefly dealt with here, shows the mountainous area on the western side of Piri Re'is' South America. This component map was added to the general map, but it was not integrated with the trigonometry of the projection. There are errors both in scale and in orientation, as shown in Figure 18.

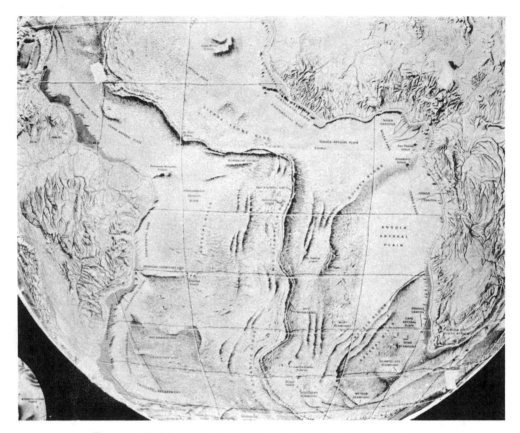

Figure 43. Modern Bathymetric Map of the Atlantic.

It seems at first glance that the mountains shown here must have been intended for the Andes. However, Kahle, one of the earlier students of the map, rejected this on the ground that the Andes were not yet discovered when Piri Re'is drew his map. On this controversial point the following considerations may be urged:

First, what is the probability that a cartographer, by pure invention, would place an enormous range of mountains on the western side of South America, *where one actually exists?*

Second, the various rivers, including both Amazons on Piri Re'is' map of South America, are shown flowing from these mountains, which is correct.

Third, the drawing of the mountains indicates that they were observed from the sea—from coastwise shipping—and not imagined.

Fourth, the general shape of the coast on the map agrees well with the South American coast from about 4° South Latitude down to about 40° S. It is between these latitudes that the Pacific cordillera of the Andes closely parallel the coast. There is even a suggestion on this coast of the Peninsula of Paracas.

Kahle adds an observation that apparently contradicts his own conclusion.

He positively identifies one of the animals shown in the high mountains as the llama, on the ground that it shows the protruding tongue peculiar to that beast. (106) Llamas are found in the Andes; but it is doubtful they could have been known to anyone in Europe in 1513.

5. A map on the main grid, from Equator II to the Peninsula of Paria.

An interesting problem on this component map is the identification of the Orinoco. The river itself is not shown, and neither is the modern delta. Instead, two estuaries extending far inland (for a distance of about 100 miles) are shown close to the site of the present river. The longitude on the grid would be correct for the Orinoco, and the latitude is also quite accurate (see Table 1). Is it possible that these estuaries have been filled in, and the delta extended this much since the source maps were made? If so, this is comparable to the extension of the delta of the Tigris-Euphrates in Mesopotamia in the last 3,500 years, since Ur of the Chaldees lay on the seacoast.

6. A map from the Gulf of Venezuela to Yucatán, omitting about 7° of coastline between the Gulf of Venezuela and the Peninsula of Paria.

A point of considerable importance here is the shape of the Atrato River. According to our grid, the river is shown for a distance of 300 miles from the sea, and its eastward bend at about 5° North Latitude corresponds to the geographical facts. This implies that somebody explored the river to its headwaters in the Western Cordillera of the Andes sometime before 1513. I have found no record of such an early exploration. Yucatán supposedly had not been discovered in 1513, either.

7. The Caribbean Islands: The Leeward and Windward Groups, the Virgin Islands, Puerto Rico on the main grid of the map; more questions about Columbus.

These islands are more accurately placed on this map, in reference to latitude and longitude, than they are on any other map of the period.

Piri Re'is wrote, in his long inscription about Columbus, that this part of the map was based on a map Columbus drew. Here the two different grids overlap to some extent: Some islands are on the special grid already discussed, and some are on the main grid. I have pointed out that one of Columbus' errors may have been due to not understanding the special grid. The Leeward and Windward Islands, which Columbus discovered, are on the main grid on this component map.

Nevertheless, it is hardly possible that he could have added them to the map, as Piri Re'is supposed. For we see them in remarkably correct latitude and longitude on the trigonometric grid of our hypothetical Alexandrian compilation. Not understanding the grid, not even dreaming of its existence, and not being able to find either correct latitude or correct longitude, how could Columbus have correctly located the islands? Piri Re'is gives names to these islands, and says that they are the names given by Columbus, yet the names are wrong! (140:408–409) It looks as if Piri Re'is here depended upon hearsay information and did not really see a map drawn by Columbus.

One group of islands on the Caribbean part of the map, the Virgin Islands, are so far out of position, so badly drawn, and so far out of scale that they might well have been added to the map by Columbus or interpolated by Piri Re'is on the basis of some contemporary report.

One of the most unusual features of this part of the map is that some features can be interpreted as two different localities, according to the grid one uses.

8. The lower east coast of South America from Bahia Blanca to Cape Horn (or Cape San Diego) and certain Atlantic islands on the main grid of the map.

Two of my students, Lee Spencer and Ruth Baraw, discovered that about 900 miles of the east coast of South America were simply missing from the Piri Re'is Map, two different source maps having apparently been erroneously put together on the general compilation. Earlier students of the map—Kahle, Goodwin, Mallery —had all assumed that the map was continuous and complete as far as it went.

Kahle's assumption of an unbroken coast required a rather forced interpretation of the map. On this assumption it was necessary to conclude that the mapmaking here was very bad. However, it seems that someone before Kahle had had the same idea. Figure 44 shows how that interpretation actually fits the oblong grid of the map. The equator is different from that of the main projection, but the length of the degree of latitude has been increased in the same way. This detail serves to support our impression of the long and complex history of this map. There is no way of knowing how many peoples of how many epochs had their fingers in the pie.

The method used by Spencer and Baraw to verify their observation of the omission of the coastline was to try identifying localities by comparison with the modern map, first from one end of the coast, and then from the other. They started first with Recife and went all the way down the coast from point to point. Everything went well as far as Cape Frio, but south of Cape Frio they thought the Piri Re'is Map ceased to correspond with the modern map at all. Then they started from the bottom, from what we assumed to be Cape Horn, or Cape San Diego (No. 74, Figure 18), and went northward identifying localities. Here again everything seemed to agree very well with the modern map until they came to a point just below Cape Frio. Farther than this they could not go. The missing coast lay in between. Our grid assisted us very much in the final verification of the break, for it gave us its value in degrees.

The omission of the coast between Cape Frio and Bahia Blanca apparently resulted in a loss of about 16° of South Latitude and about 20° of West Longitude. Therefore, in Table 1, I have added these amounts of latitude and longitude to the ones found by our grid. When this is done, the positions of the identified localities are correct to an average error of less than a degree. More important is the fact that they are correct relative to each other.

It appears significant that Piri Re'is, who stuck names taken from explorers'

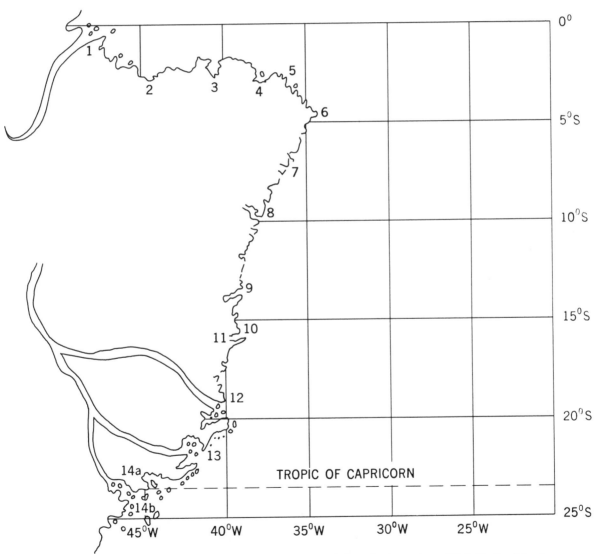

Figure 44. Alternative Grid for the coast of South America (Piri Re'is Map).
See Table 15.

accounts on much of his map (making numerous errors), did not attempt to place any names on the southern part of this coast of South America. The reason offers itself: *There were no explorers' accounts.* That coast had not been explored by 1513.

The Falkland Islands appear in this section of the map at the correct latitude relative to this lower east coast, but there is an error of about 5° in longitude. The Falklands are supposed to have been discovered by John Davis in 1592, nearly eighty years after Piri Re'is made his map (68:869).[5]

South of Cape Horn, or Cape San Diego, the coast on the Piri Re'is Map appears to continue unbroken, but here we have been able to identify another break, or rather omission.

9. The Antarctic.

Proceeding as in the case of the break in the east coast of South America, we first identified localities down to the vicinity of Cape Horn (including specifically Cape San Diego), then jumped to the next cape to the eastward, assuming as a working hypothesis that it was the Palmer or Antarctic Peninsula as claimed by Mallery. This assumption would require that the sea between the Horn and the Antarctic Peninsula had been omitted by the mapmaker. This assumption appeared to be supported by our identification of the Shetland Islands. These islands are not far off the Antarctic coast. The omission of the sea between (Drake Passage) automatically would put the South Shetlands too far north by the width of the strait, which happens to be about 9°. If the reader will compare the positions of the Falklands and the South Shetlands on a globe with their positions on the Piri Re'is Map, as we have identified them, he will see how the Antarctic coast seems to have been simply pushed northward, and Drake Passage omitted.

Interestingly enough, we find that the same mistake was made on all maps of the Renaissance showing the Antarctic. When we come, in the next chapter, to the examination of the map of Oronteus Finaeus, we shall discover the probable reason for this error.

The extraordinary implications of Captain Mallery's claim that part of the Antarctic Continent is shown on the Piri Re'is Map demand unusually thorough verification, considering that the continent was supposedly discovered only in 1818. This is no slight matter. Important questions, for geology as well as for history, depend upon it. We may begin with a brief survey of the historical background.

A good many world maps of the 16th Century show an antarctic continent.[6]

---

[5] Though some have given the credit to Amerigo Vespucci.

[6] A few of these, in addition to those discussed later in this book, are the maps of Robert Thorne (1527), Sebastian Munster (1545), Giacomo Gastaldi (1546), Abraham Ortelius (1570), Plancius (1592), Hondius (1602), Sanson (1651), Seller (1670). (See R. V. Tooley [206].)

As we shall see, Gerard Mercator believed in its existence. A comparison of all the versions suggests that there may have been one or two original versions, drawn according to different projections, which were copied and recopied with emendations according to the ideas of different cartographers.

The belief in the existence of the continent lasted until the time of Captain Cook, whose voyages into the South Seas demonstrated the non-existence of a southern continent at least in the latitudes where one appears on these maps (112). The idea of an antarctic continent was then given up, and geographers began to explain the maps as the work of geographers who had felt the need to have a land mass at the South Pole to balance off the concentration of land in the northern hemisphere. This seemed to be the only reasonable explanation, for in the first place there apparently was no such continent, and in the second place there was no reason to suppose that anyone in earlier times (Romans, Greeks, Phoenicians) could have explored those distant regions.

When we began our study of the southern sector of the Piri Re'is Map our first step was to compare it carefully, not with a flat map of the Antarctic, but with a globe. Figure 45, traced from a photograph of a globe,[7] shows a striking similarity between the Queen Maud Land coast and the coastline on the Piri Re'is Map. It should be especially noted that on the modern globe the Queen Maud Land coast lies due south of the Guinea coast of Africa, just as the coastline referred to by Mallery does on the Piri Re'is Map.[8]

This was an encouraging beginning. We went on to make a thorough examination. We asked ourselves, first, how does the coast in question on the Piri Re'is Map compare in its extent, character, and position, with the coasts of Queen Maud Land? (These coasts are named the Princess Martha and Princess Astrid Coasts.) With the gradual development of the mathematical grid we could answer two of these questions.

In the first place, we found that the Piri Re'is coast, according to our grid, extends through 27° of Longitude as compared with 24° on the modern map, a very remarkable degree of agreement. At the latitude of the coasts (about 70° S) a degree of longitude is only about 20 miles, so that the error is not great. The grid also shows the coast in good position; it is about 10°, or 200 miles, too far west.

---

[7] We took this step because flat maps distort geography in one way or another, and unless we found a map on precisely the right projection we could not be sure of a good comparison.

[8] Following publication of Mallery's views in a broadcast of the Georgetown University Radio Forum of the Air (131), the French publication, "Science et Vie" (109), published a very confused account, which was taken up in the Soviet press (28). A Soviet scientist rejected Mallery's claim, perhaps because he made the mistake of confusing the Queen Maud Land coast with the Queen Maud Mountains on the other side of the continent.

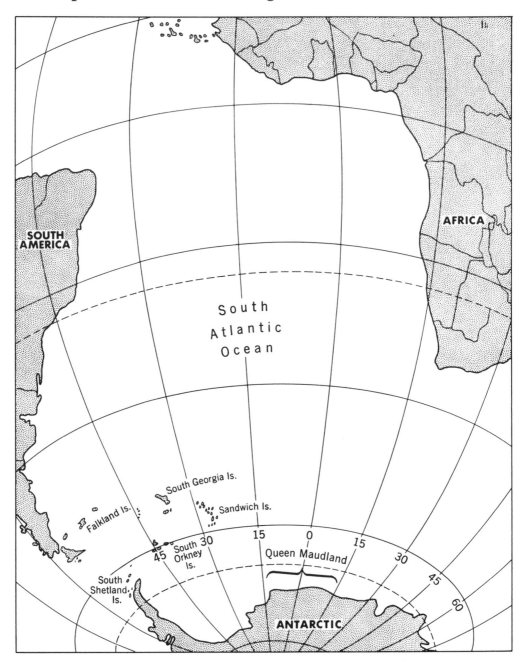

Figure 45. Relative Longitudes of the Guinea Coast of Africa and the Queen Maud Land Coast of Antarctica, for comparison with the Piri Re'is Map.

With regard to latitude, we must take account of the omissions we have noted above—part of the South American coast and Drake Passage. Together these omissions account for about 25° of South Latitude. When these degrees are added to those found by the grid for the Queen Maud Land coast, the coast appears in correct latitude (see Fig. 18).

We have noted that the omission of the South American coastline resulted in a loss of about 16° of West Longitude. The omission of Drake Passage resulted, we found, in adding about 4° to this, making 20° to be accounted for. This, with the 10° westward error of the Queen Maud Land coast, creates a deficit of some 30° between that coast and the Antarctic Peninsula. This appears to be made up for by the fact that the Weddell Sea, as we have identified it on the map, extends through only 10° of longitude, instead of 40°, as would be correct.

Now it might be argued that this result is artificial, and that we have deliberately twisted the evidence to support the conclusion, but this is not the case. My students, Lee Spencer and Ruth Baraw, had already established the omission of 900 miles of the South American coast without any thought of Antarctica. They were not interested in the bearing of their discovery on the question of the Queen Maud Land coast. We did not even see the connection until long afterwards, when the grid was worked out, and the same is true of the omission of Drake Passage. The omission is obvious from the map itself: the strait simply isn't there. In the case of both omissions we were able to measure approximately the amounts of latitude and longitude involved.

There is in addition the comparison of the character of the Queen Maud Land coast, as shown on the ancient and on the modern map. It is plain, from the modern map, that this coast is a rugged one. Numerous mountain ranges and individual peaks show up above the present levels of the ice. The Piri Re'is Map shows the same type of coast, though without any ice. The numerous mountains are clearly indicated. By a convention of 16th Century mapmaking heavy shading of some of the islands indicates a mountainous terrain (Frontispiece).

Coming to greater detail, Mallery's chief argument was the striking agreement of the map with the seismic profile across Queen Maud Land (see Figs. 46, 47, 48, and Note 10). The reader will note that the profile shows a rugged terrain, a coastline with mountains behind the coast and high islands in front. The points of the profile below sea level coincide very well with the bays between the islands on the Piri Re'is Map. This amounts to additional confirmation. The identification of specific features of the coast, as shown in Table 1, appears further to strengthen the argument.

If the Piri Re'is Map stood alone, it would perhaps be insufficient to carry conviction. But it does not stand alone. We shall shortly see that the testimony of this map regarding the Antarctic can be supported by that of several others.

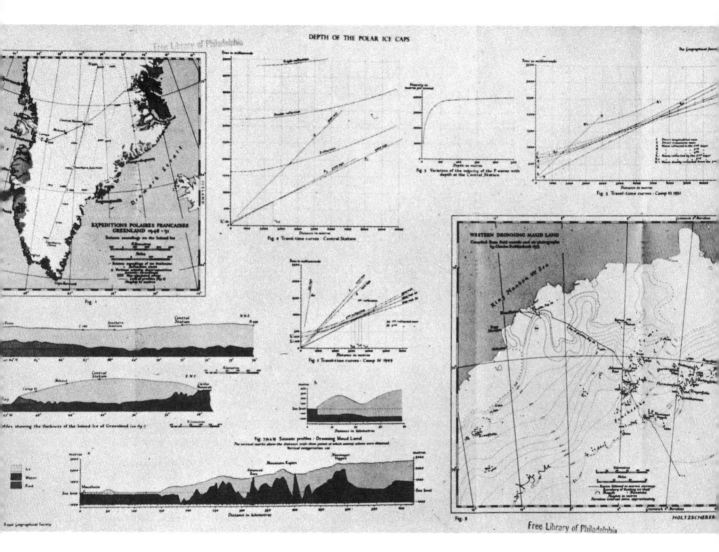

Figure 46. Cross sections of the Greenland ice cap and the Antarctic ice cap across Queen Maud Land, showing depths of the ice. (From "The Geographical Journal," June, 1954)

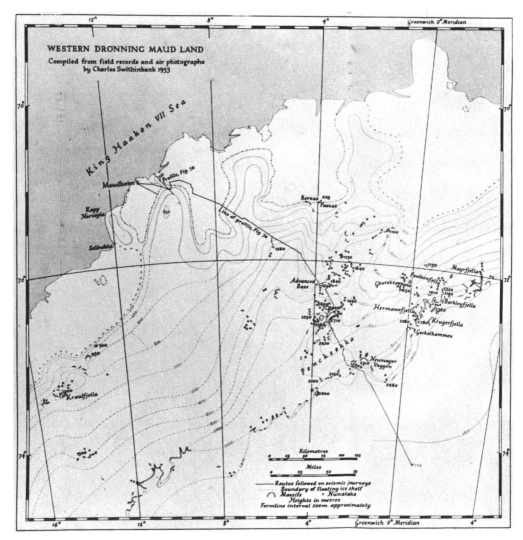

Figure 47. Route of the Norwegian-British-Swedish Seismic Survey Party across Queen Maud Land, 1949. (See Note 10)

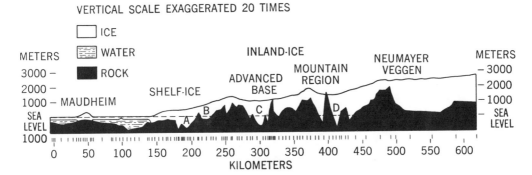

Figure 48. The profile of the Queen Maud Land ice cap: note the extensions of the ice cap below sea level, A, B, C, D. Compare with the islands and bays of the Antarctic sector of the Piri Re'is Map (Figure 18). (After Schytt)

CHAP
TER IV
THE
ANTARCTIC
MAPS OF
ORONTEUS
FINAEUS
HADJI AHMED
AND
MERCATOR

1. Oronteus Finaeus.

A part of our Piri Re'is investigation, quite naturally, was a search for other portolan charts of the Middle Ages and the Renaissance that might show Antarctica. Quite a number of these turned up, for, as we have mentioned, many cartographers of the 15th and 16th Centuries believed in the existence of a southern continent.

In the course of this investigation I arranged to spend some time in the Library of Congress during the Christmas recess of 1959–1960. I wrote ahead to the Chief of the Map Division asking if all the old maps of the periods in question could be brought out and made ready for my inspection, especially those that might show the Antarctic. Dr. Arch C. Gerlach, and his assistant, Richard W. Stephenson, and other members of the staff of the Map Division were most co-operative, and I found, somewhat to my consternation, that they had laid out several hundred maps on the tables of the Reference Room.

By arriving at the Library the moment it opened in the morning and staying there until it closed in the evening, I slowly made a dent in the enormous mass of material. I found many fascinating things I had not expected to find, and a number of portolan charts showing the southern continent. Then, one day, I turned a page, and sat transfixed. As my eyes fell upon the southern hemisphere of a world map drawn by Oronteus Finaeus[1] in 1531, I had the instant conviction that I had found here a truly authentic map of the real Antarctica.

The general shape of the continent was startlingly like the outline of the continent on our modern map (see Figs. 51 and 52). The position of the South Pole, nearly in the center of the continent, seemed about right. The mountain ranges that skirted the coasts suggested the numerous ranges that have been discovered in Antarctica in recent years. It was obvious, too, that this was no slap-dash creation of somebody's imagination. The mountain ranges were individualized,

---

[1] See Notes 11, 12, 13.

Figure 49. The Oronteus Finaeus World Map of 1532.

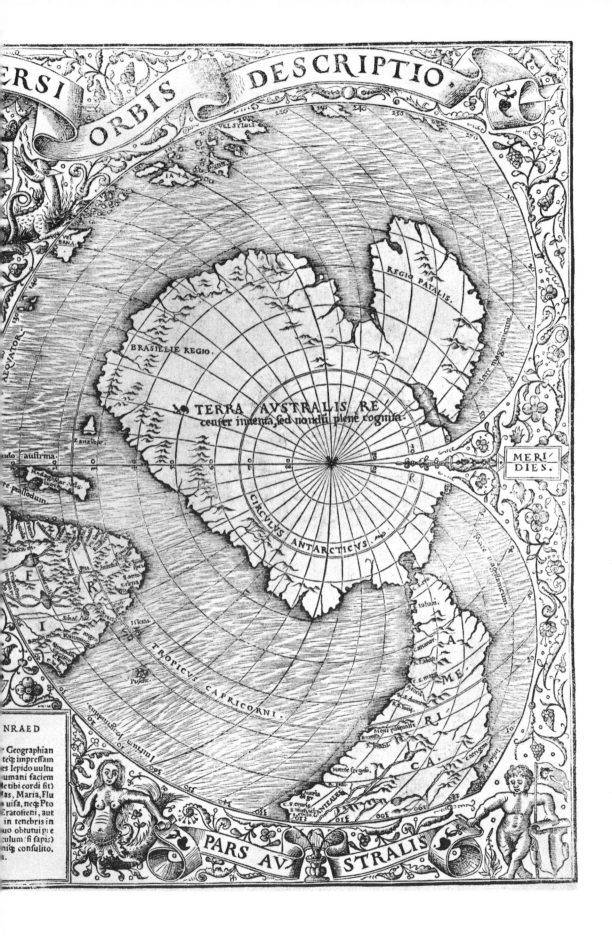

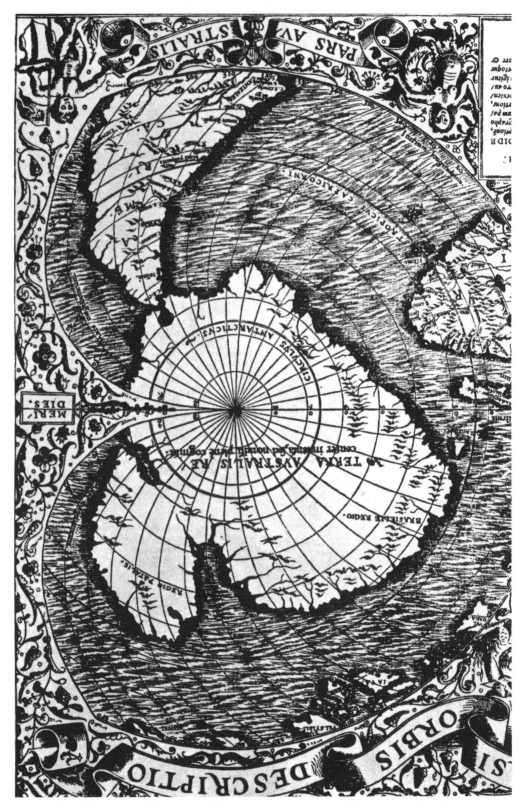

Figure 50. The Oronteus Finaeus World Map of 1532, Antarctica.

some definitely coastal and some not. From most of them rivers were shown flowing into the sea, following in every case what looked like very natural and very convincing drainage patterns. This suggested, of course, that the coasts may have been ice-free when the original map was drawn. The deep interior, however, was free entirely of rivers and mountains, suggesting that the ice might have been present there.

At the beginning of our study we made a comparison of the proportions of Antarctica as this map shows them with those shown on modern maps. I measured two traverses across the continent on modern maps and compared their ratio with the ratio of the same traverses on the map of Oronteus Finaeus. These traverses were (a) from the Antarctic (Palmer) Peninsula at 69° S and 60° W[2] to the Sabrina Coast of Wilkes Land at 66° S and 120° E; and (b) from the Ross Sea (Queen Maud Range) at 85–88° S and 180° E/W, to the Muhlig–Hofmann Mountains, in Queen Maud Land, at 72° S and 0° E/W, measuring in centimeters. On the Oronteus Finaeus Map (a small one) I measured in millimeters (since the ratios alone counted) with the following results:

The Modern Map

Palmer Peninsula to the Sabrina Coast . . . .   78.5 cm.
Ross Sea to Queen Maud Coast . . . .   38.0 cm.

The Oronteus Finaeus Map

Palmer Peninsula to the Sabrina Coast . . . .   129.0 mm.
Ross Sea to the Queen Maud Land Coast . . . .   73.0 mm.

$$\text{Thus we have: } \frac{38:78.5}{73:129} = \frac{2.06}{1.76}, \text{ or a ratio of 8:7.}$$

It is improbable that this close agreement is accidental.

Examining this map of Antarctica on the grid of latitude lines drawn by Oronteus Finaeus, we observed that he had extended the Antarctic Peninsula too far north by about 15°. At first I thought he might simply have placed the whole continent too far north in the direction of South America. Further examination, however, showed that the shores of his Antarctic Continent extended too far in all directions, even reaching the tropics! The trouble, it would seem, therefore, was with the scale. By using an oversized map the compiler was forced to crowd the Antarctic Peninsula up against Cape Horn, squeezing out Drake Passage almost entirely. Furthermore, the mistake must have been made far back, for we find the identical error in all the Antarctic maps of the period, including that of Piri Re'is.

---

[2] We measured from the beginning of the broad part of the Peninsula because study seemed to show that the upper, narrow part of the Peninsula was omitted from the Oronteus Finaeus Map, as it apparently also had been from the Piri Re'is Map.

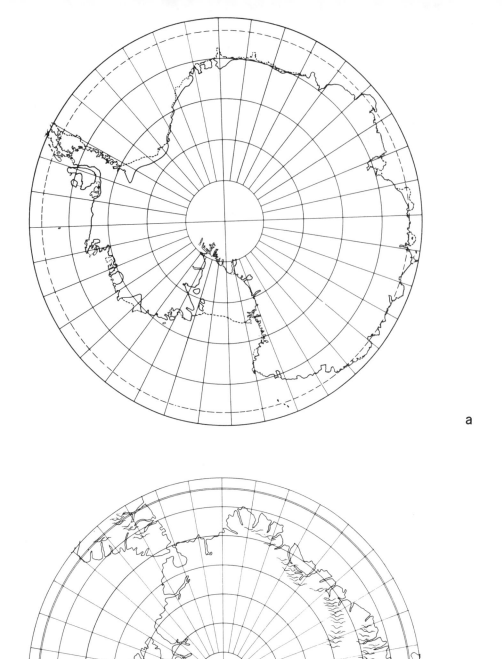

a

c

Figure 51. Four maps of Antarctica: (a) a modern map, (b) the Oronteus Finaeus
Map, (c) the Oronteus Finaeus Map redrawn on a projection similar to that of
the modern map, (d) Antarctica from the Schoner Globe of 1523–1524.

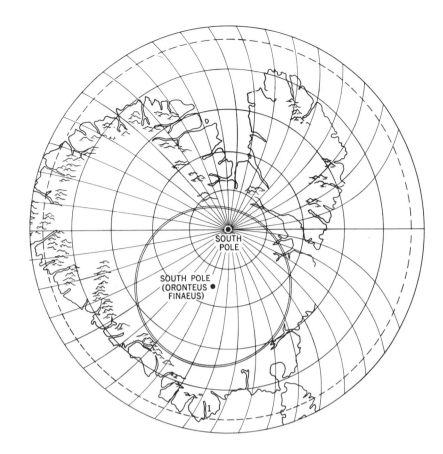

b

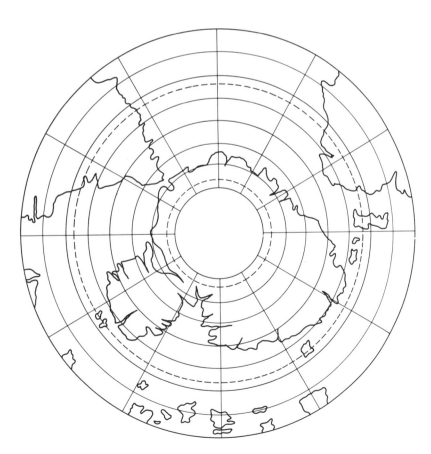

d

(a)

(b)

(c)

(d)

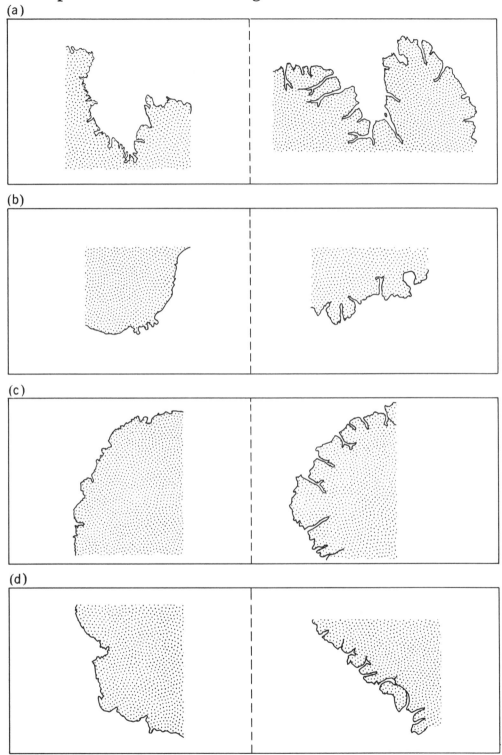

Figure 52. Antarctic coasts of the Oronteus Finaeus Map, (right), compared with those of the modern map, (left).

It is possible, indeed, that this mistake may account for the omission at some ancient period, on the source map used by Piri Re'is, of a large part of the coastline of South America: There was simply no room for it!

As our study continued, it gradually began to appear that Oronteus Finaeus' network of parallels and meridians did not fit the Antarctic as shown on his world map. Apparently a projection had been imposed by him on a source map originally drawn with a very different kind of map grid. How were we to discover the nature of this original network of parallels and meridians?

The first step seemed obvious. It was simply to remove the network of lines applied to the map by Oronteus Finaeus. We made a tracing of the map, leaving off these lines but retaining, for the moment, his position for the South Pole, and his Antarctic Circle. Since he could have had no way of knowing the position of the pole in the interior of the continent, we considered that his source map must have shown the pole.

The position of the pole looked quite correct at first glance, as I have mentioned, but, as our study and comparison of the old map with modern maps continued, we could see that the mapmaker had apparently made a mistake of a few degrees in locating the pole. We found what seemed a truer position by measuring across the continent in several directions and finding the position that would divide all the diameters of the continent in approximately the same ratio as shown on modern maps. This was, of course, an extension of our first measurement already mentioned. It was only an empirical experiment, but it seemed to give a more satisfactory result in terms of the latitudes of identifiable places.

With our adjusted pole as a center, I now constructed a grid on the supposition that the original projection might have been the equidistant polar projection, one that is said to have been known in ancient times (see Fig. 25). In this system the meridians are straight lines radiating from a pole. The parallels of latitude are circles. In order to fix the latitudes I had to find one circle at a known distance from the pole. The obvious thing to do was to locate the Antarctic Circle, which is approximately 23½° from the pole, by comparing the old map with the new. It so happens that Antarctica is circular and lies almost within the Antarctic Circle. It was comparatively easy to draw about the continent on the old map a circle that would pass at about the right distances from the various coasts, as compared with a modern map. This was, in fact, one method we used to relocate our pole.

Since the Antarctic Circle is 23½° from the pole, it was now possible to measure out one degree by dividing the distance to the pole on our draft map by 23½. With the length of the degree thus determined, we could then lay out circles 10° apart: the 80th and 70th parallels of latitude. Now we had the parallels necessary for our grid.

When it came to the meridians, we had to deal with another problem. It did not seem to us at first that the continent was properly oriented in relation to the other continents. To get correct longitude readings for our Antarctic coasts on the old map we naturally had to line it up with the meridians on the modern map. It was possible, of course, that, if we were dealing with an authentic map of Antarctica that had survived for several millennia, somebody could have placed it askew on a world map. We thought it looked as if the continent ought to be rotated about 20° to the east to bring it into correct relationship to the other continents. We selected empirically what looked like a reasonable "prime meridian" and then laid out the other meridians at five-degree intervals, thus constructing our grid.

At this point we made another vital discovery. I noticed that the circle we had drawn for the 80th parallel was almost exactly the size of the circle Oronteus Finaeus had drawn on his map and labeled *Circulus Antarcticus*—Antarctic Circle. The true Antarctic Circle follows a path in the sea off the Antarctic coasts; this Antarctic Circle of Oronteus Finaeus, on the other hand, was in the center of the continent. This suggested that Oronteus Finaeus, or a predecessor, in interpreting some old source map may have mistaken for the Antarctic Circle a circle upon the map intended to represent the 80th parallel. This mistake would have exaggerated the size of Antarctica about four times. Since every Renaissance map of the Antarctic seems to reflect this mistake, it is highly likely that the error goes back to Alexandria, or to some earlier period.

A very extraordinary aspect of this matter is that, with the correction of scale, the size of the Antarctic Continent on the map of Oronteus Finaeus is correct, by modern findings. The reader may check this matter by comparing the distribution of the land masses inside and outside the Antarctic Circle as it is shown on the ancient and modern maps (see Figs. 53 and 54).

The reader may well ask how it could happen that an ancient map, a map ancient even in classical times, could have had parallels of latitude indicated at ten-degree intervals, when this method of counting by tens and using a circle divided into 360 degrees was supposedly only applied to maps in the Renaissance. This question will be answered in connection with another matter. Meanwhile, the presumption that the ancients had a correct idea of the size of the Antarctic Continent suggests that they may also have had a correct idea of the size of the earth, knowledge that appears, indeed, to be reflected in the Piri Re'is Map.

Once we had a grid constructed, as I have described above, we tried to identify, by comparison with modern maps, as many places on this map of Antarctica as possible. The result was electrifying. All the errors of the location of places that we had identified on Oronteus Finaeus' own grid were greatly reduced. Some of the tentative identifications we had made on the basis of his grid had to be given up, but many new places were identified, so that our list of identified geographical features in Antarctica was increased from sixteen to thirty-two. For this grid we abandoned our empirically derived 80th parallel and simply used Oronteus Finaeus' own so-called *Circulus Antarcticus* as our 80th parallel. We found that by doing this we improved the accuracy of the grid. In other words, it seemed more clear than ever that the circle, misnamed by some early geographer, had originally been intended to be the 80th parallel and nothing else.

However, notwithstanding the amazing accuracy in the positions of many places, there were still numerous errors. We continued to experiment with rotating the continent a few degrees one way or the other, and changing by ever so little the position of the pole, but there were still plenty of discrepancies.

Then it appeared that this ancient map of Antarctica seemed to have been put together, like the Piri Re'is Map, from a number of local maps of different coasts, and perhaps not put together correctly. An analysis of the errors in our tables showed that, so far as longitudes were concerned, the errors differed in direction in different parts of the map. The average of longitude errors in Wilkes Land, for example, was easterly, while in the Ross Sea area and Victoria Land it was westerly. I had a transparent overlay made of the Oronteus Finaeus Map, so that we could place it over a modern map, and shift it around as we pleased. We found that the Oronteus Finaeus Map could be aligned remarkably well with a modern map, but we had to shift it around to different positions to make the individual segments of the coasts fit. It seemed impossible to make all the coasts fit at once. (See Fig. 52.) It seemed clear that we had in hand a compilation of local maps made by people who were not as well acquainted with the area as those who had originally mapped the separate coasts.

As I have mentioned, we worked for a long time on the assumption that the original projection, on which the compilation had been made, was of a sort that had meridians that were straight lines. But, on this basis, we were never able to get a satisfactory alignment of Antarctica with other continents. I was therefore forced to finally consider the possibility that the meridians might have been curved like those that actually appear on the Oronteus Finaeus Map. And so it turned out. With a grid redrawn on this basis (see Fig. 53) the identified places on the map were increased in number from thirty-two to fifty, and the averages of errors again were reduced, as shown in Table 2.[3]

At this point we should pause to consider in somewhat greater detail the obviously serious question of the ice cap which now covers the whole continent. We are not here concerned with the geological problem of accounting for a warm period in Antarctica within the lifetime of the human race. Rather, we are concerned with just what the map shows. It would appear that the map shows non-glacial conditions extending for a considerable distance inland on some of the coasts. These coasts include, it seems, the coasts of Queen Maud Land, Enderby Land, Wilkes Land, Victoria Land (the east coast of the Ross Sea), and Marie

---

[3] This finding, of course, affects very much our visual comparison of the ancient and modern maps. Since they appear to have been drawn on different projections they would naturally look different, even if they were identical. Therefore, the agreement of the two may actually be greater than it appears. Table 2 indicates this. In Fig. 51c Charles Halgren of the Caru Studios in New York has redrawn the Oronteus Finaeus Map on a projection using straight meridians. This may be compared with the modern map.

Byrd Land. Notably lacking in definite identifiable points are the west coast of the Ross Sea, Ellsworth Land, and Edith Ronne Land.

A comparison of the Oronteus Finaeus Map with the map of the *subglacial* land surfaces of Antarctica produced by survey teams of various nations during the International Geophysical Year (1958) seems to explain some of the apparent shortcomings of the Oronteus Finaeus Map, and at the same time throws some light on the question of the probable extent of glacial conditions when the original maps were drawn.

Figure 55 shows what the IGY teams discovered the actual land forms under

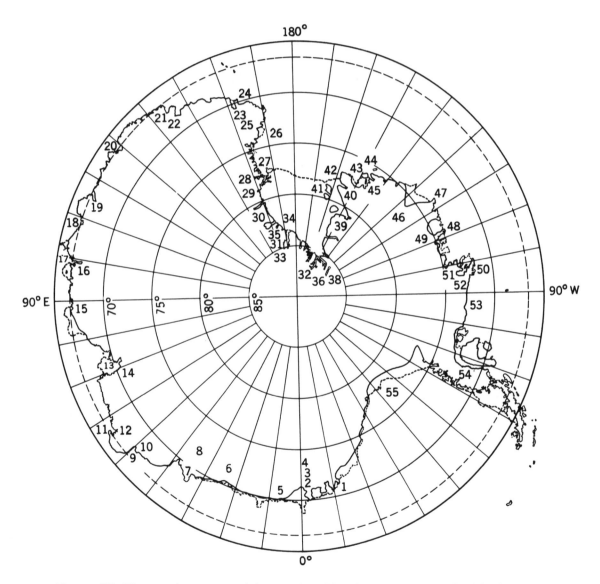

Figure 53. The modern map of Antarctica. Numbers correspond with those on the Oronteus Finaeus Map, Figure 54. See Table 2.

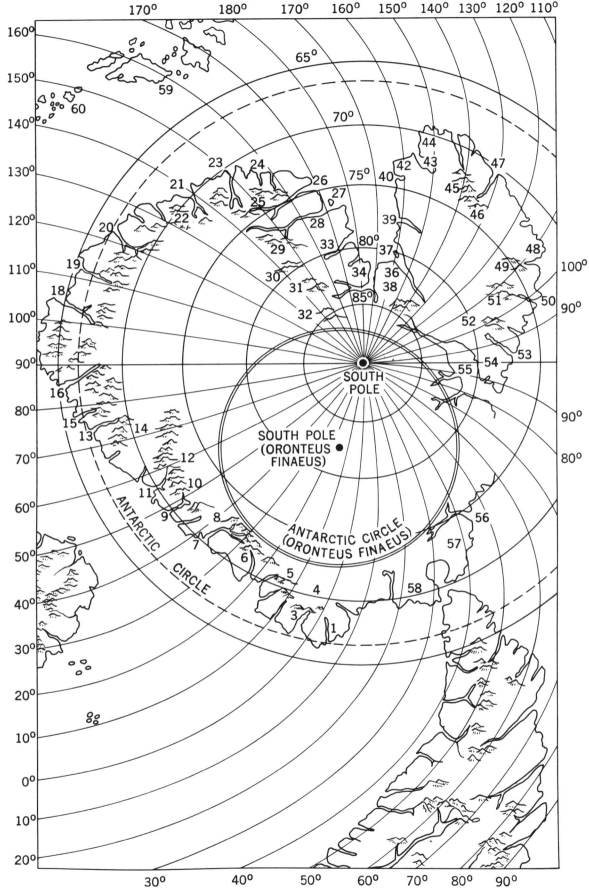

Figure 54. The Oronteus Finaeus Map of 1531: tracing with projection lines as determined by the study of the topography. See Table 2.

the present Antarctic ice cap to be. It is noticeable that, contrary to surface appearances, there is no western shore to the Ross Sea; rather, the rock surface of the continent is below sea level straight across from the Ross Sea to the Weddell Sea, and most of Ellsworth Land is also below sea level. If the ice cap melted, all these areas would be shallow sea—not land.

It is plain, of course, that if the western coast of the Ross Sea and the coast of Ellsworth Land are, in fact, non-existent, the absence of definite physical features in these sections of the Oronteus Finaeus Map is well explained. But, it seems that the ice cap may already have been in existence at least in West Antarctica when the original maps were drawn, for the interior waterways connecting the Ross, Weddell, and Amundsen Seas are not shown.

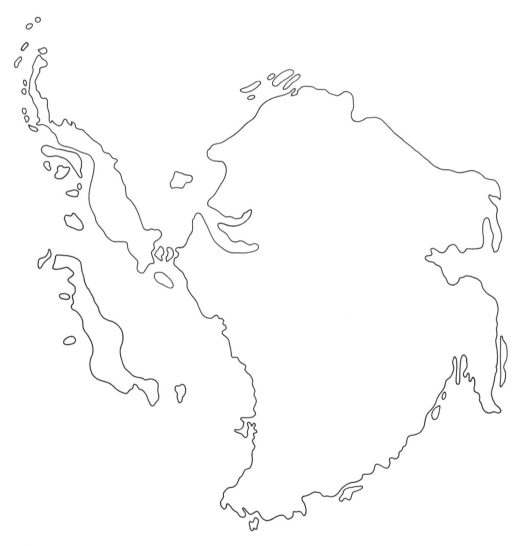

Figure 55. Recent map of sub-glacial topography of Antarctica, as shown by seismic surveys carried out during the International Geophysical Year.

The Antarctic (Palmer) Peninsula presents a point of special interest. As already noted, only the base of the Peninsula can be identified on the Oronteus Finaeus Map. The upper part of the Peninsula is missing. We find, now, from the results of the IGY investigations, that there is, in fact, no such peninsula. There is, in fact, what would be an island if the ice cap melted. It would seem, then, that even if a great deal of ice was already in existence when the original map of this portion of the Antarctic was drawn, the ice cap had not yet covered the area of shallow sea between the continental shore and this island.

It must, of course, be remembered that thousands of years may have elapsed between the drafting of the earliest and latest of the original maps of different parts of Antarctica. We cannot therefore draw the conclusion that there was a time when there was a great deal of ice in East Antarctica and none in West Antarctica. The maps of East Antarctica may have been drawn thousands of years later than the others.

Another very extraordinary map may serve to throw some light on this. Buache, the 18th Century French geographer already referred to, left a map of Antarctica that may show the continent at a time when there was no ice at all (see Fig. 56). Compare this with the IGY map of the land masses (Fig. 55). If an apparent error in the orientation of the continent to other land masses is disregarded, it is quite easy to imagine that this map shows the waterways connecting the Ross, Weddell, and Bellingshausen Seas.

When we discovered that the meridians of the original map were curved, essentially as Oronteus Finaeus had constructed them, it was no longer necessary to rotate his map of the Antarctic eastward in order to bring it into agreement with the other land masses. Instead, it became apparent that his source map of South America and his source map of Antarctica probably came to him in one piece. Their relative longitudes were correct.

The eastern hemisphere on the Oronteus Finaeus Map of 1531 in no way compares with the Antarctic and South American parts. He seems to have based his Mediterranean, for example, on the inaccurate Ptolemy maps rather than on the portolanos.[4]

Among the most remarkable features of the Oronteus Finaeus Map is the part we identify as the Ross Sea. The modern map indicates (dotted lines in Fig. 53) the places where great glaciers, like the Beardmore and Scott Glaciers, pour down their millions of tons of ice annually to the sea. On the Oronteus Finaeus Map (Fig. 50), fiord-like estuaries are seen, along with broad inlets and indications of rivers of a magnitude that is consistent with the sizes of the present glaciers. And some of these fiords are located remarkably close to the correct positions of the glaciers (see Table 2).

---

[4] For further discussion of the projection of the Oronteus Finaeus Map see Notes 11, 12, 13, 22.

Figure 56. Sub-glacial topography of Antarctica, suggested by the 18th century French geographer Buache.

CARTE
DES TERRES AUSTRALES,
Comprises entre le Tropique
du Capricorne et le Pôle Antarctique.

Où se voyent les Nouvelles découvertes
faites en 1739. au Sud du Cap de
Bonne Esperance. Par les Ordres de
M.rs DE LA COMPAGNIE DES INDES.

Dressée sur les Mémoires et
sur la Carte Originale de
M.r de Lozier Bouvet
chargé de cette Expedition.

Par Philippe Buache
de l'Académie R.le des
Sciences, Gendre de
feu M.r Delisle P.er
Géographe du ROY
De la même Acad.
Augmentée
de diverses vues
Physiques &c.
1754

PLAN
et VUE des Terres
DU CAP DE LA
CIRCONCISION

Situé à 54 degrés
de Latitude Merid.le
Et environ à 28 deg.
30 min de Longitude.

TERRE qui s'étend
Ras-ti a l'E N E. et à o li.
au S E.
Circoncision.

Cette Variaõn a été Observée
de 9° 15' sur un Compas
et de 8° sur un autre le
10 dec.1738 à la lat de
51° 26' et à 25° 45'
de Longitude.

Cap de la

Glaces vûes en Janvier 1739.

Variation 6° 45' au M.di.

Glaces.

Ces Glaces ont paru avoir 2 à 300 pieds de haut. Et
depuis une demie lieue jusqu'à 2 ou 3 li de tour.

Lieues

d'Amsterdam
Roterdam
Capricorne

AFRIQUE
LES HOTTENTOTS
C. das Voltas
B. S. Elene
B. de Saldana
B. de la Paltate
C. DE BON E DE BON
ESPERANCE
Pic
L. de Goetin

Roy du Sofala
MONOMOTA
L. Bourbon
I. de France
MER DES INDES

Terre des
Perroquets

Terre Comore

ANTARCTIQUE

AUSTRAL

Baye de Frederic Henry
C. du Sud
L. de Schouten
Terre
de Diemen
découverte
par Tasman
I. Maria
Vandieven

NOUVELLE HOLLANDE

Terre de
Leuwin
Terre d'Eendracht
Terre de la Concorde
Terre de Nuits
Terre de Wit
I. S. François
I. S. Pierre

sous le Privilège de l'Académie R.le des Sciences le 3 Septembre 1739.

The open estuaries and rivers are evidence that, when this source map was made, there was no ice on the Ross Sea or on its coasts. There had also to be a considerable hinterland free of ice to feed the rivers. At the present time all these coasts and their hinterlands are deeply buried in the mile-thick ice cap, while on the Ross Sea itself there is a floating ice shelf hundreds of feet thick.

The idea of a temperate period in the Ross Sea in time so recent as is indicated by this map will, at first acquaintance, be incredible to geologists. It has been their view that the Antarctic ice cap is very ancient, perhaps several million years old, although, curiously enough, it seems that previously in the long history of the globe the climate of Antarctica was often warm and sometimes even tropical (85:58–61).[5]

In answer to this possible objection I can cite, in addition to the map itself, only one further piece of evidence, but it is a very impressive piece of evidence indeed. In 1949, on one of the Byrd Antarctic Expeditions, some sediments were taken from the bottom of the Ross Sea, by coring tubes lowered into the sea. Dr. Jack Hough, of the University of Illinois, took three cores to learn something of the climatic history of the Antarctic. The cores were taken to the Carnegie Institution in Washington, D.C., where they were subjected to a new method of dating developed by the nuclear physicist Dr. W. D. Urry.[6]

This method of dating is called, for short, the ionium method. It makes use of three different radioactive elements found in sea water. These elements are uranium, ionium, and radium, and they occur in a definite ratio to each other in the water. They decay at different rates, however; this means that when the sea water containing them is locked up in sediments at the bottom of the ocean and all circulation of the water is stopped, the quantities of these radioactive elements diminish, but not at the same rate. Thus, it is possible, when these sediments are brought up and examined in the laboratory, to determine the age of the sediments by the amount of change that has taken place in the ratios of the elements still found in the sediments.

The character of sea-bottom sediments varies considerably according to the climatic conditions existing when they were formed. If sediment has been carried down by rivers and deposited out to sea it will be very fine grained, more fine grained the farther it is from the river mouth. If it has been detached from the earth's surface by ice and carried by glaciers and dropped out to sea by icebergs, it will be very coarse. If the river flow is only seasonal, that is if it flows only in summer, presumably from melting glaciers inland, and freezes up each winter, the sediment will be deposited somewhat like the annual rings in a tree in layers or "varves."

All these kinds of sediments were found in the cores taken from the Ross Sea

---

[5] See Chapter VII.

[6] Not to be confused with another nuclear physicist, Dr. Harold D. Urey, of the University of Chicago.

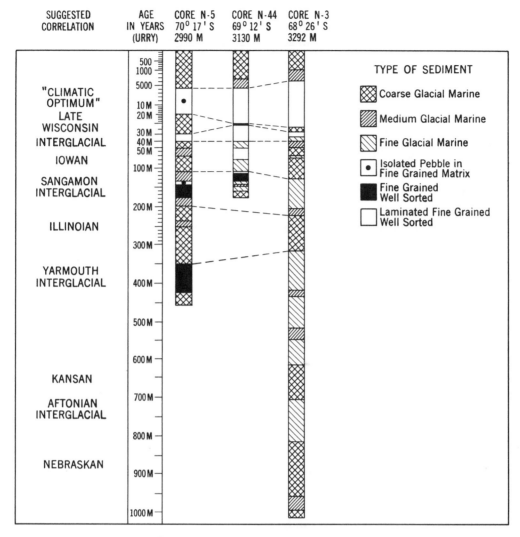

Figure 57. The Ross Sea cores.

bottom. As you will see from the illustration (Fig. 57) there were many different layers of sediment in the coring tubes. The most surprising discovery was that a number of the layers were formed of fine-grained, well-assorted sediments such as is brought down to the sea by rivers flowing from temperate (that is, ice-free) lands. As you can see, the cores indicate that during the last million years or so there have been at least three periods of temperate climate in Antarctica when the shores of the Ross Sea must have been free of ice.[7]

This discovery would indicate that the glacial history of Antarctica may have been roughly similar to that of North America, where we have had three or more ice ages in the last million years. Let us remember that, if most geologists cannot imagine how Antarctica could have had warm climates at short and relatively

---

[7] See Note 14, for Hough's interpretation of the cores.

recent geological intervals, neither can they explain how North America could have had arctic conditions at equally short intervals and just as equally recently. Ice ages remain for geologists an unsolved mystery (85:35).

The date found by Dr. Urry for the end of the last warm period in the Ross Sea is of tremendous interest to us. All three cores agree that the warm period ended about 6,000 years ago, or about 4000 B.C. It was then that the glacial kind of sediment began to be deposited on the Ross Sea bottom in the most recent of Antarctic ice ages. The cores indicate that warm conditions had prevailed for a long time before that.

An important fact about the Oronteus Finaeus Map is that all the rivers on it are shown flowing from mountain ranges near the coasts, except those near the southern tip of South America. No rivers are shown in the deep interior. This suggests that, very possibly, when the source maps were made, the interior was already covered by the ice cap. In that case, the ice cap was an advancing continental glacier that had not yet brimmed the encircling mountain ranges to reach the sea, nor had it yet stopped the flow of rivers on the seaward side of the mountains.

Let us connect this situation for a moment in regards to the Princess Martha Coast as we have identified it on the Piri Re'is Map. It would seem that the ice cap had not yet crossed the mountains that stretch along behind that coast. Supposing the ice cap to have advanced from the direction of the South Pole, which area would it have reached first—the Princess Martha Coast or the Ross Sea? It would have reached the Ross Sea first, and the shores of that sea would no doubt have been glaciated quite a good deal earlier than the Princess Martha Coast, in fact, possibly some thousands of years before. If this was the case the ancient voyages to the Princess Martha Coast that may be reflected in the Piri Re'is Map may have been made as recently as about 1000 B.C. While this may go a little way to relieve the historian of the problem of accounting for the mapping of that particular coast, it does nothing to help him in the Ross Sea area, for there it seems that the mapping would have to have been done at least 6,000 years ago.

So far, then, we find that this map of Oronteus Finaeus seems to be based on an authentic ancient source map of Antarctica compiled from local maps of the coasts drawn before the Antarctic ice cap had reached them. The individual maps of the different coasts are fairly accurate, taking account of the differences that may be attributed to the presence of the ice cap now over the coasts. In addition, the general compilation, which successfully placed the coasts in correct latitudes and relative longitudes and found a remarkably correct area for the continent, reflects an amazing geographical knowledge of Antarctica such as was not achieved in modern times until the twentieth century. The minor error in the location of the pole was perhaps subsequent to the date of compilation of the general map.

This map appears to confirm our impression as to the presence of a part of the Antarctic coast on the Piri Re'is Map. We have been successful, it would seem, in the quest for supporting evidence.

2. The remarkable map of Hadji Ahmed.[8]

In some respects this Turkish map of 1559 is one of the most remarkable I have seen (see Fig. 58). There is a striking difference between the drawing of the eastern and western hemispheres. The eastern hemisphere seems to have been based on the sources available to geographers of the time, mostly Ptolemy, and to be somewhat ordinary. The map of the Mediterranean is still evidently based on Ptolemy instead of on the much better portolan maps. The African coasts do not compare in accuracy with the same coasts on the Piri Re'is Map of 1513 or on other maps to be discussed shortly.

But if this is true of the eastern hemisphere, it is an entirely different story in the west, and here it is evident that the cartographer had at his disposal some most extraordinary source maps. The shapes of North and South America have a surprisingly modern look; the western coasts are especially interesting. They seem to be about two centuries ahead of the cartography of the time. Furthermore, they appear to have been drawn on a highly sophisticated spherical projection. The shape of what is now the United States is about perfect.

This remarkable accuracy of the Pacific coasts of the Americas, and the difficulty of imagining how they could have been drawn in the middle of the 16th Century,[9] adds significance to another detail of the map: the suggestion of a land bridge connecting Alaska and Siberia. This land bridge actually existed in the so-called Ice Age. The map suggests that the land bridge was a broad one, perhaps a thousand miles across (Fig. 59).

In case the reader is drawing back at this moment, in a state of amazement mingled with horror, I am forced to remind him that this bit of evidence is only a link in a chain. We have completed a study of the Piri Re'is Map of 1513, and have concluded that it may contain a representation of part of the Antarctic coast drawn before the present ice cap covered it. We have examined the 1531 Oronteus Finaeus Map of Antarctica and have come to much more far-reaching conclusions. We cannot estimate, of course, the lapse of time implied by these remarkable maps of Antarctica. But we have presented evidence that the deglacial or unglaciated period in the Antarctic cannot have come to an end later than 6,000 years ago and must have existed for a very long time before that. The warm period in the Antarctic may, then, have coincided with the last glacial period in North America. If this is true it follows that this map need be based on maps no older than the maps already discussed.

A more detailed examination reveals further interesting facts. The grid drawn on the map enables us to check accuracy. This particular projection has all the meridians curved except one, which we refer to as the prime meridian. The reader

[8] For a discussion of this map see Marie Armand d'Avesac-Macaya, "Note sur un Mappemonde Turke" (19).

[9] More than two centuries before the solution of the problem of longitude.

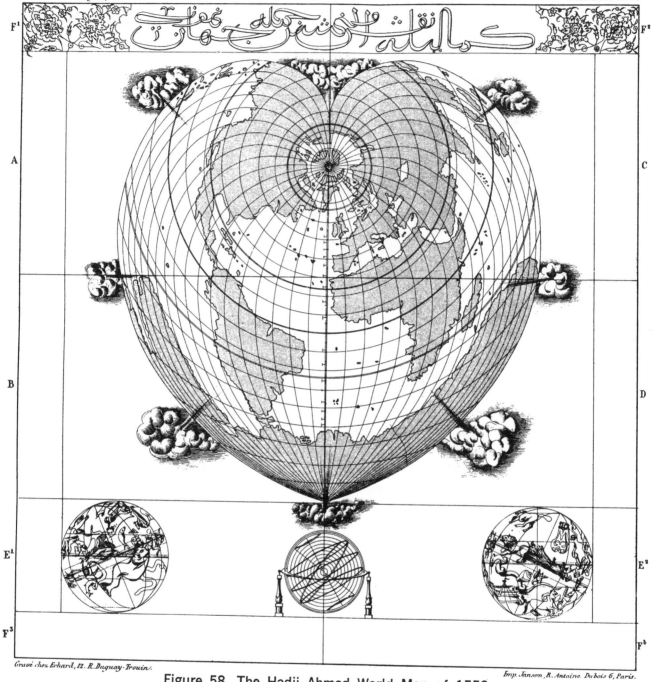

Gravé chez Erhard, 12. R. Duquay-Trouin.

Imp. Janson, R. Antoine Dubois 6, Paris.

Figure 58. The Hadji Ahmed World Map of 1559.

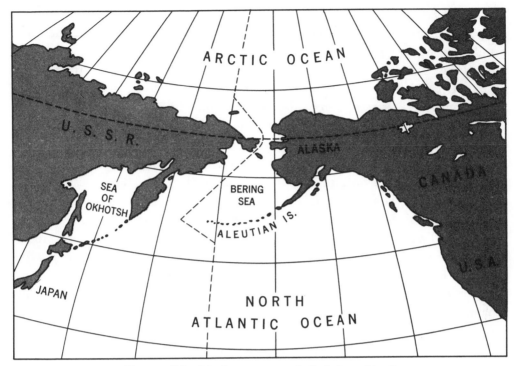

Figure 59. Modern map of Behring Strait.

can see the prime meridian on this map, running from the North to the South Poles, and passing near the coast of Africa. The other meridians are all spaced ten degrees apart, as are the parallels of latitude from pole to pole. The prime meridian on this map appears to coincide closely with the 20th meridian of West Longitude on modern maps. Thus, to find the longitude of any place, we will start with this line as 20° West and count by tens, adding West Longitude westward and subtracting it eastward. In Table 3 I have listed a number of places and compared their positions on this map and on modern maps. Table 3 is in two parts; the first part deals with places that are fairly close to, and the second part with places that are far from, the prime meridian. Note that both latitude and longitude are surprisingly accurate for places near the meridian, the accuracy of longitude being especially noteworthy. But that accuracy declines rapidly with distance from the meridian.

This increased inaccuracy with distance from the prime meridian indicates an error in the projection, but not necessarily an error in the drawing of the coasts that seem too far off. It may merely be another case of imposing a projection on a map that was originally drawn on an entirely different projection.

Some of the apparent exaggerations of the size of Antarctica on the map of Hadji Ahmed can, of course, be attributed to the same error we found on the Oronteus Finaeus Map, namely the confusion of the 80th parallel of latitude with the Antarctic Circle. But even considering this, the continent on this map seems too large, and its shape is hardly recognizable.

To understand the cause of the extreme distortion of the map, let us consider the polar regions on any Mercator map. It is difficult to find a modern Mercator world map showing Antarctica, but anyone who has seen a Mercator world map cannot have failed to notice how the projection exaggerates the northern polar regions. On such maps Greenland, for example, appears to be about the size of South America. This results from the fact that on this projection the meridians are parallel straight lines that never meet. The whole line across the top or bottom of such a map represents the pole, and the geography is distorted accordingly.

What I am suggesting is that some of the ancient source maps of Antarctica may have been drawn on a projection resembling the Mercator at least in this respect of having straight meridians parallel to each other. Such a projection existed in Greek times and was, according to Ptolemy, the projection used by Marinus of Tyre (39:69). If ancient source maps survived on two different projections—some on a circular projection such as we have apparently found on the Oronteus Finaeus Map and some on a straight-meridian projection like that of Marinus of Tyre or Mercator—the appearance of this map would be readily explained.

3. Mercator's maps of the Antarctic.

Gerhard Kremer, known as Mercator, is the most famous cartographer of the 16th Century.[10] There is even a tendency to date the beginning of scientific cartography from him. Nonetheless, there never was a cartographer more interested in the ancients, more indefatigable in searching out ancient maps, or more respectful of the learning of the long ago.

I think it is safe to say that Mercator would not have included the Oronteus Finaeus Map of Antarctica in his *Atlas* if he disbelieved in the existence of that continent. He was not publishing a book of science fiction. But we have further reason to know he believed in its existence: He shows Antarctica on maps he drew himself. One of his maps of the Antarctic appears on Sheet 9 of the 1569 *Atlas*. (See Fig. 60.) At first glance I could see little relationship between this Mercator Map and that of Oronteus Finaeus, and I had little reason to suspect that it could be a good map of the Antarctic coast. But careful study showed that a number of points could be clearly identified (see Fig. 61). Among these were Cape Dart and Cape Herlacher in Marie Byrd Land, the Amundsen Sea, Thurston Island in Ellsworth Land, the Fletcher Islands in the Bellingshausen Sea, Alexander I Island, the Antarctic (Palmer) Peninsula, the Weddell Sea, Cape Norvegia, the Regula Range in Queen Maud Land (as islands), the Muhlig–Hofmann Mountains (as islands), the Prince Harald Coast, the Shirase Glacier (as an estuary) on Prince Harald Coast, Padda Island in Lutzow–Hölm Bay, and the Prince Olaf Coast in

---

[10] For a short biography of Mercator, see Note 15.

Figure 60. Mercator's Map of the Antarctic (1569).

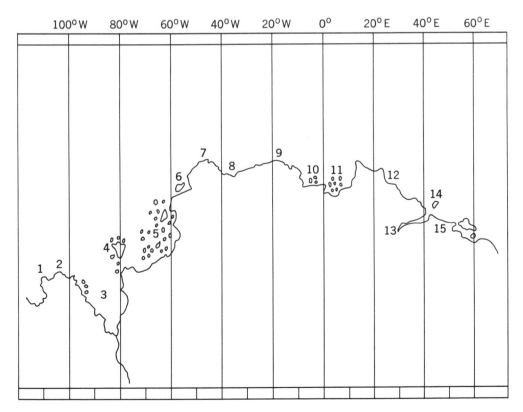

Figure 61. Mercator's Map of the Antarctic: tracing, with straight meridians according to Mercator's Projection. See Table 4.

Enderby Land. In some cases these features are more distinctly recognizable than on the Oronteus Finaeus Map, and it seems clear, in general, that Mercator had at his disposal source maps other than those used by Oronteus Finaeus.

The projection on Mercator's map of Antarctica is the one that is named after him. It has straight meridians that run parallel from pole to pole, and these, of course, enlarge the polar regions very much, as already explained.

I thought at first that Mercator might have drawn his map of Antarctica to fit his projection; in which case its large size might be thus explained without recourse to any other considerations. To test this, I traced the map and drew parallel meridians on it at ten-degree intervals, converting his longitude figures into East and West Greenwich Longitude (Fig. 61). Mercator, accepting the cartographical convention of his time, counted 360 degrees from a meridian off the west coast of Africa, approximately in what we now call 23° West Longitude. This is indicated on his world map of 1538, which also shows the Antarctic (Figs. 63, 64).[11]

In order to correlate his system with ours, it is necessary to find a point common to both. I picked the meridian of Alexandria as this common point because I saw that Mercator's 60th meridian[12] passed through Alexandria, which, in our system, is the 30th meridian of East Longitude. Thus his 30th meridian, we might suppose, should be equivalent to our zero meridian (the meridian of Greenwich). To convert his longitude to our system, accordingly, it seemed that we should merely have to subtract 30° going eastward. His zero meridian and his 360th meridian coincided, and should be, according to this, equivalent to our meridian of 30° West. But, as we have seen, this is not the case. His zero/360th meridian actually coincides fairly closely with our meridian of 23° West. The discrepancy amounts to about 7°.

I should perhaps explain at this point that the exact location of Mercator's zero/360th meridian depends on the accuracy of his placing of the Cape Verde Islands, the Canary Islands, and the Azores in longitude. He has his zero meridian running through the easternmost of the Cape Verde Islands, missing the westernmost Canaries by a degree and a half, and passing through the easternmost Azores, so that he has the easternmost islands of the Cape Verdes and of the Azores on the same meridian. But they are actually not on the same meridian. For this reason I thought it best to take a definite point, like Alexandria, as a common point for converting his system into ours. However, we have just seen that this will not precisely do, either. There is a discrepancy.

What is the matter? This brings up a most important point. Mercator is rightly regarded as a great cartographer. We forget that he had to work within the limitations of his age. Since he did not know the true circumference of the

---

[11] Mallery maintained that this map of Mercator must have been based on an authentic ancient source map (131).

[12] On his world map of 1538.

earth, he had to take the best guess going. We have seen that when we start counting degrees from Alexandria by his system and by our own we do not end up in the same place. We find ourselves 7° too far west. On the other hand, if we start counting from our 23rd meridian, converting that to Mercator's zero meridian, we are going to find Alexandria at 37° East Longitude, 7° too far east. The actual longitude difference between the meridian of 23° West and Alexandria is 53°. A simple calculation shows that 7° is 13 per cent of the longitude difference. So Mercator used a circumference of the earth 13 per cent too short.

With regard to Mercator's 1569 map, my first step was to pick a reference point for longitude. It seemed to me that our zero meridian, which intersects the Queen Maud Land Coast between the Regula Range and the Muhlig–Hofmann Mountains, might be a good point to start with, experimentally. I took no account of the difference in the length of the degree, but drew my meridians the same distance apart as Mercator did, numbering them after our modern system. The errors in the longitudes thus found for the various points convinced me that Mercator had not redrawn his source map; apparently he had simply taken a map constructed on quite a different projection and transferred it bodily to his own map.

I conjectured that the original projection may have been a polar type of projection with straight meridians. In this case the parallels of latitude would be circles. To test the matter, I listed the identifiable points on the map with their latitudes. As the reader can see (Fig. 61), the localities are distributed in a semicircle. They are, nevertheless, closely in the same latitudes, averaging about 70° South, as this list shows:

Cape Dart	73.5 S
Cape Herlacher	74.0 S
Amundsen Sea	72.0 S
Thurston Island	72.0 S
Fletcher Island	73.0 S
Alexander I Island	69–73 S
Bellingshausen Sea	71.0 S
Antarctic Peninsula (truncated)	70.0 S
Weddell Sea	72.0 S
Cape Norvegia	71.0 S
Regula Range	72.0 S
Muhlig–Hofmann Mts.	71–73 S
Prince Harald Coast	69–70 S
Shirase Glacier	70.0 S
Padda Island	69.0 S
Casey Bay	67.5 S
Edward VIII Bay	67.0 S

The indication is that the parallels of latitude on this map were originally circles. I found it possible with a pair of compasses to draw a circle that would pass close to all of them. A series of experiments finally located a point for the

South Pole that gave me a satisfactory parallel of 70° South Latitude, with respect to most of the localities. Having the pole and the 70th parallel, it was a simple matter to find the length of the degree of latitude and then measure out the 80th, 75th. 70th, and 65th parallels. We could then check the latitudes and longitudes of the various points as shown in Table 4; we found our grid fairly well confirmed (Fig. 62).

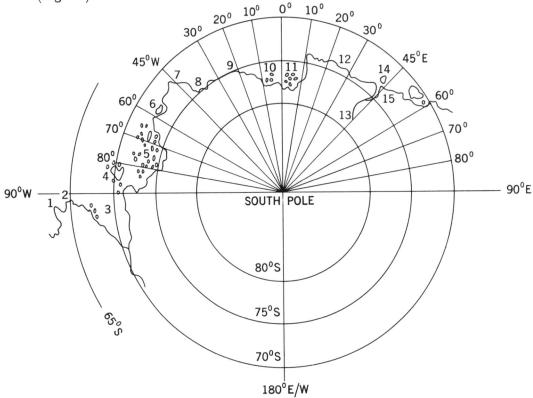

Figure 62. Mercator's Map of the Antarctic: tracing, with superimposed polar projection. See Table 4.

These findings indicate that Mercator had a real map of the Antarctic, though he was unable to transfer the points on it to his own projection. The errors of longitude are less than they seem, since, as we have already mentioned, the degree of longitude is very short at the high latitudes of Antarctica.

Earlier, in 1538, Mercator drew a world map that also showed Antarctica, as we have seen. The similarity to the map of Oronteus Finaeus is obvious, but there are important differences. Mercator has the Antarctic Circle inside the continent, as Oronteus Finaeus does, but not at the same distance from the pole. In other words, Mercator seems to have changed the *scale*. On the Oronteus Finaeus Map, as we have seen, the so-called *Circulus Antarcticus* seemed to be a mistaken interpretation of the 80th parallel of the source map, as confirmed by the agreement of the geography with the grid drawn on that assumption. By shifting this, Mercator destroyed the original scale. Therefore, it is impossible for us to reconstruct a grid of latitudes on this map, as we did for the other map. Longitudes, however, are remarkably accurate (Table 5).

It seems that Mercator made constant use of ancient source maps available to him. What eventually happened to these maps we do not know, but we are able to distinguish, in a number of cases at least, where he depended on them and where he was influenced by contemporary explorations.

For the Antarctic, of course, he had to depend on the ancient sources.[18] The source maps here may have come to him through Oronteus Finaeus, who may have found them in the library of the Paris Academy of Sciences, now part of the Bibliothèque Nationale; or he may have had others of his own. For Greenland he used the Zeno Map of the North, with the mountain ranges conventionalized. So far as his 1569 map of South America is concerned (Fig. 41), a number of interesting points emerge.

First, with regard to the northern coast, it is clear that he depended on ancient maps as well as modern explorations. He has the Amazon misplaced with regard to the equator, just as it appears on the Piri Re'is Map, but the course of the Amazon is conventionalized with a number of snakelike meanders. The Island of Marajo, correctly delineated on the equator of the mathematical projection of the Piri Re'is Map, is here confused with the Island of Trinidad, off the mouth of the Orinoco. Trinidad is therefore shown as much too large. The southeast coast of South America, from the Tropic of Capricorn to the Horn, is very badly drawn, evidently from the accounts of the explorers, while the west coast is completely out of shape.

Oddly enough, in his map of 1538, thirty years earlier, Mercator had represented the west coast of South America much more correctly (Fig. 63). How is this explained? I suggest that in his first map he depended on the ancient sources, while on his map of 1569 he depended on the modern explorers, who, since they could find no accurate longitude, could merely guess at the trends of coasts.

---

[18] **Since the very continent was not "discovered" until 250 years later.**

Figure 63. Mercator's World Map of 1538.

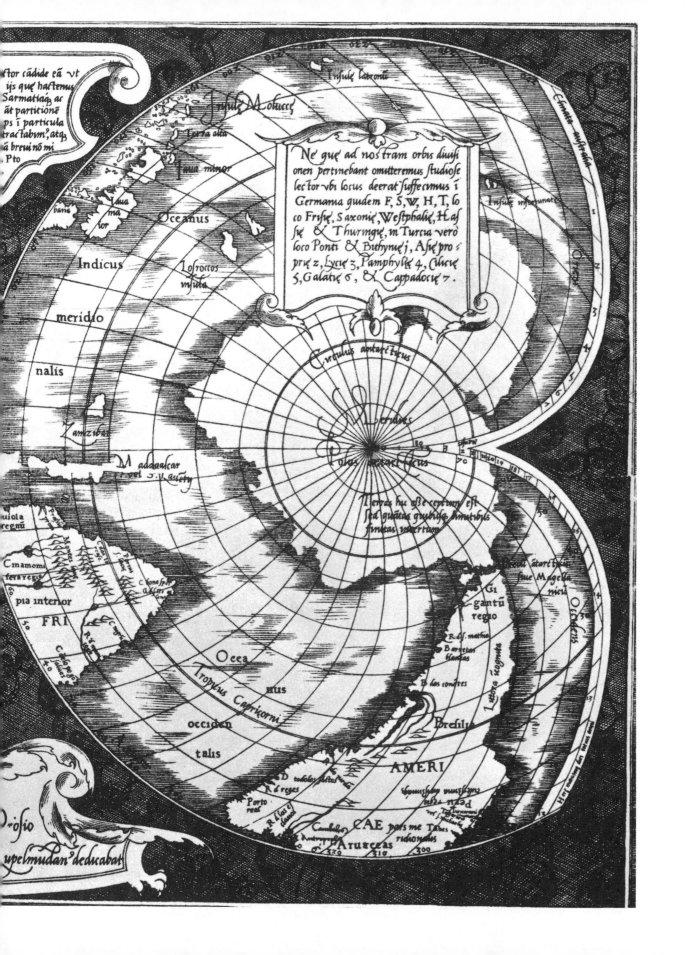

Insulæ latronu

Insulæ Moluccæ

Terra alta

Iaua minor

bana

Iaua maior

Oceanus

Indicus

Iosroctos insula

meridio

nalis

Zanzibar

Maddaalcar vel S. V. auety

uiola regnu

Cmamomi feraregio

pia interior

FRI

Ne quæ ad nostram orbis diuisi
onem pertinebant omitteremus studiose
lector vbi locus deerat suffecimus i
Germania quidem F, S, W, H, T, lo
co Frisiæ, Saxoniæ, Westphaliæ, Has
siæ & Thuringiæ, in Turcia verò
loco Ponti & Bithyniæ j, Asiæ pro
priæ z, Lyciæ 3, Pamphyliæ 4, Ciliciæ
5, Galatiæ 6, & Cappadociæ 7.

Insulæ infortunatæ

Circulus antarcticus

Berides

Polus antarcticus

Terras hic esse certum est
sed quátas quibusq́ finitibus
finitas incertum

Circul. antarcticus
siue Magella
nicia

Gi
gantũ
regio

R d.s. mathia

Barreras
blancas

B los conctes

Littora incognita

Occeans

Ocea

nus

Tropicus Capricorni

occiden

talis

Bresilia

AMERI

CAE pars me Tdies
ridionalis

Porto
real

Aruareas

stor cadide ea vt
ijs quæ hastenus
Sarmatiaæ ac
at partitione
ps i particula
tractabim̃, atq̃
a breui nõ mi
Pto

Prosio
upelmudan̄ deducabat

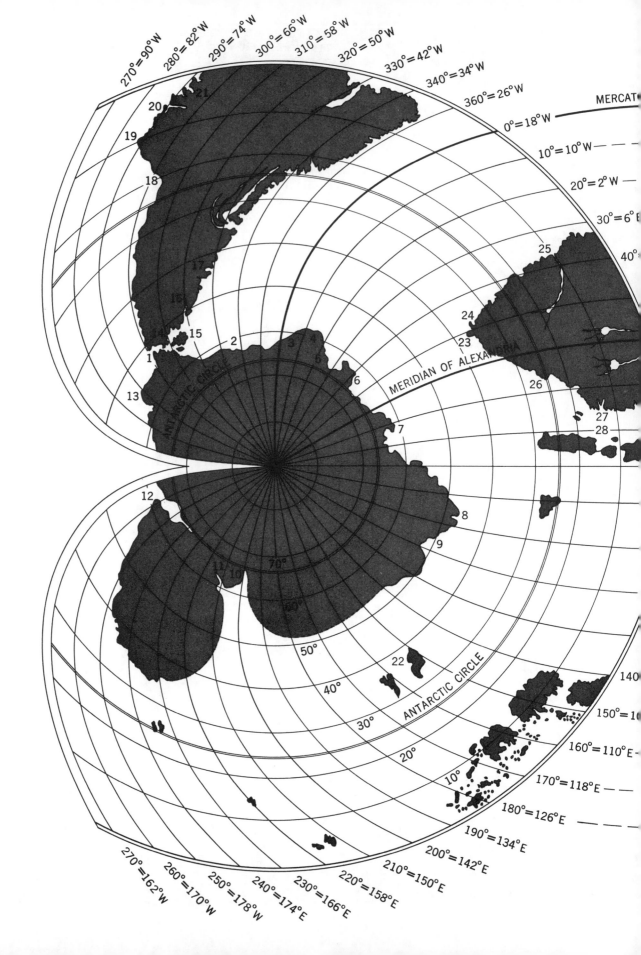

Figure 64. Mercator's World Map of 1538, converted to Greenwich Longitude.
See Table 5.

CHAP
TER V
ANCIENT
MAPS OF
THE EAST
AND
WEST

The Dulcert Portolano of 1339. The De Canerio Map of 1502. A Venetian Chart of 1484. The Mathematical Basis of these Three Charts. The Twelve-Wind System in the Venetian Chart, reflected also in the De Canestris Map of 1335–37. The earliest Portuguese Chart of the Indian Ocean, and the discovery that it shows Australia. A Twelfth Century Map of China.

We have already noted that Nordenskiöld in his essay on the portolan charts stated that they were too accurate to have originated in the Middle Ages. He found evidence that they probably existed in classical times, alongside the inferior maps of Eratosthenes, Pomponius Mela, and Ptolemy. He even hinted that he thought they were of Carthaginian origin. It is our purpose now to examine a number of these charts, to see how accurate they really are, and how far they may be related to a possible worldwide system of sophisticated maps deriving from pre-Greek times.

1. The Dulcert Portolano of 1339.

The Dulcert Portolano of 1339 is an early version of the "normal portolano"—the highly accurate map that appeared suddenly in Europe in the early 14th Century, seemingly from nowhere. This kind of map did not evolve further but was simply copied and recopied during the rest of the Middle Ages and during the Renaissance (see Fig. 3).

In Figure 65 I have worked out the grid of this map. I began with the assumption that the grid would be a square grid. I identified a number of geographical points around the map and from these discovered how much latitude and longitude was covered by the map. Dividing the number of degrees into the millimeters of the draft map, I found the length of the degree. It did appear that there was a square grid.

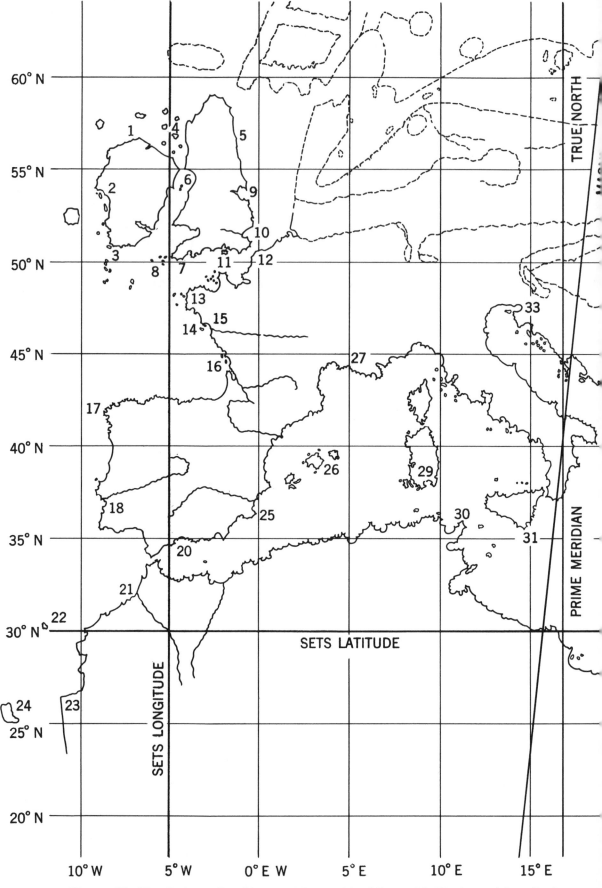

Figure 65. The Dulcert Portolano, with a grid of line of latitude and longitude constructed empirically from the geography. See Table 6a.

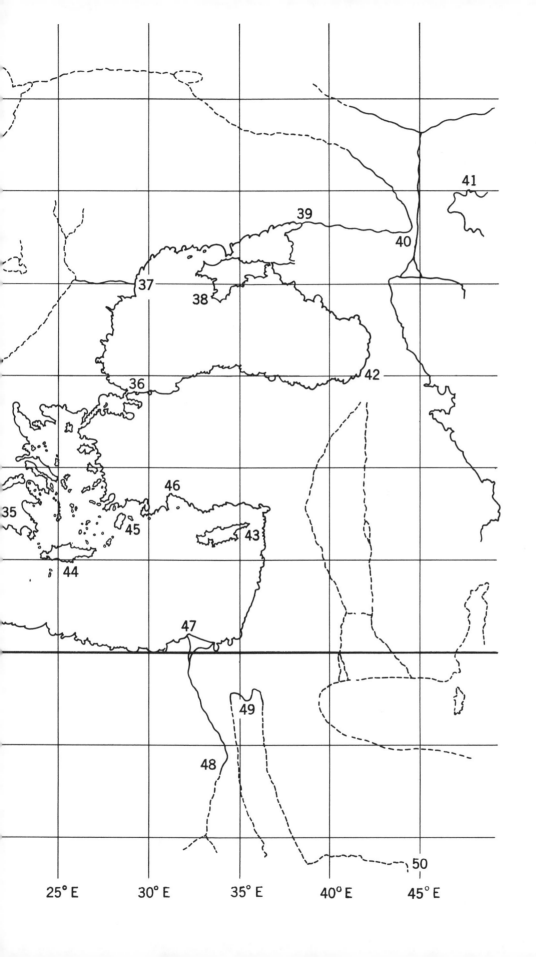

35

37

38

39

40

41

36

42

46

45

43

44

47

48

49

50

25° E   30° E   35° E   40° E   45° E

It was necessary to lay out this grid from some definite point. For a first experiment I selected Cape Bon, in Tunisia, close to the ancient site of Carthage. I was influenced here by the idea that perhaps this map had been drawn in ancient times by the Carthaginians, using Carthage as a center. I drew my first grid on the assumption that the vertical line (or prime meridian) through the center of the portolan projection was drawn on True North (see Fig. 3). The resulting table revealed errors that indicated that the map was not oriented to True North, but about 6° to the east. It appeared that 6° was probably just about the amount of the compass declination in the Mediterranean at that time.[1] A grid drawn on this basis, however, revealed yet further errors, which seemed to indicate that Alexandria, not Cape Bon, might be a better reference center for the map. A new grid, based on Alexandria, proved to be very satisfactory with respect to latitude, but about 2° off with respect to longitude. A final grid was drawn with latitude based on the parallel of Alexandria, and longitude based on the meridian of Gibraltar, and this proved extremely satisfactory (see Table 6).

The grid applied to this portolano reveals some very interesting facts. First, it appears that the geographical information contained in the chart is much greater than can reasonably be expected from medieval sailors and cartographers. The map falls into three parts: a very accurate map of the Mediterranean and Black Sea regions and of the coasts of Europe as far north as the Hebrides; a very inaccurate map of the Baltic region; and a very inaccurate map of the eastern regions, embracing the Persian Gulf and the Indian Ocean. It seemed that the inaccurate parts of the map were simply tacked onto the portolano proper. They would seem to reflect the true state of medieval geographical knowledge. The portolano proper, on the other hand, is a remarkably scientific work. It is evident from the table, for example, that the latitudes of all places, except perhaps for the Black Sea region, are too accurate to have been determined by medieval sailors. The remarkable degree of accuracy as to latitude extends all the way from the northern tip of Ireland to the Sinai Peninsula and to Assam in southern Egypt.

The accuracy of longitude is far more remarkable. The total longitude of the Mediterranean and Black Seas is correct to within half a degree on this grid. That means that the mapmaker achieved highly scientific accuracy in finding the ratio of latitude to longitude. He could only have done this if he had had precise information on the relative longitudes of a great many places scattered all the way from Galway in Ireland to the eastern bend of the Don in Russia. Nordenskiöld would seem to have been clearly right when he said that no medieval mapmaker could have drawn this map. Not even Mercator in the 16th Century could have done so.

Another point calls for mention. How could it have been possible to draw so

---

[1] Compass declination in the Mediterranean was 6° E in 1599 (89:14) and 6.1° in 1296 (50). It may have remained about the same in the intervening period.

accurate a map of the vast region covered by the Dulcert Portolano (one thousand miles north and south, almost three thousand miles east and west) without the aid of trigonometry? Let us remember that the mapmaker's problem was to transfer points on the spherical surface of the earth to a flat plane in such a way as to preserve correct distances and land shapes. For this, the curvature had to be calculated and transferred to a plane by trigonometry.[2] That this probably was done for the Dulcert Portolano will be shown a little later.

In conclusion, we may remark that, since the Dulcert Portolano represents essentially Nordenskiöld's "normal portolano," we have evidence here that all the portolanos stemmed from a common origin in remote times.

2. The De Canerio Map of 1502.

The complete Piri Re'is Map, before it was torn in two, had included the whole continent of Africa as well as Asia. In view of this, and in view of the probability that other copies or versions of the source map Piri Re'is used for Africa (or that were used by the Alexandrian compilers) might have survived, we continued a search for a map of Africa drawn on the same projection. We finally found what we thought was such a map.

My first glance at the De Canerio Map of 1502 (see Fig. 66) gave me a feeling that our search had been successful. The South African part (from the equator southward) looked astonishingly modern. I felt fairly confident that this was an authentic ancient map in Renaissance dress.

The abundance of easily identifiable points on the coasts made it easy to work out the scale and construct an empirical grid, and their positions with reference to the grid indicated that the mapmaker had achieved considerable accuracy in both latitude and longitude. The errors of latitude averaged only 1.6° and those of longitude only 1.4°. We thought it remarkable that longitudes seemed more accurate than latitudes.

For some time it was not possible either to connect the map directly with the Piri Re'is Projection, or to solve the mathematical structure that, I still confidently felt, underlay it. Finally, the discovery of the magnetic orientation of the Dulcert Portolano furnished the key.

In the center of Africa, there appears a very large wind rose, obviously the center of the portolan design. It had not appeared to me to lie on any significant parallel or meridian, and I had therefore been unable to link it up with the projection of the Piri Re'is Map until it occurred to me to find out whether this map was oriented to Magnetic North. Experiment indicated that the map was, in fact, oriented about 11¼° east of north. It was a simple matter to rotate the map, on its center, westward to True North. Figure 67 shows how this was done and

---

[2] See Note 7.

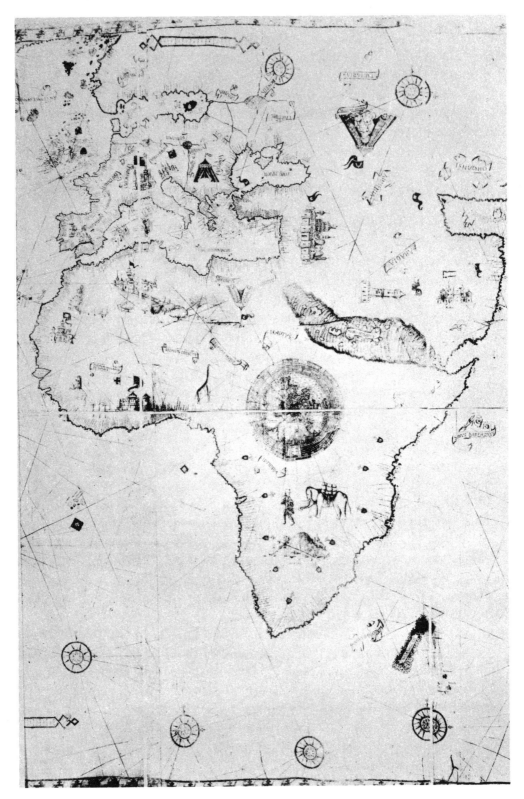

Figure 66. The Nicolo De Canerio World Map of 1502.

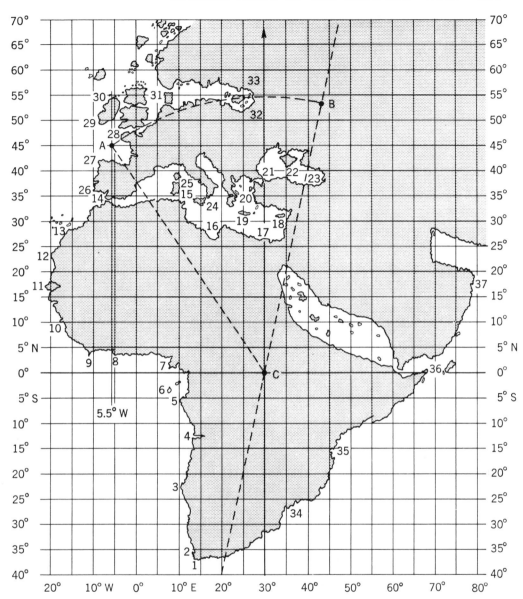

Figure 67. The De Canerio World Map of 1502, with an oblong grid constructed by spherical trigonometry. See Table 7.

how, with this shift, the center of the map turned out to lie on the equator—and *on the meridian of Alexandria!*

This was an extraordinary discovery. It constituted as good proof as might be necessary to establish the Alexandrian derivation of the map. It demonstrated, too, that the original map had been drawn on True North, and that the magnetic orientation was probably introduced by De Canerio or some other geographer of the relatively modern period. Why that geographer gave the map an orientation more than twice too far to the east is difficult to imagine. It would, of course, have rendered all compass courses hopelessly wrong. The same error appears in numerous other portolanos.

Now that the exact center of the map had apparently been established, it occurred to me that it might be possible to solve its mathematical structure and to construct a grid based on trigonometry.

This proved easier than I had expected. A number of minor projection points appeared at equal intervals on the map, obviously arranged on the perimeter of the circle of the portolan projection. The trigonometric solution would depend on finding the exact length of the radius of the circle. For this all that was needed was the exact position of one of these minor projection points in latitude and longitude. Fortunately, one of these points lay just off Land's End, England, as near as we could estimate in Latitude 50° North and Longitude 5.5° West. We now had the two co-ordinates necessary for a trigonometric calculation.

From our experience with the Piri Re'is Map, we assumed at first that plane trigonometry would be involved, and found the length of the radius to be about 61.3°. This gave us the latitudes of the two projection points located on the perimeter, where the latter was intersected by the prime meridian, and we thus obtained the length of the degree of latitude. Since we had not then discovered the oblong grid of the Piri Re'is Map, we assumed that the lengths of the degrees of latitude and longitude would be the same, and drew a square grid for the map. Some remarkable facts now emerged.

The most sensational development was what the grid revealed in the Mediterranean and Black Seas. It was obvious, by comparison with the Dulcert chart, that the De Canerio Map was based on the "normal portolano," though it did not show such fine detail. This was natural enough, considering the much vaster area covered. This part of the De Canerio Map, however, was evidently an integral part of the map of Africa; it had not been just tacked on. It would seem that it had been drawn originally on the same trigonometric projection. This is shown by what follows.

A table of thirty-seven geographical points, as found by our square grid, revealed the extraordinary accuracy of the map as to latitude and longitude. We found, for example, that the average error in the latitudes of eleven places (Gibraltar, the northern coasts of Sardinia, Sicily, Cyprus, and Crete; Cape Bon, Bengazi, Lesbos, the Bosphorus, Sevastopol, and Batum) was only one half of one degree.

The longitudinal distance between Gibraltar and Batum was correct, proportionate to the latitude, suggesting that there may have been no considerable error in the original source map, as to the size of the earth. It seemed that the trigonometric solution of the De Canerio Map carried with it the implication that trigonometry underlay the normal portolano and, in fact, the whole group of portolan charts.[3]

The other parts of the De Canerio Map were not as accurate as the Mediterranean and Black Sea areas. The eastern section (including the upper part of the African coast of the Indian Ocean and Arabia) was evidently plastered onto the accurate source map by De Canerio or somebody else. It did not fit the grid, and it seemed to have been derived from Ptolemy. Another section in the far north, covering the Baltic, also appeared to have been originally a separate source map which at some time had been incorrectly compiled with the principal part of the map.

Other errors appeared within the limits of the trigonometric chart itself. Points on the west coast of Africa from the Cape of Good Hope to the delta of the Niger averaged about 4° too far south. Points from Freetown to Gibraltar averaged about 3.6° too far north. The total latitude error from the Cape of Good Hope to Gibraltar was 5.5°, implying an error in the length of the degree of latitude of about 8 per cent. Latitude errors continued to increase northward on the coasts of Europe as far as northern Ireland.

At first I supposed this might imply an error in the scale of the source map, but corresponding longitude errors were not found. An error in scale would carry with it proportional errors in both latitude and longitude. There were larger longitude errors, it is true, along the African and European coasts than in the Mediterranean, but they did not suggest an error in the length of the degree of longitude. From the Cape of Good Hope to Walvis Bay, on the west coast, the average error was 3.5° West. From the Congo to Cape Three Points it was 3.5° East. From Cape Palmas to Gibraltar longitude errors were negligible. On the European coasts, from Cape St. Vincent to Londonderry, they averaged 3.5° East. There was no indication here of any error in scale, and, in view of the distribution of latitude errors in the Mediterranean, very little suggestion of any error in the orientation of the continent. We did, however, change the orientation later, making the shift from the magnetic orientation 12° instead of 11¼°.

The apparent increase of latitude errors with distance from the equator gained

---

[3] An exception must be made for the earliest of the maps called portolanos, the Carta Pisana. This apparently dates from the 13th Century. In this case the typical portolan design was applied to an extremely inferior map, such as might have been drawn in the Middle Ages or very sloppily copied from an accurate portolano. The latter supposition is supported by the fact that the mapmaker made a botch of the portolan design. This consisted of two circles, but the mapmaker made them of different diameters, and hardly a line in the design is straight.

added significance with our discovery of the oblong grid on the Piri Re'is Map. If no error in scale was responsible, perhaps it was a question of an original projection that might have taken account of the curvature of the earth by spreading the parallels with distance from the equator, as in the modern Mercator Projection. Hints were found, although not confirmed, of a possible knowledge of the principle of the Mercator Projection in medieval Europe and in ancient China. Accordingly, we decided to find out whether there could be any truth in this. Charles Halgren, of the Caru Studios, was kind enough to construct a Mercator grid for the map, and this was then examined by William Briesemeister. Unfortunately, it turned out that there was very little basis for supposing that the original source map had been drawn on anything resembling Mercator's projection.

We now came back to the point from which we had started: the question of the alternatives of plane versus spherical trigonometry. I decided to draw a grid based on spherical trigonometry to see whether that would solve our problem. Three different persons—Richard Strachan, Professor E. A. Wixson, of the Department of Mathematics of Keene State College; and Dr. J. M. Frankland, of the Bureau of Standards—independently used spherical trigonometry to calculate the length of the degree, and agreed on essentially the same result: 58.5° for the radius of the projection. The diagram in Figure 67 shows that, by this calculation, the degrees of latitude and longitude *differ* and that, as a result, we have an oblong grid, as we found empirically to be the case with the Piri Re'is Map. This grid, based on spherical trigonometry, solved our problem of latitude errors, as can be seen by an examination of Table 7. The following paragraphs summarize the general results:

1. Longitude in the Mediterranean and Black Seas: The average of the errors of longitude of twelve places from Gibraltar (5.5° W) to Batum (42° E) is about one-fifth of a degree or about 12 miles. Over a total longitudinal distance of 47½° (about 3,000 miles) between Gibraltar and Batum, we find an error of only 1°, equal to about 2 per cent of the distance.

2. Latitude on the Atlantic Coasts: From the Cape of Good Hope (35.5° S) to Londonderry, Ireland (55° N), over a total latitude distance of 90½°, the error is 1°, about 1 per cent of the distance. There are larger latitude errors at many points in between, but these may represent distortions of local geography introduced by careless copyists. The accuracy of longitude east and west in the Mediterranean, and of latitude north and south in the Atlantic, suggests the basic accuracy of the grid based on spherical trigonometry.

3. Latitudes in the Mediterranean and Black Seas: There seems to be a regular error of about 3° applying to this whole area, which is thus placed too far south. The relative latitudes of places, however, are good. Deviations from the regular, or standard, error of 3° average less than 1°. It would seem probable that the general error was introduced by the compilers who originally combined maps of the Mediterranean and of the Atlantic coasts on the trigonometric projection.

These findings with regard to the De Canerio Map affect rather deeply our views with regard to the Piri Re'is Map and other maps to be considered later. It would now seem that the original source maps used by Piri Re'is for Africa and Europe, and perhaps also for the American coasts, as well as all the portolanos, may have been based on spherical trigonometry.

The De Canerio Map of 1502, showing, as it does, both the Atlantic and the Indian Ocean coasts of Africa, raises another problem, especially for those who are anxious to attribute its origins to the Portuguese and other explorers of the 15th Century. An investigation of the history of the discovery of the African coast in the century before the drawing of this map reveals no solid basis for believing that the explorers could have drawn the map or even supplied cartographers at home with the data necessary for drawing it.

To begin with, it appears that by 1471, only thirty-one years before the map was drawn by De Canerio, the Portuguese had not even reached the mouth of the Niger, four degrees north of the equator on the west coast. The Portuguese scholar Cortesao (54) says:

. . . The whole of the Gulf of Guinea was discovered by the Portuguese during the third quarter of the 15th Century, and Rio de Lago, where the present Lagos, the capital of Nigeria, lies, not far from Ife, was reached for the first time in 1471. . . .

Lagos is in 6° North Latitude and 3.5° East Longitude, and there is 100 miles or more of coast between Lagos and the mouth of the Niger. Boies Penrose, in his scholarly account of the Age of Discovery, gives a chronology of the discovery of the African coast and states that by 1474 the Portuguese had just reached Cape St. Catharine, two degrees below the equator (162:43). It is plain from this that only a quarter of a century before the De Canerio Map was compiled the Portuguese had not even begun the exploration of the west coast between the equator and the Cape of Good Hope, to say nothing of exploring the eastern coast.

To understand how impossible it would have been for Portuguese or other western explorers to have accurately mapped these coasts, even if they had explored them, we have only to understand that sea charts with graduated scales of degrees, subdividing the multiples of them into equal smaller units, were not in use by navigators until after 1496. Until then, therefore, even if the navigator could have found longitude—which was impossible—he could not have entered any notations of longitude on the charts, and the same is true for latitude. Penrose describes the state of nautical science just before 1502 in the following passage:

King John [of Portugal] was very interested . . . in cosmography and astronomy, and he had a committee of experts—the Junta—headed by the brilliant Jews, Joseph Vizinho and Abraham Zacuto, to work on the problem of finding position at sea. Zacuto had written in Hebrew in the previous decade his Almanach Perpetuum, the most advanced work yet to appear on the subject, and one containing full tables of the sun's declination. But its technical nature, coupled with the fact that it was written in a language but little understood by the average skipper, rendered it quite impractical. Vizinho, therefore, translated it into Latin (printed at Leiria, 1496) and later made an abridged version. . . . One result of this technical research was the expedition of Vizinho in 1485 along the Guinea Coast as far as Fernando Po, for the purpose of determining the declination of the sun [the Latitude] throughout Guinea. . . . The observations made by the Vizinho expedition led to the introduction of graduated sailing charts into Portugal. . . . (162:44–45)

The story of the exploration of the coast from 1496 when graduated sailing charts were introduced to 1502 gives no basis for supposing that the De Canerio Map of 1502 resulted from it. An important explorer, Diogo Cão, discovered the Congo, reaching a latitude of 13° S, and returned to Portugal in 1484 (162:45–47). On his next voyage he explored the coast for nine degrees farther south and returned to Portugal in 1487. This was five years before the drawing of the De Canerio Map, and there were still about 800 miles of unexplored coast lying between the point reached by Diogo Cão and the Cape of Good Hope.

It is true, of course, that Bartholomew Diaz rounded the Cape of Good Hope in 1488, but his was not a mapping expedition. He did not follow the coast down to the Cape. Instead, just south of Caboda Volta (Lüderitz) in 27° South Latitude, he was blown off course and around the Cape, making his landfall 250 miles to the East! He returned to Portugal in 1489 (162:47).

After Diaz, the next expedition was that of da Gama, who left Portugal in 1497 and returned in 1499. This expedition may have carried graduated sailing charts, for it was very carefully planned. Penrose says:

Four ships were constructed under the supervision of Bartholomew Diaz. . . . Bishop Diogo Ortiz supplied the fleet with maps and books, and Abraham Zacuto provided astronomical instruments, and made tables of declination, and trained the ships' officers in the art of making observations. . . . (162:50)

This fleet might have produced some accurate observations of latitude along the coast, but this was not its purpose. Its destination was *India*. Therefore, da Gama plotted his course *to avoid the coast*. He followed it a short way, and then made for the Cape Verde Islands. From there he steered a "circular course far out into mid-ocean to the southwest, to escape the doldrums and the currents of the Gulf of Guinea" (162:51). He reached St. Helena, on November 8, 1497, a few days later set to sea again, and rounded the Cape of Good Hope without touching at any other port. His first landfall after rounding the Cape was at Mossel Bay, 300 miles to the east of the Cape. He touched at a few other points before heading out across the Indian Ocean to India, but the African coast was out of sight most of the time and therefore could not have been mapped. He might have found the latitudes of his ports of call, but he could not, at any point, have determined longitudes.

We can conclude that neither da Gama, nor Diaz, nor any of their predecessors, could have done the accurate mapping of the west and east coasts of Africa that we find on the De Canerio Map.

3. The Venetian Chart of 1484.

Among the most noteworthy of the portolan charts is one drawn, or at least found, in Venice in 1484 (see Fig. 68). This chart is remarkable for its accuracy and because it was based both on trigonometry and the so-called "twelve-wind system" known to the ancients. In the last particular it appears to be unique among the known portolan charts. We will consider these points in reverse order.

The usual portolan design, with which the readers of this book have now become familiar, is one in which the circle is bisected a number of times to make angles at the center of 180°, 90°, 45°, 22½°, and 11¼° (and occasionally with still another bisection into angles of half of 11¼°). This has already been explained (see Fig. 9). There also was in antiquity the so-called "twelve-wind system." My student, Alfred Isroe, who illustrated the eight-wind system, has also illustrated the more sophisticated twelve-wind system (Fig. 69). Instead of requiring only the bisecting of angles this calls for the trisecting of the hemisphere, which, in turn, requires a knowledge of the ratio of the circumference of the circle to its diameter. This system produces angles of 60°, 30°, 15°, and 5° and appears related to the 360° circle, known from ancient times but not used, at least for navigation, in the Renaissance.

Various writers refer to the use of the twelve-wind system among the ancients. According to one (199:54), it was employed by the Greek geographer Timosthenes, an immediate predecessor of Eratosthenes. The latter is said to have abandoned it in favor of the eight-wind system, because it was too difficult for mariners (39:124–125).[4] The system continued to be the one preferred by the Romans, who were not much interested in the sea. It was known in the Middle Ages[5] and is said to have been used in the earliest editions of Ptolemy's maps when they were recovered in the 15th Century.

When I first examined the Venetian chart, what struck me most was that, more distinctly than any other chart I had seen, it showed a square grid, dominating the portolan design, which appeared to be drawn on True North. Only after long examination did I discover that it was, in fact, oriented about 6° to the

---

[4] The Venetians, however, apparently made an attempt to use it with the compass. See Note 16.

[5] See the discussion of the De Canestris Map below.

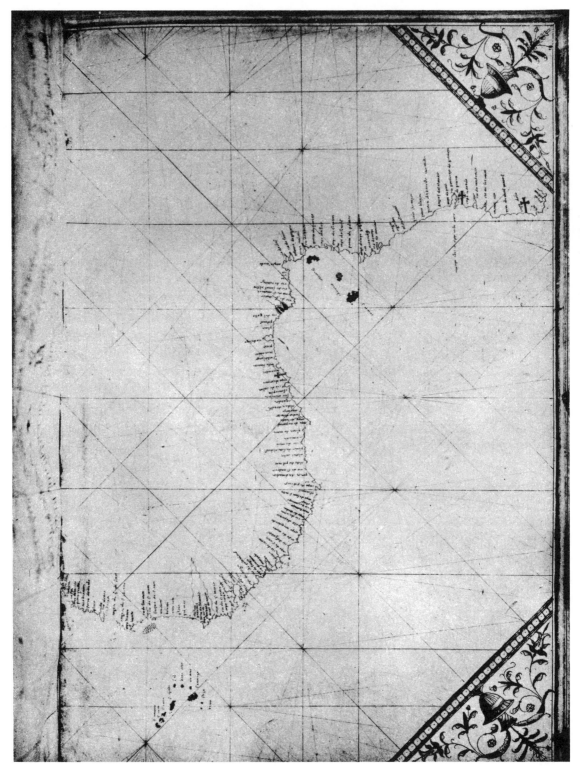

Figure 68. A Venetian Map of the African Coast (1484).

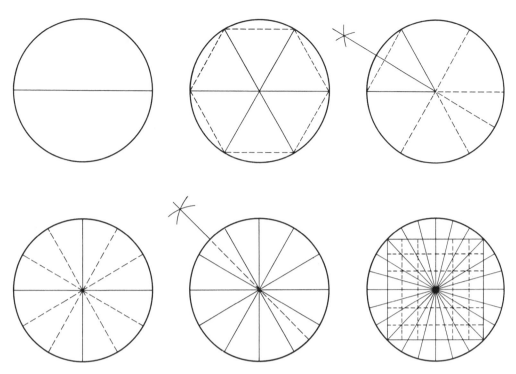

Figure 69. Diagram of the Twelve-Wind System found in the Venetian Map (by Alfred Isroe).

east. Obviously this map has a grid of lines of latitude and longitude. The diagonal lines were less emphasized than on most other portolanos. Examination now suggested the possibility of finding a solution of this map by trigonometry.

The first step was to make a careful comparison of this map with a good map of the African coast. This revealed that previous scholars, who seem to have assumed that the map showed the coast from the Strait of Gibraltar to the Cape of Good Hope, or to a point near the Cape, were apparently in error. It appeared that the map extended on the north only to about 26° or 27° North Latitude, while on the south it extended only a few degrees below the equator.

The intervals of the twelve-wind system made it simple to draw an equilateral triangle with its apex on the 27th parallel and its base on the equator, and solve for the length of the degree with trigonometric tables. A square grid, based on the length of the degree found in this way, seemed to give very good results, at least so far as latitude was concerned. It seemed that the latitudes of all identified points on the coast were accurate to within one-third of one degree, or about 20 miles.

Longitude findings, however, were not as accurate. Errors averaged about one degree. This was not very bad, excepting that they were distributed in such a way as to imply an error of some kind in the projection. The easternmost points were too far east, the western points too far west, so that it was a question of the length of the degree of longitude. Were the degrees of latitude and longitude really equal, as we had assumed?

In this situation it seemed best to set aside the trigonometry and try to work out a grid empirically to see whether the degrees of latitude and longitude were equal. The reorientation of the map to True North revealed that the top of the map was not at 26° or 27 ° North, but at 24° North. On this basis, measurements showed that the degrees were not quite equal. The degree of latitude was, it appeared, slightly *shorter* than the degree of longitude. Surprisingly, however, the length of the degree of longitude found empirically now turned out to be precisely the same as the length of the degree found by trigonometry (see Fig. 71 and Table 8). It is one thing to work out the grid of a map that has already been drawn, but quite another to draw the map in the first place. Our work indicated that the map must have been originally drawn on a plane trigonometric projection. The fact that the apex of the triangle was found to be at 24° North also was interesting, in view of the fact that the Greek geographers (Eratosthenes, Hipparchus, and their successors) accepted that as the Tropic, for the sake of simplicity even though they knew better. It seems then, that the map was intended to be fixed astronomically between the Tropic of Cancer and the equator.

There is evidence that at the southern end of the map some 15th Century navigator added some coastline. The errors of latitude increase sharply from Cape Lopez southward to the Congo and Benguela; they are of the sort to be expected from 15th Century navigators.

Another detail should attract our attention in passing. An extra island appears near São Tomé on the equator. The fact that the second island (No. 19, in Fig. 70) has the same relationship to the equator of the projection oriented to Magnetic North that the other island we have identified as São Tomé has to the true equator suggests that No. 19 is an addition by somebody exploring Africa's equatorial coast with the map already oriented to 6° E. This would mean, of course, that the original explorers were using True North, not Magnetic North. The 15th Century navigator, sailing by the compass, may have had with him this map already showing the island, but at its correct place on the sidereal grid. And so he added the second island.

But why weren't these explorers honest enough to admit they were exploring these coasts with the help of maps many times better than they could draw for themselves? Or if the Portuguese were using trigonometry and the twelve-wind system, and had a means of finding longitude, why didn't the facts leak out? King John II of Portugal must have had a very efficient security system!

4. The De Canestris Map of 1335–37.

Our discovery of the twelve-wind system in the Venetian Map of 1484 led us into a search for other such maps. Various persons collaborated in this search.

Richard W. Stephenson, of the Map Division of the Library of Congress, went through the map collections in that library; Dr. Alexander Vietor, Curator of Maps at the Yale University Library, also made a search for us, without success. Finally, Alfred Isroe detected the twelve-wind system, in a very dilapidated form, in the De Canestris Map of 1335–37[6] (see Fig. 72).

At first glance this looks like many medieval maps, presumably originating in the peculiar ideas and limited knowledge of the time. Most ingenious work was done in adapting the geography to human forms—including those of a man and a woman (who are seen in lively dispute). Other human heads are observable. This anthropomorphism appears to have been accomplished without distorting the geography to any noticeable extent.

Among the various irregular lines on this map (many of them introduced to complete the human forms), Isroe noticed a few straight lines that suggested the survival of parts of an original pattern resembling that of the portolanos. Measurements with a protractor showed that, while the angles between them were not precisely those of the twelve-wind system, they were much closer to those than to the angles characteristic of the eight-wind pattern.

Taking this suggestion of Isroe's, I thought I would try to reconstruct the possible original pattern. I straightened the two lines emanating from the projection center at the left of the map, on the assumption that they might have been intended originally to represent one straight line. This involved only a slight change. With this change all the other angles of the intersections of the lines traced by Isroe from the photograph of the original map fell into agreement with the twelve-wind system (see Figs. 74 and 75).

In addition to these indications of an original twelve-wind system on this map, I discovered a straight line in the Mediterranean that suggested a parallel of latitude of the original source map. Comparing this line with the present geography of the Mediterranean, I observed that it indicated an orientation of the whole map about $11\frac{1}{2}°$ or 12° east of True North, as on so many of the maps recognized as portolanos. On rotating the map to take into account this apparent magnetic orientation and get it back on True North, I found that the parallel in question was the parallel of Alexandria (Fig. 76). It would seem, then, that this map, and a whole family of other maps from this period of the Middle Ages, are, in fact, not so much original productions of the Middle Ages as degenerated versions of ancient maps, very possibly drawn by the geographers of the School of Alexandria.

---

[6] Isroe had left Keene State College to transfer to the University of Amsterdam and was pursuing a research project there at my suggestion. I wanted very much to locate, if possible, the source maps used by Mercator for his "Atlas" of 1569. Despite co-operation extended by the Dutch Government, the source maps were not found. However, Isroe did make a significant discovery, anyway, as we see here.

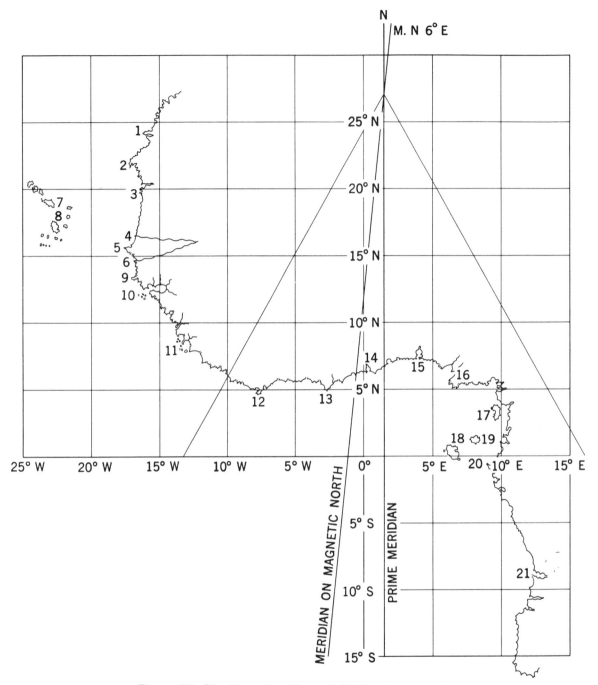

Figure 70. The Venetian Map of 1484, with a modern grid
constructed empirically from the geography and confirmed
by plane trigonometry. See Table 8.

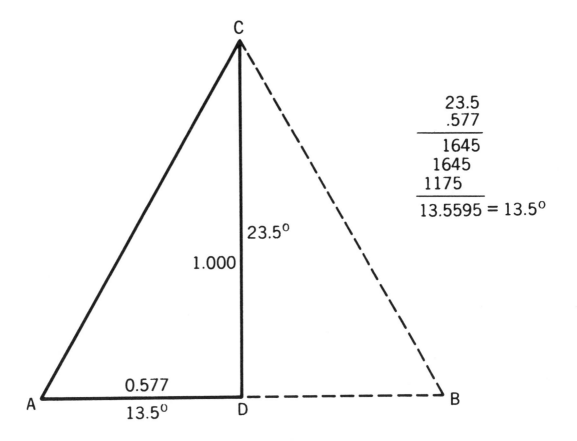

$$\begin{array}{r} 23.5 \\ .577 \\ \hline 1645 \\ 1645 \\ 1175 \\ \hline 13.5595 = 13.5^{\circ} \end{array}$$

Figure 71. The trigonometry of the Venetian Map. Given an equilateral triangle, A-B-C-D, on the earth's surface between a point at 23½° North Latitude and the Equator, A-B, with the length of C-D given (23½°). Problem: to find the length of A-D in degrees, to find the length of the degree of longitude on the draft map. (The ratio 1.000:0.577 is taken from "Natural Trigonometric Functions," in Chemical Rubber Company Tables, C. D. Hodgman, editor. Cleveland: 1956, p. 107.)

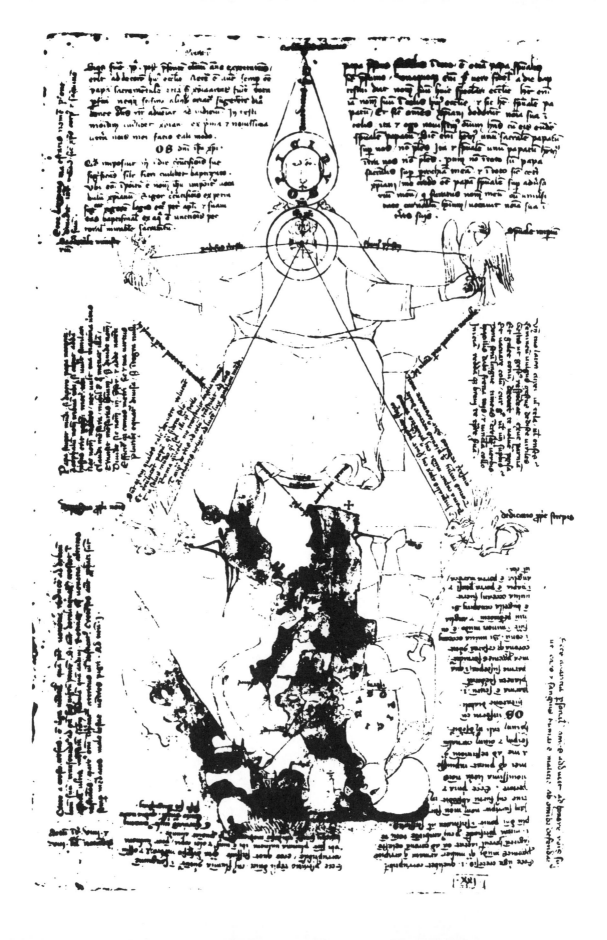

Figure 72. The Opicinus di Canestris Map of 1335–1338. The square grid suggests one of the projections attributed by Nordenskiöld to Ptolemy or Marinus of Tyre. One of the meridians and one of the parallels intersect at the site of Alexandria (see redrawing, Fig. 73). Assuming the interval to be about 5°, the map shows surprising accuracies in the latitudes and longitudes of many geographical localities.

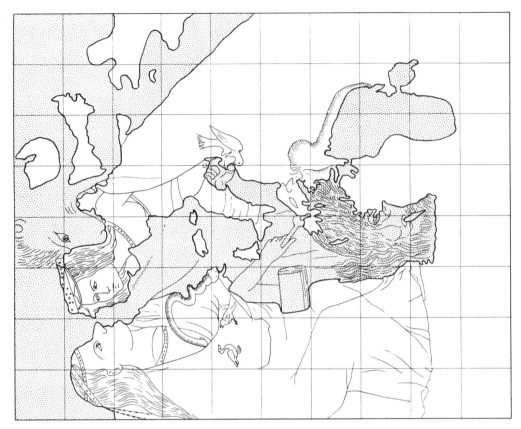

Figure 73. The Opicinus di Canestris Map, section (redrawn).

5. The Reinel Chart of the Indian Ocean.

I felt it important to see whether, having found a map of Africa that seemed to be based on ancient mathematical cartography, it might not be possible to extend the system to Asia. Thus, I might be able to determine whether the ancient cartographers of a vanished race had extended their system farther east. With this in mind, I examined what is considered to be the earliest Portuguese chart of the Indian Ocean (see Fig. 77). I attempted an empirical solution of the grid of this map, along the lines of the investigations of other maps, and made some surprising discoveries. One was the extent of the geographical knowledge contained in the map. I found that it showed a number of islands in the Atlantic, as well as a remarkable knowledge of the archipelagoes of the Indian Ocean itself (see Table 9). As I worked out a grid from the identifiable geographical localities on the map itself, I was astounded to find that this map apparently shows the coast of Australia, the first and only portolano to do so. The map also appeared to show some of the Caroline Islands of the Pacific. Latitudes and longitudes on this map are remarkably good, although Australia is shown too far north.

As I continued to examine this map, I saw that the trend of the Australian coast was wrong, as was its latitude. This reminded me of the Caribbean area of the Piri Re'is Map. Was it possible that we had here another example of a satellite grid, with a different north, integrated with the Piri Re'is World Projection? A comparison of the map with the world map drawn by the Air Force centered on Cairo (see Fig. 25) was extremely thought-provoking. A glance at a tracing of this map with the Piri Re'is projection superimposed on it (Fig. 27) showed that the design of the Piri Re'is projection was capable of being used to cover this area just as well as it was used to cover the Caribbean.

It seemed evident to me that this map showed much more geographical knowledge than was available to the Portuguese in the first decade of the 16th Century, and a better knowledge of longitudes than could be expected of them. The drawing of the coasts, however, left much to be desired. The map looked much like a map, once magnificently accurate, that had been copied and recopied by navigators ignorant of the methods of accurate mapmaking.

6. A Twelfth Century Map of China.

In the effort to see whether the system of ancient maps extended farther east than the Indian Ocean, I examined the available Chinese and Japanese maps. Despite the splendid co-operation of the staff of Japan's great Diet Library (the equivalent of our Library of Congress), which sent me many old Japanese charts, I was not able to discover any maps that bore apparent relationships to the western portolanos, except maps of a comparatively late date which might have been influenced by western cartography.

I had much better luck in China. This was owing entirely to the availability of Needham's great work on *Science and Civilization in China* (145). In Volume III of that work he reproduced a very remarkable map that had been carved in stone in China in the year 1137 A.D. (see Fig. 79). Although the map was carved in 1137, it is known to have been in existence for an indefinite period before that. Its real date of origin is unknown. Therefore it is wrapped in the same mystery as are the portolanos of the West. A comparison of the river system shown on this map with that on a modern map of China shows a remarkable accuracy (see Figs. 80 and 81). This map was evidently drawn with excellent information as to longitudes, such as we find on the portolanos, but do not find on the classical maps of Greece and Rome, and which was certainly not typical of the cartography of medieval China or Japan.

Needham, and presumably the Chinese scholars who have studied this map, apparently assumed that its square grid was the original grid on which it had been drawn; this was a perfectly natural conclusion for them. On the other hand, I had just recently discovered that the square grid inherent in the plane trigonometry of the portolan projection was evidently *not* the original grid on which some of

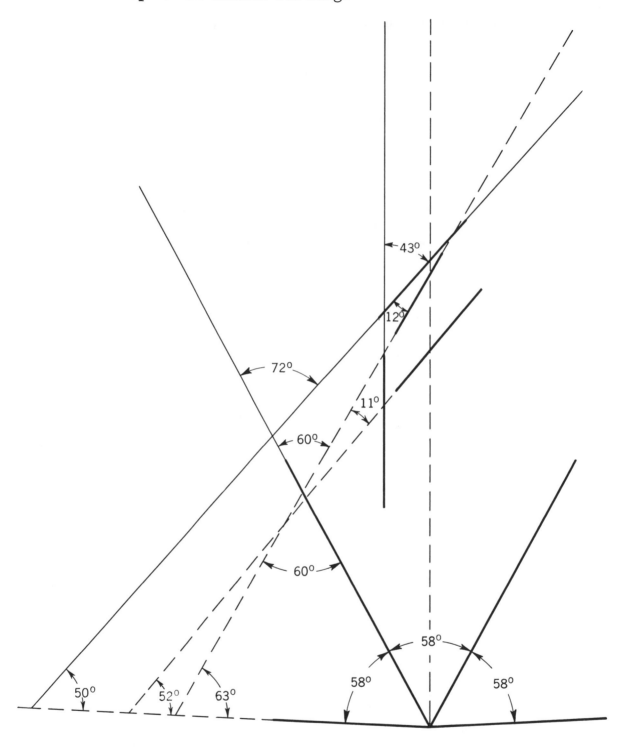

Figure 74. Opicinus di Canestris Map, tracing of projection lines.

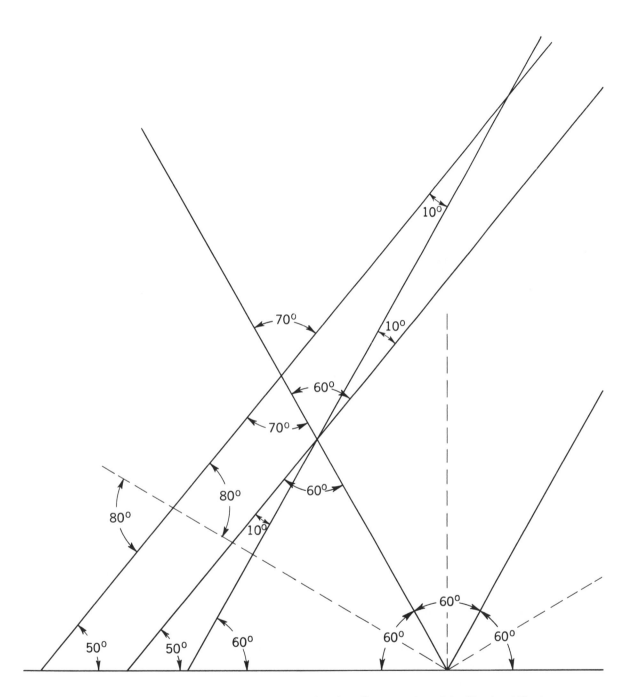

Figure 75. Opicinus di Canestris Map, projection lines restored to Twelve-Wind System.

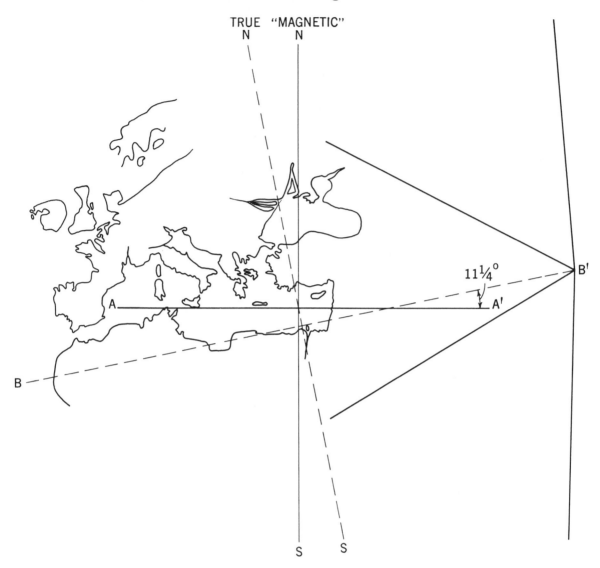

Figure 76. Opicinus di Canestris Map reoriented.

the source maps used by Piri Re'is and other mapmakers had been drawn. I had just come to believe, on the contrary, that Piri Re'is' source map had originally been drawn with an oblong grid of some kind. Therefore I decided to test the grid of this Chinese map.

I began by trying to find the length of the degree of latitude on a tracing of the map. As before, the procedure was to pick a number of geographical features that were easily and clearly identifiable and find their latitudes on a modern map. These were distributed from the northernmost to the southernmost parts of China. I extended lines from these points to the margins of the tracing and found the length of a degree of latitude by dividing the number of millimeters on the tracing from north to south by the number of degrees of latitude between the identified points.

Then I repeated the process to find the length of the degree of longitude. I supposed it would probably come out the same, but nevertheless a sense of excitement gripped me as I noted the longitudes of identifiable places across the map and drew lines from these to the bottom of the tracing. I used a number of geographical points in each case, for finding the lengths of the degrees of latitude and longitude, to ensure against the risk that any one of the chosen points might be out of place because of a local error in the map. Thus, if I depended only upon two positions, at either extreme of north and south or east and west, an error might be made in the length of the degree.

When I finished the measurement of the degree of longitude on the map I was truly electrified, for I found that it was unmistakably shorter than the degree of latitude. In other words, what revealed itself here was the oblong grid found on the Piri Re'is Map, found on the Ptolemy maps, and found, through spherical trigonometry, on the De Canerio Map. The square grid found on the map was, then, clearly something superimposed on the map in ignorance of its true projection. This, together with the fact that the square grid was similar to the square portolan grid, created an altogether astonishing parallel, a parallel that suggested an historical connection between this map and the maps of the West. If I may be allowed to speculate here, I may suggest that perhaps we have here evidence that our lost civilization of five or ten thousand years ago extended its mapmaking here, as well as to the Americas and Antarctica.

The square grid imposed on the map is evidence of the same decline of science we have observed in the West, when an advanced cartography, based on spherical trigonometry and on effective instruments for determining latitudes and longitudes, gave way to the vastly inferior cartography of Greece—and when, later in the Middle Ages, even the geographical science known to the classical world was entirely lost. In China, the square grid was apparently imposed on the map by people who had entirely forgotten the science by which it was drawn.

There are other indications that the map was drawn in its present form in an age of the decline of science in China. Despite the extraordinary accuracy of the geographic detail of the Chinese interior, the coasts are hardly drawn in at all; they are only schematically indicated. This suggests to me that the map was carved in stone in an age when China had no interest in the outside world, but an enormous interest in the great river system that carried the internal commerce of the fabulously rich empire. The original map may have shown the coasts in detail; but in the 12th Century they were apparently of interest to nobody.

The map shows some of the rivers flowing in directions different from those of the modern map. This does not necessarily mean that there were inaccuracies in the ancient map. The rivers of China—particularly the Hwang Ho, or Yellow River—have the habit of changing their course, with the most disastrous conse-

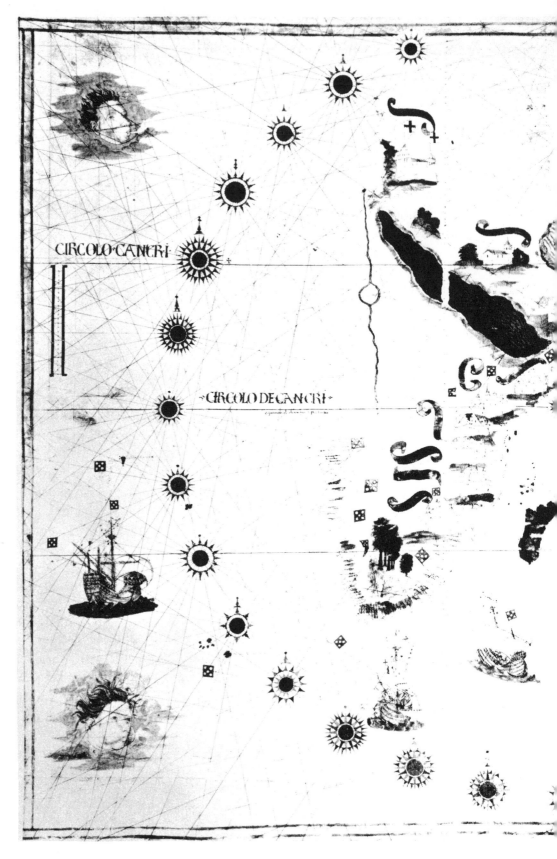

CIRCOLO·CANCRI

·CIRCOLO DE CANCRI·

Figure 77. Portuguese Map of the Indian Ocean, Jorge (?) Reinel, 1510.

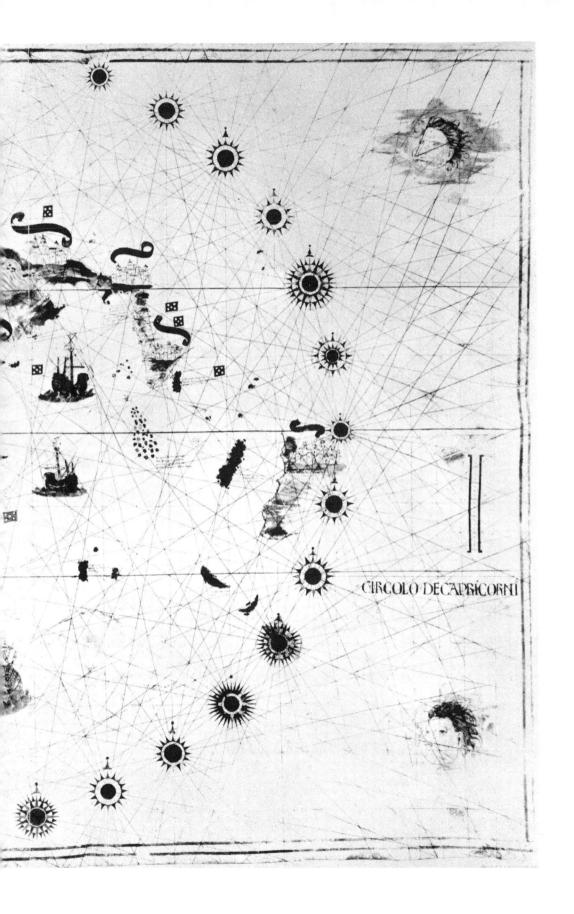

CIRCOLO DE CAPRICORNI

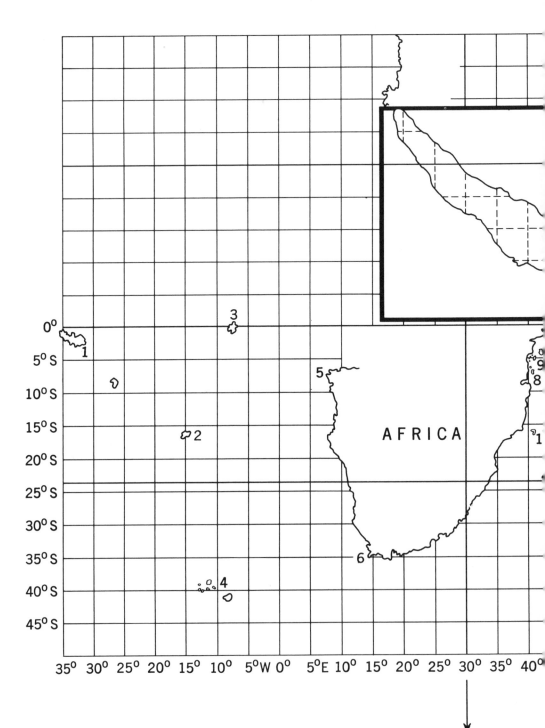

0°
5° S
10° S
15° S
20° S
25° S
30° S
35° S
40° S
45° S

35° 30° 25° 20° 15° 10° 5°W 0° 5°E 10° 15° 20° 25° 30° 35° 40°

3

1

2

5

AFRICA

6

4

8

9

1

1

(MERIDIAN OF
ALEXANDRIA)

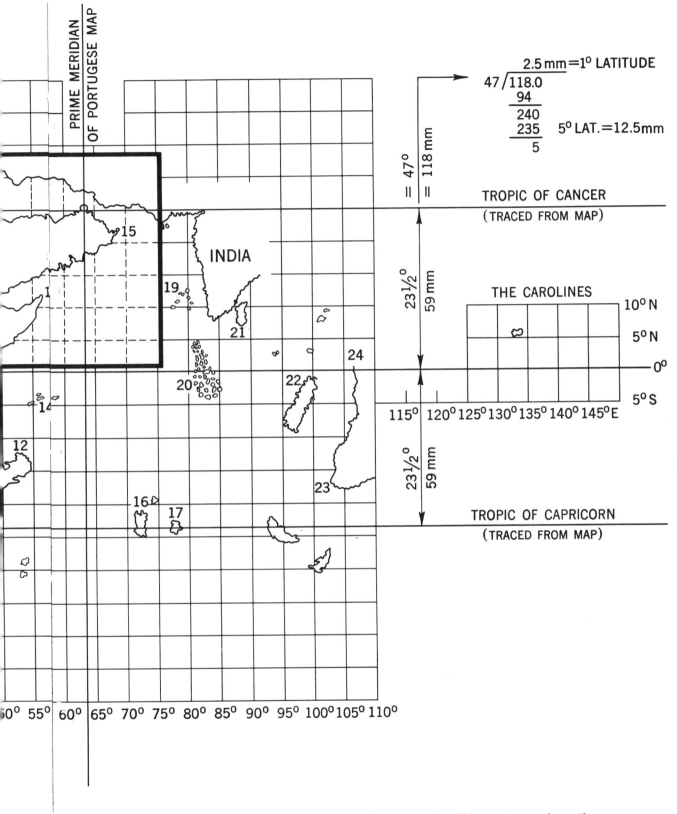

Figure 78. Portuguese Map of the Indian Ocean, with grid constructed on the basis of the length of the degree of latitude as measured between the Tropics of Cancer and Capricorn, and on the assumption of equal lengths of the degrees of latitude and longitude. See Table 9.

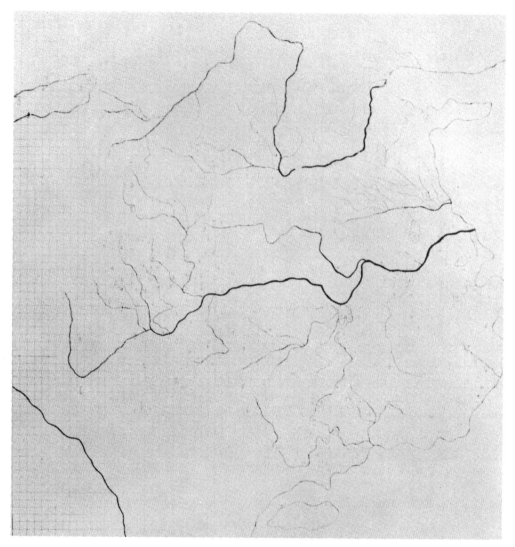

Figure 79. The Chinese map of the 12th Century.

quences. The Yellow River is, in fact, called "China's Sorrow." It has changed its course three times in a century and a half. The ancient map shows it following a course to the north of its present course, but its course, in one of the northern valleys, is perfectly reasonable.

I subjected the grid I had constructed for this map to the most rigorous testing. Using the grid, I identified a large number of additional geographical localities, mostly the intersections of major rivers, rejecting any that appeared in the least dubious. I have listed these localities, with the discrepancies in their positions, in Table 10, a, b, c. I grouped the localities in the northwest, northeast, southwest, and southeast quadrants of China. In each quadrant in turn I averaged the discrepancies, or errors, in the latitudes and longitudes of places, with the following results.

Table 10a (summarized)

Quadrant	Number of Localities	Average Errors
1. Northwest	8	0.4° Lat. 0.0° Long.
2. Northeast	10	0.0° Lat. 0.0° Long.
3. Southwest	9	1.3° Lat. 1.2° Long.
4. Southeast	7	0.8° Lat. 1.2° Long.

Here we have evidence that when this ancient map of China was first drawn, mapmakers had means of finding longitude as accurately as they found latitude, exactly as was the case with the portolan charts in the West. The accuracy of the map suggests the use of spherical trigonometry, and the form of the grid, so like that of the De Canerio Map, suggests that the original projection might have been based on spherical trigonometry.

As a further test of the grid I had drawn for the map, I listed separately all the northernmost and southernmost places identified on the map and averaged their errors in latitude. I also listed all the easternmost and westernmost places and averaged their errors in longitude (Tables 10b and 10c). The average error of latitude on the north was less than one-half of one degree (or 30 miles!), and the average error on the south balanced out to zero (with four localities 1° too far south and four 1.2° too far north). So far as longitude was concerned, the errors both on the east and on the west balanced out to zero. There was no indication, therefore, that the grid constructed for the map was seriously in error.

It seems to me that the evidence of this map points to the existence in very ancient times of a *worldwide* civilization, the mapmakers of which mapped virtually the entire globe with a uniform general level of technology, with similar methods, equal knowledge of mathematics, and probably the same sorts of instruments. I regard this Chinese map as the capstone of the structure I have erected in this book. For me it settles the question as to whether the ancient culture that penetrated Antarctica, and originated all the ancient western maps, was indeed worldwide.

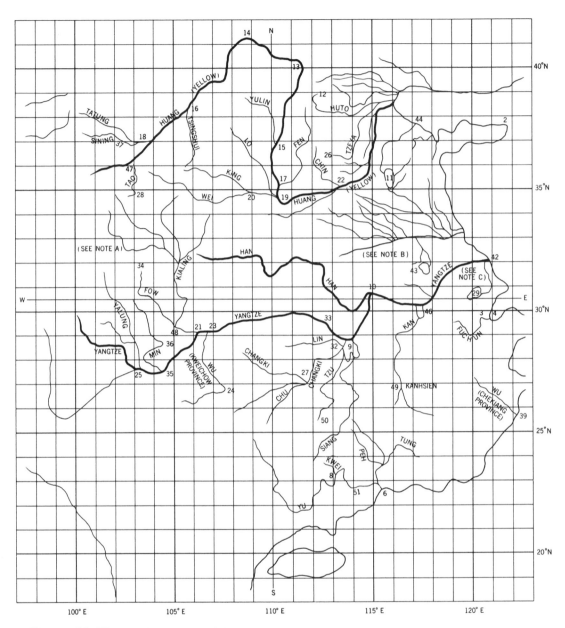

Figure 80. The ancient map of China with grid of lines of latitude and longitude constructed empirically from the geography. Numbers indicate identified localities. See Table 10a.

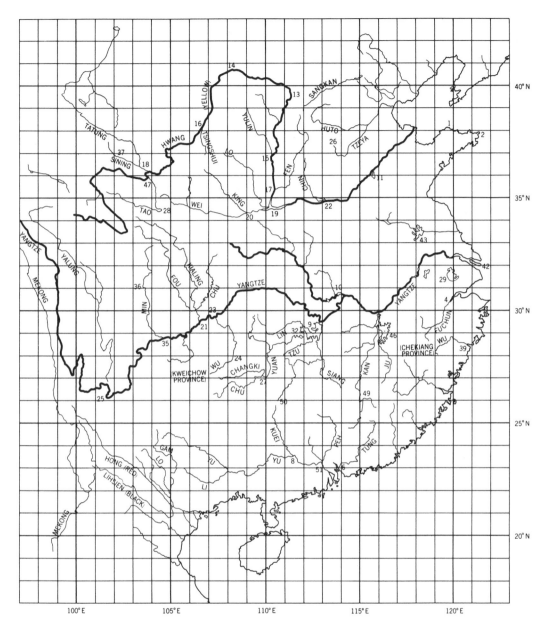

Figure 81. A modern map of China, with numbers indicating the same localities as Figure 80.

CHAP
TER VI

THE
ANCIENT
MAPS
OF THE
NORTH

The Zeno Map of 1380. The Ptolemaic Map of the North. The Map of Andrea Benincasa. The Portolano of Iehudi Ibn Ben Zara of Alexandria.

We have seen that the analysis of a number of maps has led to surprising conclusions and implications. They appear to call in question not only the accepted ideas of ancient history, and especially the history of cartography, but also fundamental conceptions of geology. We have seen, in the maps already considered, suggestions of voyages to America and Antarctica that must have occurred in times preceding the oldest of our historical records— voyages accomplished by a people or peoples whose memory has not survived. The Oronteus Finaeus Map appears to document the surprising proposition that Antarctica was visited and perhaps settled by men when it was largely, if not entirely, non-glacial. It goes without saying that this implies a very great antiquity. In the next chapter we shall consider in greater detail the reasons for supposing that the evidence of the Oronteus Finaeus Map takes the civilization of the original mapmakers back to a time contemporaneous with the end of the ice age in the northern hemisphere.

Facts do not stand alone. A given statement may mean more or less depending upon the context. The maps we are now to consider cannot be considered *in vacuo*. They should be evaluated with some reference to what we have already learned. If they suggest an enormous antiquity for the cartographic tradition their evidence cannot be so easily dismissed as it might be if it stood alone. I cannot claim that the evidence is so strong as to compel acceptance. I admit frankly that it is doubtful and may be interpreted in more than one way, but I do not apologize for suggesting my own interpretation of it.

1. The Zeno Map of 1380.

The Zeno Map of 1380 was supposedly drawn by two Venetians, Niccolo and Antonio Zeno, who made a famous voyage to Greenland and perhaps Nova

Figure 82. The Zeno Map of the North (1380 A.D.).

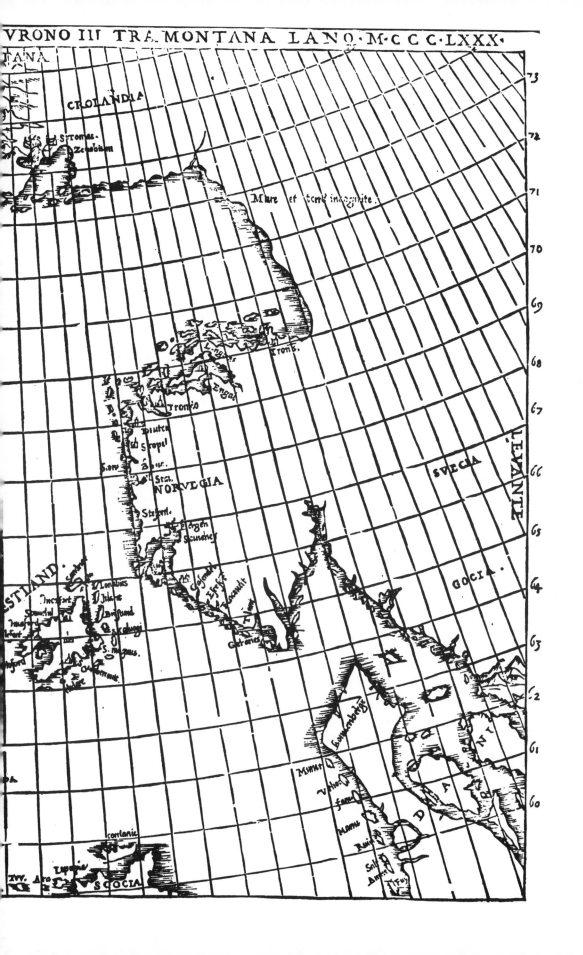

Scotia in the 14th Century (see Fig. 82). Two hundred years later, a descendant of these Zenos—an important family in Venice—found the map mouldering away among the family papers, and copied it. He appears to have removed the lines of the ancient projection, most probably, it would seem, of portolan type, and attempted to substitute one of the new projections that were being developed in the Renaissance.

A study of the map itself shows that it was probably not drawn by the Zeno brothers. In the first place, though they are supposed to have visited only Iceland and Greenland, the map also shows the coasts of Norway, Sweden, Denmark, Germany, and Scotland, and, in addition, the Shetland Islands and the Faroes. In the second place, a polar type of projection applied empirically to the map (Fig. 83) shows that the latitudes and longitudes of many places scattered all over the map are amazingly correct (see Table 11a). It is unbelievable that anyone in the 14th Century could have found accurate latitudes for all these places, to say nothing of accurate longitudes. Nevertheless, when the map is assumed to be based on a correct measurement of the longitude across the Atlantic, then the latitudes of places turn out also to be closely correct, so that the original mapmakers must have known the correct comparative lengths of the degrees of latitude and longitude in the North Atlantic. This suggests they had sound information on which to base their map.

A third consideration is that when we select the 30th meridian of East Longitude—the meridian of Alexandria—as our prime meridian, and use two norths, as in the Piri Re'is Map (but in this case at right angles to each other), we can draw a typical portolan grid for this map, which is almost as accurate as the polar projection (see Fig. 84 and Table 11b). It seems probable that this is the result of conscious intention. It looks very much as if early mapmakers, somewhere and sometime, found themselves with an excellent map of the North Atlantic and the Arctic, drawn on a projection based on spherical trigonometry, and very ingeniously applied their flat projection to it. Their motive in doing this eludes us. They must have possessed all the necessary information on the latitudes and longitudes of the geographical features, which could have been derived from the original grid of the map. Did they work out their map with two norths as a sort of game? There are a good many examples of crude medieval maps that seem to have originated as cryptograms or puzzles.

The suggestion of a vast antiquity behind this map is conveyed by a feature to which Captain Mallery first drew attention. He pointed out that the Zeno Map shows Greenland with no ice cap (130). The interior is filled with mountains. Rivers are shown entering the sea, in some cases at the points where at present great mountain glaciers are moving down through the mountains to the coast. He called attention to a flat area stretching across Greenland, interrupted by some mountains halfway across, and stated that seismic expeditions in recent years found that the under-ice topography agreed with the Zeno Map. It is quite true that the

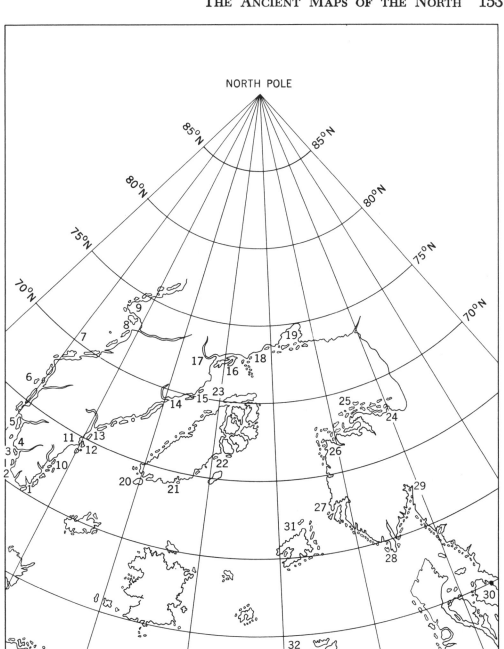

Figure 83. The Zeno Map with a reconstructed polar projection.
See Table 11a.

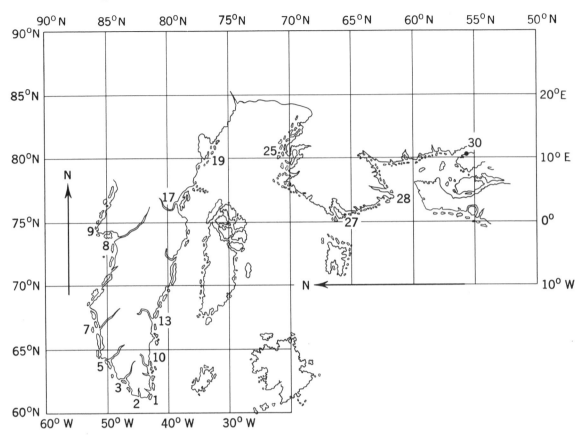

Figure 84. The Zeno Map with a square portolan projection.
See Table 11b.

Paul-Emile Victor French Polar Expeditions of 1947–49 did cross the Greenland ice cap and make a seismic profile of the thickness of the ice. (For the profiles, see Fig. 46.)

It might be objected that the flat territory shown extending across Greenland in the Zeno Map is above water, while what the Victor Expedition revealed was a strait cutting Greenland into two or three islands. This, however, might be explained. It is not impossible that if the original map was made when Greenland was ice-free, the region was then standing higher relative to sea level. Later, with the imposition of a very thick ice cap (more than a mile thick) the land might be expected to sink under the weight enough to bring it below sea level.

Captain Mallery placed great emphasis on this evidence that the mapping was done a very long time ago. In my opinion this aspect of the map needs to be considered in connection with the entire body of evidence derived from the ancient maps, even though, by itself, it may be easy to disregard it. The mountains, valleys, and rivers shown in Greenland, it might be argued, represent the medieval cartographic imagination, and the low area stretching across Greenland may be mere coincidence. As I have already said, the reader is entitled to his own conclusions.

Each of the maps presented new and special problems, but the Zeno Map was especially difficult. Since my solution was achieved by a step-by-step process, and can only be clearly understood if followed chronologically, I will try to reconstruct each step in the process in turn, as it actually occurred. I must emphasize that when I started on the analysis of this map, I had not the slightest idea what it would reveal. I was not optimistic that it would reveal anything. I had been discouraged by the fact that the portolan type of design, which I supposed must have originally been on the map, had apparently been replaced by the 16th Century Zeno with a more modern sort of grid, ending the possibility of a solution along the lines of the Piri Re'is Map. My students had failed to solve the problem, and so had the cartographers of the Air Force. Therefore, we just put the map away, and I intended to ignore it.

The map had been studied by Mallery in his book, *Lost America*, but we were not in agreement with his conclusions. He had noted the accuracy of the ancient map, with respect to many points in Greenland, by drawing a grid of his own, based on the geographical points themselves, but not extending to the whole map. He assumed further that the large island to the east of Greenland on the Zeno Map was not meant to be Iceland. In his opinion this island represents Gunnbiorn's Skerries, islands that reportedly existed in medieval times along the Greenland coast, but now are partly subsided beneath the sea and partly covered by the Greenland ice cap. We could not agree with this.[1]

In March, 1964, during the preparation of the manuscript of this book, I decided to take one last look at the map and to review carefully just what Mallery had done, in order to see whether it was really as accurate as he claimed. I had run across an article by the geologist William H. Hobbs (93), who knew his Greenland and who said that the map was remarkably accurate. Therefore, I got out the map, looked at it, and collected and laid out a number of modern maps of the same general area—maps of the North Atlantic and of the Arctic.

First I noticed that the grid actually on the map was not the portolan kind, but a sort of circular, polar one. By comparison with modern maps, I could see the sense of this. After all, this was a polar area. It seemed that the square or oblong grids of most of the other ancient maps would never do here. Meridians in Greenland pointing north could not be parallel to meridians in Norway pointing north. The meridians had to converge at the poles. It did not occur to me then that the problem might be solved by using two norths.

Was it possible that this was, after all, the original projection, as the 14th Century Zenos had found it?

I analyzed the projection farther. Assuming that each space between the

[1] Although we did not agree with Mallery's conclusions, we wish to emphasize that he deserves the credit for pioneering the study of this map, as he pioneered that of the Piri Re'is Map.

meridians and the parallels represented one degree, I counted the degrees to see how much longitude there was across the Atlantic. I picked two recognizable places, Cape Farewell at the tip of Greenland and Cape Lindesnës at the tip of Norway. These places were nearly at the same latitude. Since Cape Farewell is at Longitude 44° West, and Lindesnës is at Longitude 5° 30′ East, the total longitude difference equaled 49½°. On the Zeno Map, however, counting each meridian as one degree, the longitude difference was only 30°, obviously very far off. With latitude similarly, starting with the true position of Cape Farewell at 60° North Latitude and counting northward, the northernmost identifiable point in Greenland, Cape Atholl on the west coast, would lie at only 67° N, instead of 77° (though the map starts by erroneously putting Cape Farewell nearly 6° too far north, as the reader can see).

I played with the idea that perhaps the 16th Century Zeno had made a mistake. He might have misinterpreted the grid. Perhaps each interval equaled two degrees of latitude or longitude, instead of one. This idea was not really satisfactory either, because it would give 60° across the Atlantic, instead of the correct 49½°. Furthermore, the curvature of the parallels of latitude across the Atlantic did not seem to me to be sufficient for the high latitude of Greenland. A comparison with modern maps of the polar region showed the difference in curvature. It seemed to me that the degree of curvature on the Zeno Map would be appropriate to a much lower latitude.

Finally, it was evident from the geography itself that the grid did not accurately represent north for either Greenland or Norway; there should be a much sharper convergence of the meridians. If the meridians as shown on the map are projected to the point of meeting, the pole so found is much too far north of Greenland (the northernmost point of which is actually only about six degrees from the pole) and the island thus is pushed much too far south.

I concluded that somebody had made a mistake, at some time or other, in applying this sort of projection to the map. It might well have been the sixteenth century Zenos, but hardly the earlier Zeno brothers, for no one in the 14th Century drew grids of latitude and longitude on circular projections.

There was nothing to do but start afresh and draw a projection to fit the map—not just to fit Greenland, but to fit the map as a whole. The first problem was to find the right location for the North Pole. My first step was to find two localities on opposite sides of the Atlantic in about the same latitude. As stated, I found Cape Farewell in Greenland at 60° North, and Cape Lindesnës in Norway at 58° North. These are both very clearly shown on the Zeno Map. I now sought to draw a curved line, to represent the 60th parallel of north latitude, from Cape Farewell to a point just north of Cape Lindesnës. To do this I began by finding what seemed the direction of north in Greenland and also in Norway on the Zeno Map and drawing lines running due north until they met at a point representing the pole. The first experiment did not work because, when I described a circle with this point as a center and with a radius to Cape Farewell, it did *not* pass just north of Cape Lindesnës. I then experimented in raising or lowering the polar point, and moving it slightly this way and that, until I found a point from which I could describe a circle that would intersect Greenland and Norway at the *same* latitude (60° N), that is, the latitudes of Cape Farewell and Cape Lindesnës. Then it was a simple matter to subdivide the radius by thirty (the number of degrees between the 60th parallel and the pole) to find the length of the degree and draw a grid. This grid, it is true, did resemble very much the grid of the original map, though it was differently oriented and the parallels curved more sharply. Meridians could now be drawn easily at five-degree intervals from this pole, starting with Cape Farewell in 44° West Longitude.

The first grid drawn in this way, when tested, did not prove sufficiently accurate. It indicated that my pole was too low, because of some mistake in finding the direction of north in Greenland and Norway. There were twenty-one and a half degrees of longitude too many between Cape Farewell and Cape Lindesnës. Since I had started counting my degrees from Cape Farewell, assuming correct longitude for that point (44° W), I found Cape Lindesnës to be 21½° too far east. At the same time, however, the latitudes of all the points seemed remarkably accurate. I was afraid that if I corrected the grid to get correct longitude across the Atlantic, the latitudes would be thrown way out. And I reflected that the ancient map was more likely to have been wrong on longitude than on latitude.

Nevertheless, I thought it worthwhile to assume that the ancient mapmaker knew the relative longitudes of Norway and Greenland precisely, and I therefore raised the pole enough so that the meridians from Cape Farewell and Cape Lindesnës would meet there at precisely the angle of 49½°—the correct longitude difference. This meant a very considerable change in the length of the degree of latitude as well as in that of longitude. It was therefore with real anxiety that I drew a new grid on the map (as shown in Fig. 83) and tabulated the positions of the localities. To my astonishment I found that with the revision of the degree of longitude, the accuracy of latitude of the whole map was notably increased. How accurately the ancient mapmaker knew the latitudes and longitudes of points in the North Atlantic is evident from Table 11a. The navigators who gave him his data must have had good instruments for finding both latitude and longitude. It is highly probable also that they knew the distances they traveled, and had an accurate knowledge of the size of the earth.

It goes without saying that this map was not drawn by the Zeno brothers. They traveled over only a small part of the area mapped; they had no way of finding longitude, and in the late 14th Century they could not have found accurate latitudes for the points on the Greenland coast. So far as we know, they had no idea at all of the size of the earth—even the estimate of Ptolemy, wrong as it was, was not yet available to them.

The map, then, was very probably a copy of an ancient map. A version may have reached Venice from Constantinople, perhaps soon after the Fourth Crusade

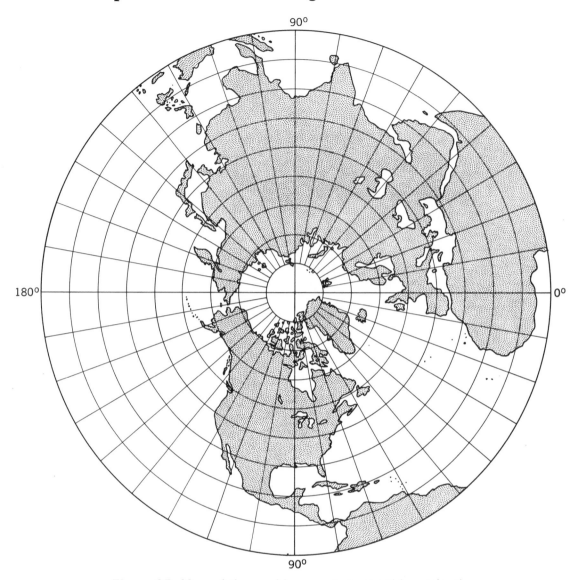

Figure 85. Map of the world on a stereographic projection.

of 1204 A.D., in which the Venetians captured Constantinople. It is interesting to speculate on the nature of its original projection. In reviewing the different projections used by ancient and modern mapmakers (60) I noted that on a map of the world drawn on the stereographic projection Greenland is shown at an angle to Norway similar to its angle on the Zeno Map (see Fig. 85). The stereographic projection dates from ancient times.

2. The Ptolemaic Map of the North.[2]

One of the great events of the 15th Century was the recovery of the works of Claudius Ptolemy, the last great geographer and cartographer of classical antiquity, who lived in the 2nd Century A.D. The works included a treatise on geography, still of great interest, tables of latitudes and longitudes of known geographical localities, and a large body of maps.

The maps published in the 15th Century, although attributed to Ptolemy, are not considered to have been actually drawn by him. Some authorities have considered that they were reconstructed from the tables sometime during the Middle Ages, or even in the 15th Century. Others, on the other hand, feel that no one in the Middle Ages (or 15th Century) was capable of reconstructing the maps in such detail from the tables left by Ptolemy. Among the latter is the Danish scholar Gudmund Schütt, author of a treatise on Ptolemy. Schütt writes:

It is well known that the study of geography decayed lamentably after the close of the Roman period, or even earlier. How, then, could ignorant copyists in medieval times have undertaken the enormous task of constructing a detailed atlas on the base of the Ptolemaic text, and have carried it out so remarkably well? Such an idea cannot be entertained. The manuscript atlases, as we have them, at the first glance are proved to be copies of a classical original, executed by an expert who . . . represented the highest standard of geographical science of the classical era. (186)

Schütt adds more evidence to support his conclusion, showing in some detail that the manuscript atlases of Ptolemy recovered in the 15th Century are closer in style to other surviving works of the 4th Century A.D. than they are to those of the 5th and 6th Centuries. This would suggest that the maps we have were the work of someone who lived within two centuries of the lifetime of Ptolemy even if they were not drawn by him. It is entirely possible that they were good copies of maps he drew.

Ptolemy himself worked at the Library of Alexandria, and had at his disposal not only the contemporary information on the geographical features of the known world (see his World Map, Fig. 6) but also the works of preceding geographers, such as Marinus of Tyre, and the maps that had been accumulated during the library's five centuries of existence. It can be considered likely that he saw the prototypes of some of the maps we have been studying in this book, though he may not have realized the aspects in which they were superior to the cartography of his own time.

---

[2] I have seen three versions of this map. The one reproduced in Fig. 26 is from the "Monumenta Cartographica Vaticana" (Vatican Atlas), Vol. I, Plate LII, and is entitled "Carta Dei Paesi Settentrionale Dell' Europa Contenant la Geographia di Tolemeo."

Figure 86. A Ptolemaic Map of the North.

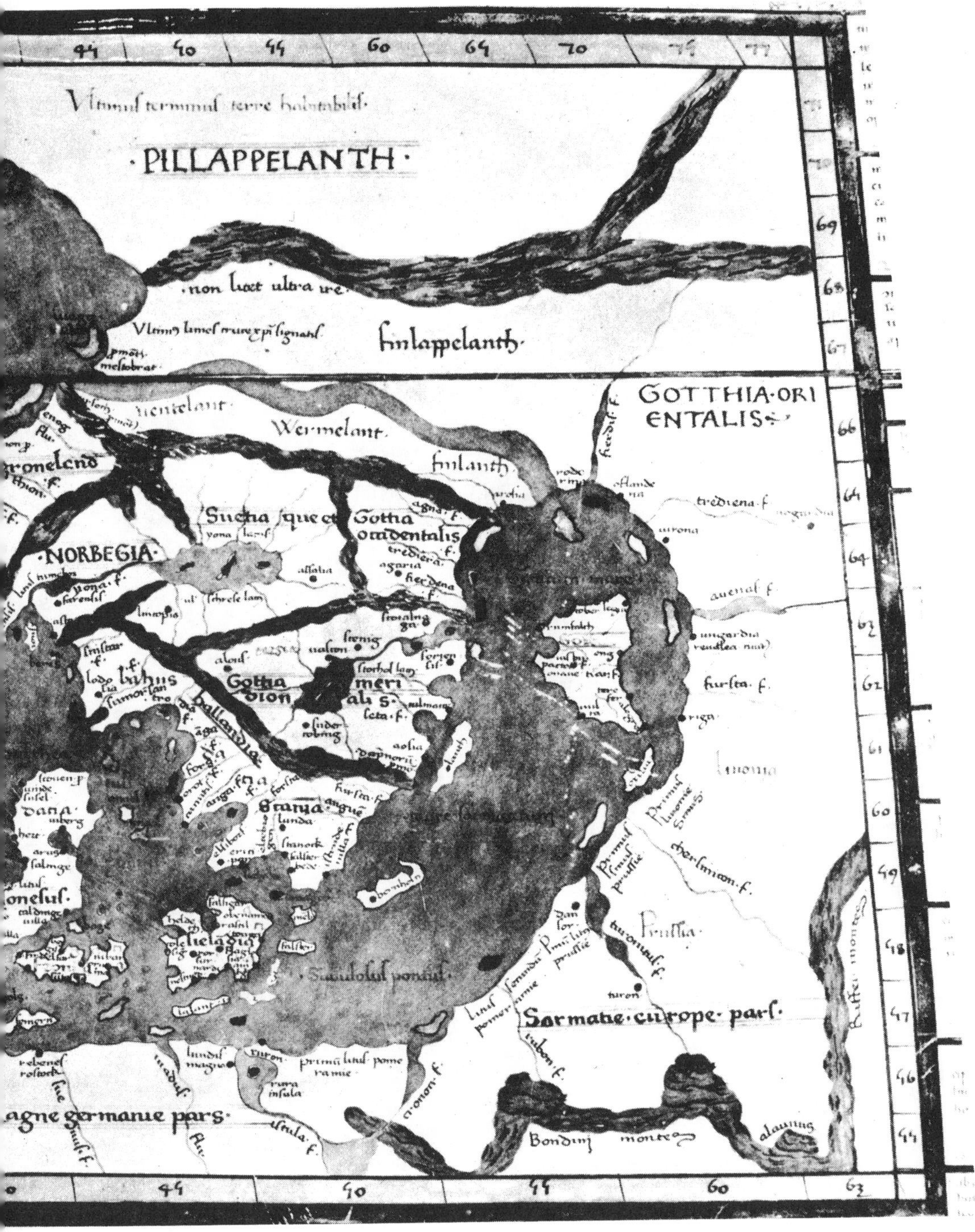

·PILLAPPELANTH·
non licet ultra ire
finlappelanth
GOTTHIA·ORIENTALIS
Wermelant
finlanth
NORBEGIA
Prussia
Sarmatie·europe·pars
agne germanie pars
Bondini montes

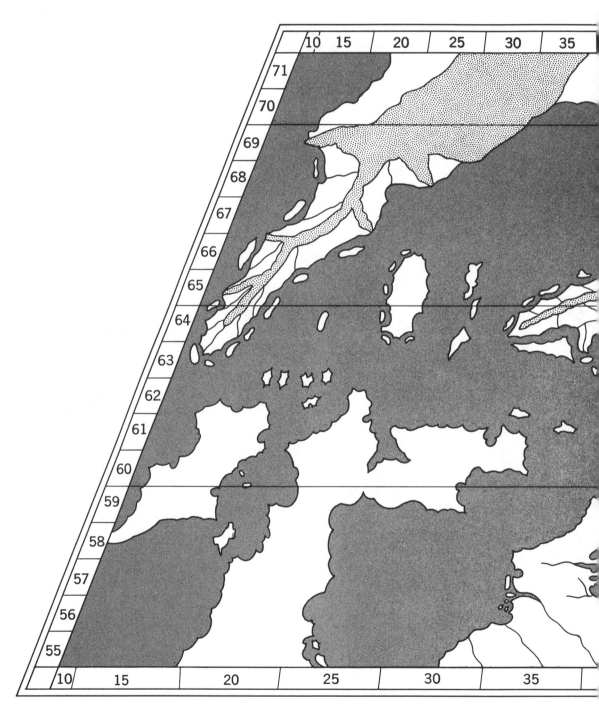

Figure 87. Redrawing of the Ptolemaic Map of the North.

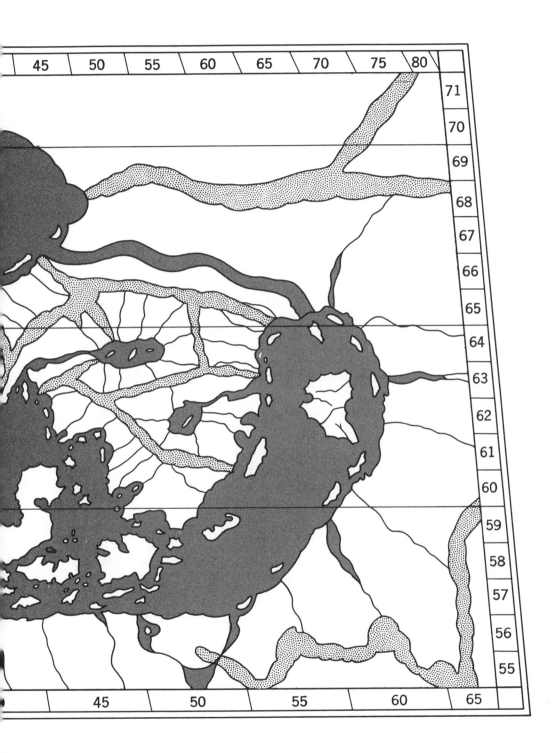

The map we are now to consider is similar in style to those published in all the Ptolemy atlases in the 15th Century. It reflects, as they do, considerable information on the latitudes of places but exceedingly poor conceptions of longitude. Ptolemy had to depend on travelers' itineraries and similar information for his estimates of distances in the Roman Empire, for there was little geographical information derived from stellar observations. What there was dealt with latitudes only, since there was no scientific way of determining the longitudes of places. As a result, the shapes of countries and seas were sadly distorted on the Ptolemy maps, as we saw in the comparison Nordenskiöld made of his Mediterranean map with the Dulcert Portolano (Fig. 4).

In Ptolemy's Map of the North (Figs. 86, 87) we see these characteristics. Our version of the map seems to be the work of two different copyists. The part including Britain and Ireland is Ptolemaic in outline, but it is decorated with an artistic scroll device of geometrical character and has no internal details. The rest of the map is more typical of Ptolemaic maps in general, and shows a number of authentic geographical details, such as the lakes of southern Sweden.

The most remarkable detail of the map is the evidence it appears to contain relative to glaciation. It shows Greenland largely, but not entirely, covered by ice. The shape of the island suggests that of the Zeno Map and may come from the same ancient source. The ice is artistically suggested—there even seeming to be a sheen such as might be produced by the reflection of sunlight from the ice surface. There is a suggestion that when the map was drawn the ice cap was much smaller than it is now (such as it is indeed supposed to have been during the early Norse occupation).

If we turn our attention to southern Sweden we see further evidence of what seems to be glaciation. Although there are still glaciers in Scandinavia, there is none in this part of Sweden. But the map shows features drawn in the same style as the Greenland ice cap. Unbelievable as it may appear, they actually do suggest the remnant glaciers that covered this country at the end of the last ice age, about 10,000 years ago. Some fine details strengthen the impression. Lakes are shown suggesting the shapes of present-day lakes, and streams very much suggesting glacial streams are shown flowing from the "glaciers" into the lakes. To me, there is a strong suggestion here of the rapid melting of the glaciers during the period of the withdrawal of the ice. It goes without saying, of course, that no one in the 15th Century, no one in earlier medieval times, and no one in Roman times ever had any suspicion of the former existence of an ice age in northern Europe. They could not have imagined glaciers stretching across southern Sweden—and they would not have invented them.

Additional details deserve notice. Features of the same type—some of them following the ridges of mountain chains, but some not—can be observed on this map behind the German and Baltic coasts. They begin in the Erz Gebirge, or Hartz Mountains, in Germany, in correct longitude relative to southern Sweden,

Figure 88. Physiographic Map of southern Sweden.

and stretch eastward across the Riesen Gebirge (the Sudeten Mountains) to the main range of the Carpathians, where they turn sharply southward, in the direction of the axis of the mountain range. Then the map shows the glacier turning northward, where it seems to follow quite accurately the highlands of Western White Russia (bordering Poland on the east) and ending in the Livonian Highlands in about 57° N, in correct latitude relative to southern Sweden.

I do not think that these features should be dismissed as merely representing mountains, even although the 15th Century copyist can only have assumed they were mountains. It is natural enough that the glaciers of the end of the ice age should have lingered longest in the mountainous areas, but there are no mountains in southern Sweden, and there are no mountains in Poland or Livonia. Figure 88 shows the present topography in southern Sweden.

Comparing the Ptolemy Map of the North with the Zeno Map, we can see

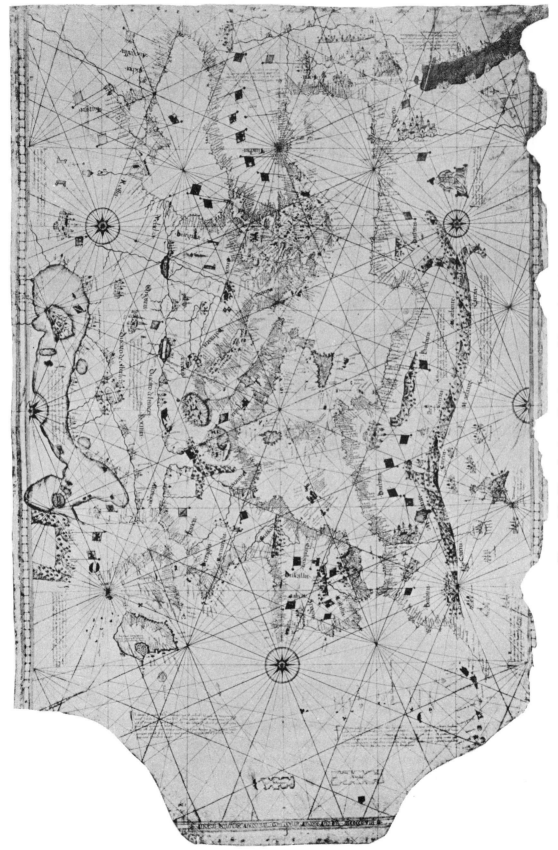

Figure 89. The Andrea Benincasa Map of 1508.

that they are related, but that they must have derived from sources dating from different times. If the original source of the Ptolemy Map came from the end of the ice age, that of the Zeno Map may have originated much earlier.

3. The Andrea Benincasa Map of 1508.[3]

This is one of the best of the portolan charts (see Fig. 89). Examination revealed that it was oriented to Magnetic North, about 6° E.

To draw a grid for this map we first found the length of the degree of longitude by measuring the distance on the map in millimeters between known points at either end of the map—in this case Gibraltar and Batum—and dividing the millimeters by the number of degrees of longitude between them. The length of the degree of latitude was found in the same way separately by using points on the Atlantic coast from Cape Yubi to Ireland, and points on the east from Cairo to Yalta. On our draft map, we found the longitude degree to be 7 mm, the latitude degree on the Atlantic to be 9 mm, and the latitude degree on the east to be 5 mm. As the longitude degree was intermediate, we took this as the basis for our grid. There were not enough points (and not enough total latitude) on the east to give us a reliable measure for the degree (Fig. 90).

The grid had to be tied to some geographical reference point, and for this we chose Cape Bon, near the former site of Carthage, as it was central and well delineated on the map. Subsequently we discovered an error of about one degree affecting the whole map, and therefore moved our meridians one degree to the east. The resulting grid indicated an amazing accuracy for the map as a whole. Table 12 shows that longitude errors, on the basis of the length of the degree assumed by us, average less than a degree in a total of forty-seven degrees (or nearly 3,000 miles) between Batum and Gibraltar. This is an indication that the mapmaker could find precise relative longitudes of places, something which, as we have just seen, Ptolemy could not do.

In order to evaluate this map's most remarkable feature it is necessary to lay emphasis upon the fact that it is one of the most accurate of all the portolanos in its delineation of the details of the coasts. At the same time it shows in its accuracy of latitude and longitude that, like some of the other maps, it can only have been drawn originally with the aid of spherical trigonometry. It is therefore a scientific product in the true sense of that term.

The feature in question is at the north, and looks at first glance like a very bad representation of the Baltic. A comparison with a modern map shows that the Baltic runs nearly north and south, while on this otherwise so accurate map it runs east and west. There is no evidence of either the existence of the upper

---

[3] "Carta Nautica de Andrea Benincasa" (Ancona, 1508) Borg. VIII, in "Monumenta Cartographica Vaticana," Vol. I, Plate XX.

Figure 90. The Andrea Benincasa Map with square grid constructed empirically from the geography. See Table 12.

Baltic or of the Gulfs of Bothnia and Riga. What could be the reason for this? This map is dated 1508, when the Baltic was, in fact, very well known. For nearly three centuries before this date it had been a highway of commerce, dominated by the merchant ships and navy of the Hanseatic League. Furthermore, its shape was better known to Ptolemy, as is shown by the map we have just discussed.

As we look at this feature on the Benincasa Map we note details that differ considerably from other representations of bodies of water on maps of the 15th and 16th Centuries. Is this large feature really the Baltic—*or is it a mass of ice?* Are those blobs along the southern edge supposed to be harbors along the Baltic coast of Germany, or are they run-off lakes from the melting glacier? Are those apparent islands really islands, or are they deglaciated tracts in the middle of the retreating ice cap? I was greatly intrigued by these possibilities, and considered the evidence very strong indeed when I observed that the general contour of the southern side of this "Baltic" followed very precisely the shape of the southern side of the Scandinavian continental ice cap as it stood about 14,000 years ago. Unfortunately, further observation showed that this ice cap did, in fact, closely follow the shape of the Baltic coast (see Fig. 91).

In all the maps that show this erroneous shape of the Baltic there also appears a break in the Baltic coast. It seems that the accurate portolano in each case extends to Britain, and to the coast of the Netherlands. Then an entirely different source map has been used, and this map has been misinterpreted so that the coast of the Netherlands is mistaken for the coast of Denmark, which is thus placed about 250 miles too far west, much coastline being omitted. This apparently distorted map of the Baltic may have been circulating in southern countries— Portugal or Italy—where the true shape of the Baltic was less known. It is my suggestion that cartographers in those countries may have happened upon this old map, along with the others that had somehow reached Europe from Constantinople or elsewhere, and combined it with the normal portolano.

Another possibility is provided by the consideration that there was a time in the post-glacial period when the Baltic may very well have had the shape shown on these maps. The northward extensions of the sea—the Gulfs of Bothnia and Riga—probably were covered by the ice long after the lower Baltic had become ice-free. Since they are both very shallow, they may even have been above sea level when the sea level was several meters lower.

4. The Portolano of Iehudi Ibn Ben Zara of Alexandria.[4]

We have mentioned that Nordenskiöld considered all the portolan charts to be copies of one original. It seems to me that the portolano of Iehudi Ibn Ben Zara of Alexandria (Fig. 92) may stand very close to this original.

---

[4] "Carta Nautica di Iehudi Ben Zara," Alessandrio d'Egitto, 1497. Borg. VII, "Monumenta Cartographica Vaticana," Vol. I, Plate XIX.

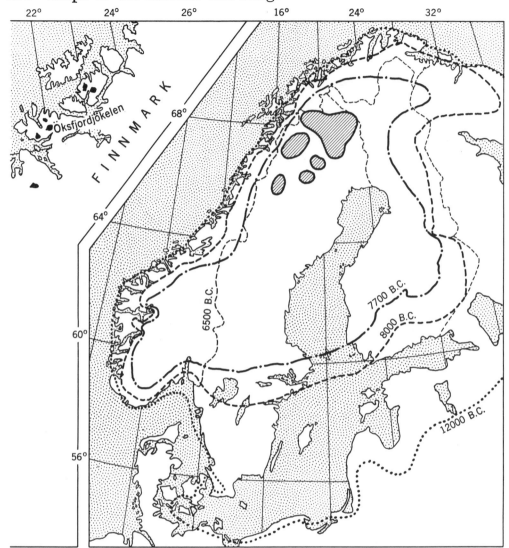

Figure 91. Glacial map of the Baltic.

I had been attracted to the study of this portolano because it seemed definitely superior to all the other portolan charts I had seen in the fineness of its delineation of the details of the coasts. As I examined these details in comparison with the modern maps, I was amazed that no islet, no matter how small, seemed too small to be noted. For example, on the French coast, along with the principal features, I found the mapmaker had drawn in the tiny islets, the Ile de Re and the Ile d'Oleron, north of the mouth of the Gironde River. North of the mouth of the Loire he included Belle Ile and two other small islets. Off Brest, he drew in the Ile d'Ouessant. Similar fine details can be found all around the coasts.

The grid worked out for the map (Fig. 93) revealed, indeed, a most amazing accuracy so far as relative latitudes and longitudes were concerned. Total longitude

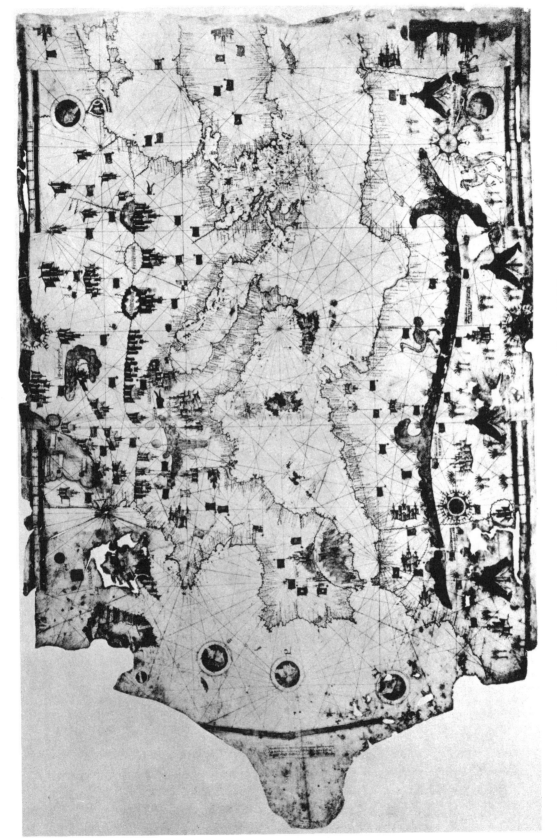

Figure 92. The Map of Ibn Ben Zara, 1487.

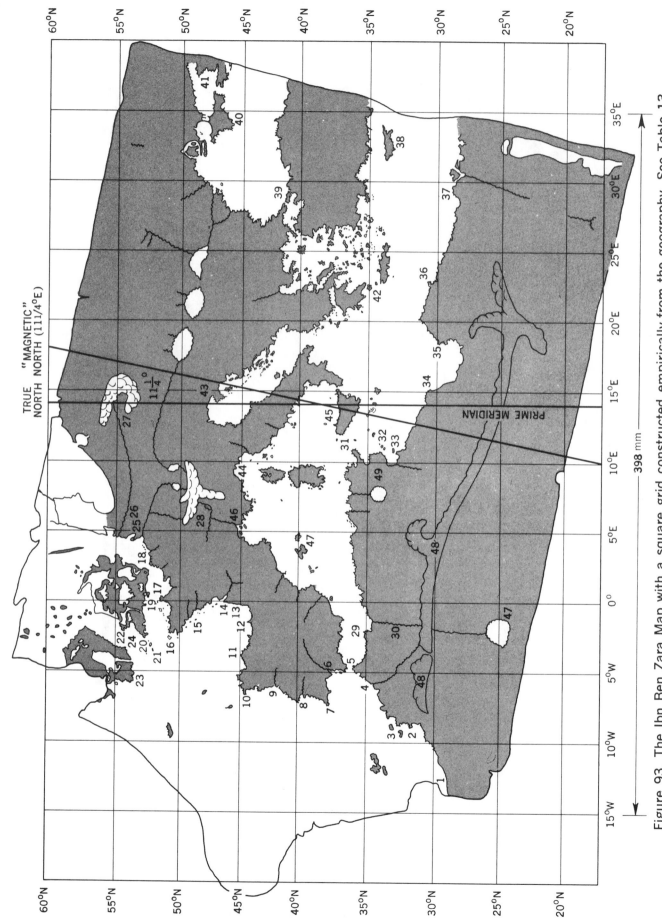

Figure 93. The Ibn Ben Zara Map with a square grid constructed empirically from the geography. See Table 13.

Figure 94. The Faces of the Ibn Ben Zara Map.

between Gibraltar and the Sea of Azov was accurate to half a degree, total latitude from Cape Yubi to Cape Clear, Ireland, was accurate to a degree and a half. Average errors of latitude for the whole map were less than one degree; average errors of longitude amounted to little more (see Table 13). As far as the map as a whole was concerned, there was no evidence of an oblong grid. Like the Benincasa Map, it seemed to have been drawn for a square grid. Yet a complication was to appear, as we shall see below.

Alfred Isroe, my student, drew my attention to one of the most remarkable features of the map. These were five tiny faces in medallions in the corners of the map, where mapmakers of the Renaissance followed the custom of placing faces symbolizing the winds. Usually such faces are not found on the portolan charts.

In Renaissance maps they are usually shown with their cheeks puffed out, obviously blowing vigorously in the appropriate direction.

The faces on the Ben Zara Map are not typical of the cartography of the Renaissance. Their cheeks are not puffed out. The faces are calm and aristocratic in mien, and the clothing indicated does not seem characteristic of the contemporary styles.

Isroe suggested at first that the faces resembled faces found on icons of the Greek Orthodox Church, such as were, he said, produced by the famous iconographic school of Parnassus in the 7th and 8th Centuries A.D. This was exciting indeed. Could it be that we had here an accurate copy of an ancient portolano that had come through a Greek monastery of the 8th Century? Ibn Ben Zara had,

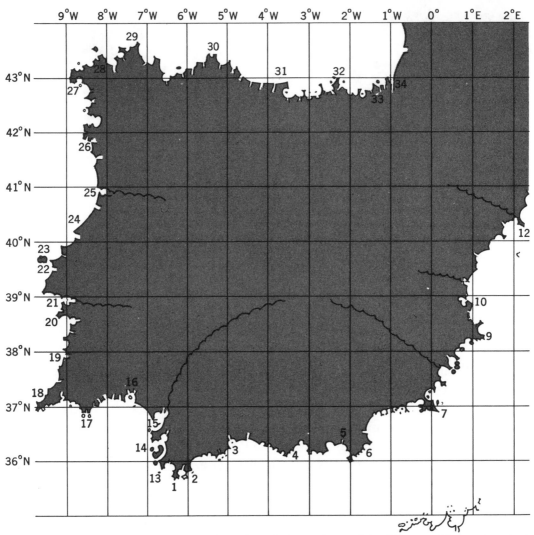

Figure 95. The Ibn Ben Zara Map, Spanish section, with oblong grid constructed empirically from the geography.   See Table 14.

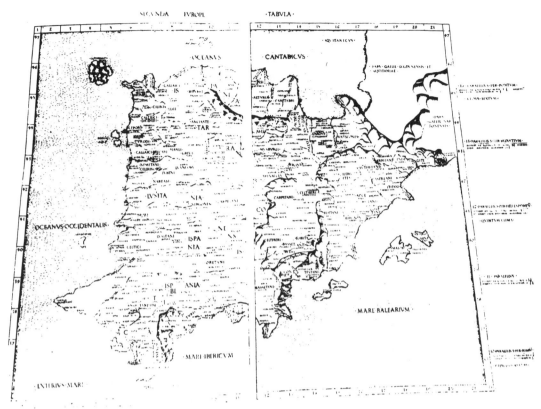

Figure 96. Ptolemy's Map of Spain.

of course, added modern names to his chart, but perhaps he had made no other changes in his ancient source map. This matter required intensive examination.

I took the matter up with my aunt, Mrs. Norman Hapgood, who is a scholar and a translator with a knowledge of Russian and other eastern tongues. She said that, to her, the faces looked Coptic. I investigated Coptic art in Harvard's Fogg Museum and was well rewarded. A number of treatises gave me some light on the subject. Among these, two in particular were useful (80, 4).

Gruneisen, one of these scholars, describes the Coptic art born in Alexandria before the rise of Christianity as *"frivole, spirituel, profondement raffiné et aristocratique par excellence."* [5] The reader may judge for himself how far the little faces agree with this description (Fig. 94). As I have already pointed out in connection with the Piri Re'is and other maps there is reason to believe that the portolan charts did in fact come through Alexandria and were copied and arranged by the geographers of Alexandria, who may, indeed, have originated the flat portolan projection itself. In view of this, the Hellenistic-like faces may be quite significant.

I have mentioned that the grid worked out for this map indicated that a

---

[5] **"Frivolous, intellectual, deeply refined, above all aristocratic."**

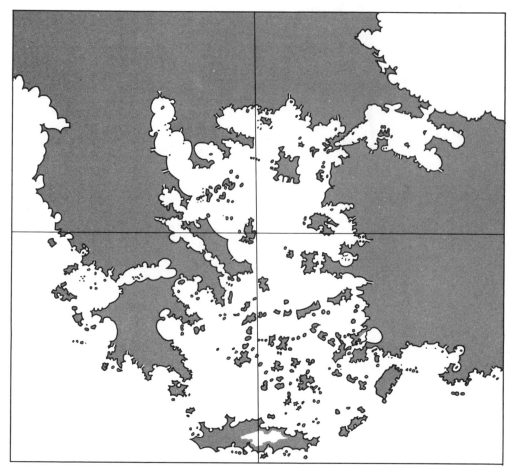

Figure 97. The Ibn Ben Zara Map, Aegean section.

square grid, not an oblong one, was probably the sort of grid used in drawing the map, or at least in compiling it from local maps. Table 13 strongly suggests this.

I was profoundly surprised, therefore, when one of my students, Warren Lee, discovered that, in regards to Spain, the grid indicated by the topography was oblong, and not square (see Fig. 95 and Table 14). This is, indeed, astonishing. How can it be explained? Are we to suppose that the mapmaker who compiled the whole map, and did it so very well, made use of separate maps of different countries, and among these used a map of Spain that had been drawn earlier, perhaps on the same projection as the De Canerio Map?

Warren Lee is responsible for another interesting observation. In his study of the Spanish sector of the map he observed that it showed a large bay at the mouth of the Guadalquivir River. At this point the modern map shows a large delta, composed of swamps, about thirty miles wide and fifty miles long. The bay on the Ben Zara Map might seem thus to represent the coastline before the growth of the delta of the Guadalquivir. Since the Guadalquivir is not an enormous

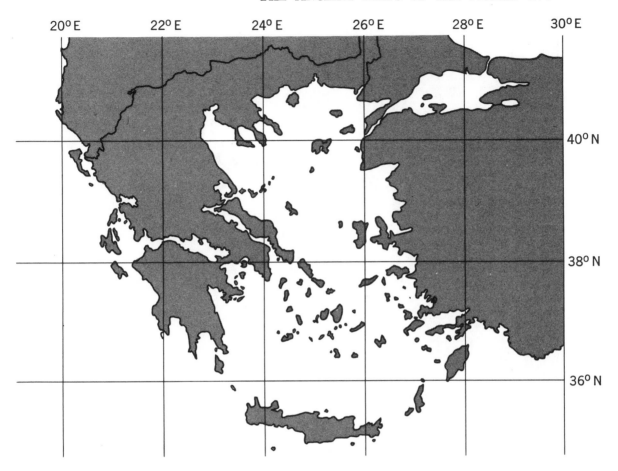

Figure 98. Modern map of the Aegean.

river, and does not carry huge loads of sediment, it would have taken a considerable time indeed to build the delta. Several other maps we have examined carry indications of delta-building since they were originally drawn, but in no other case is the evidence so clear as this.

Another matter of importance to us was to determine whether, in the remarkably detailed representation of the islands, especially in the Aegean Sea, we might not have evidence of a change of sea level since the original map was made. Comparison of the Aegean Sea on this map (Fig. 97) with the Aegean on a modern map (Fig. 98) suggests that many islands may have been submerged. There are many fewer islands on the modern map, and many of these are smaller than they are shown to be on the old map. One may ask, if the mapmaker was so conscientious in drawing in the smallest islands, and showing all the features of the coast with the greatest accuracy possible, why in the Aegean should he suddenly take leave of his senses, and fill the sea with imaginary islands, *while still showing the real islands in their correct positions?*

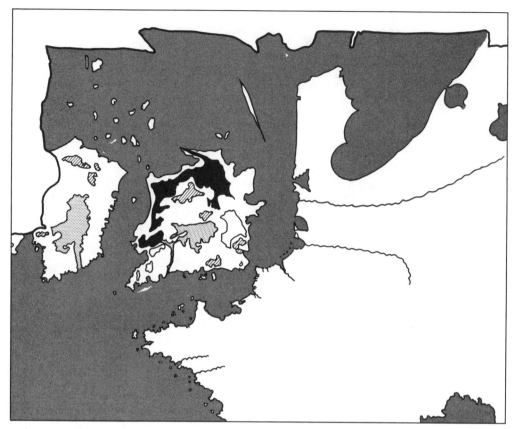

Figure 99. The Ibn Ben Zara Map, tracing of the western section.

The reader will note that a corner of this map apparently shows the same feature we have discussed as a possible ice cap in the Benincasa Map. We had to admit that the evidence in that case was equivocal. Here, in this map, we have more of the same sort of evidence, in the form of features suggesting glaciers in central England and in central Ireland (see Fig. 99).

Here, at least, we can say the evidence all hangs together: the great lapse of time required for building the delta of the Guadalquivir, the evidence of a lower sea level at the time the map was drawn (which we know did exist at the end of the ice age), and now these remnant glaciers in the British Isles.

CHAP
TER VII

THE
SCIENTIFIC
IMPLICA
TIONS OF
THE
MAPS

1. Spherical Trigonometry.

When we started our investigation of the ancient maps we had a limited goal: We wanted to find out simply whether the Antarctic Coast appeared on the Piri Re'is Map of 1513. This led us to a probable solution of the Piri Re'is Projection. We found that the projection of the source map used by Piri Re'is for his Atlantic coasts (both of the Old and of the New World) had apparently been based on plane trigonometry, and on the computation of the circumference of the earth ascribed to Eratosthenes. We found an agreement between the length of the Greek stadium implied in this projection, and its length as found by Dr. George Sarton. We found the center of the projection to lie at the intersection of the Tropic of Cancer with the meridian of Alexandria, and we found that the use of the meridian of Alexandria accorded with the procedure adopted by ancient Greek geographers generally. We had the statement of Piri Re'is himself that some of his source maps descended from the time of Alexander the Great.

Thus we had strong support for the conclusion that the Piri Re'is source map was of ancient vintage. It appeared to reflect an unexpected level of scientific achievement in Alexandrian science—that is, it suggested that the geographers of the great "Museum" or Academy attached to the Library of Alexandria might have solved the problem of applying mathematics to mapmaking—something that all known geographers from Eratosthenes to Ptolemy knew ought to be done but were unable to accomplish. It would seem that this had been accomplished, however, before the time of Ptolemy (in the 2nd Century A.D.). It can hardly have been done later. The only era of brilliance in science between the days of Ptolemy and the modern period of the Renaissance was the great era of Arab science from the 10th to the 13th Centuries, but the Arab maps reflect no application of trigonometry to mapmaking. They are only beautiful pictures.

In Alexandrian times we found that not only had Hipparchus in the 2nd Century B.C. discovered or rediscovered plane and spherical trigonometry, but he had also developed one or more types of mathematical map projections based on spherical trigonometry. We do not know whether he drew maps—perhaps he did—but the truth is that in his day he could not possibly have applied his projections to the globe because the necessary data, in the form of correct findings of latitudes and longitudes of a very large number of places over the known areas of the earth, were not available. This was the weakness of all Greek cartographic science. In Greek times mathematics was in advance of mechanical instrumentation: There was no instrument for easily and correctly determining the longitude of places. However, the Piri Re'is and the other maps we went on to study, seemed to suggest that such an instrument or instruments *had* once existed, and had been used by people who knew very closely the correct size of the earth. Moreover, it looks as if this people had visited most of the earth. They seem to have been quite well acquainted with the Americas, and to have mapped the coasts of Antarctica.

A particularly difficult problem—a very strange and even seemingly contradictory sort of enigma—was presented by the apparent use of plane trigonometry at a time when, according to the historical record, spherical trigonometry was known. *Why*, in the Piri Re'is Projection, do we find evidence of the use of plane trigonometry, as repeated experiments have indicated, when the Greek geographers after Hipparchus were familiar with the theory of spherical trigonometry in map-making? (As far as we know, though, they could not make practical use of the theory.) It does not seem very likely that the map was drawn at precisely that period in the life of Hipparchus when he had completed the development of plane trigonometry, but had not yet developed spherical trigonometry.

The way to the solution of this particular enigma was pointed out first by a curious fact that emerged from the Tables of the Piri Re'is Map. We found a very remarkable accuracy of longitude on this map; but, strangely, latitude seemed rather less accurate. That is, for example, there was a regular and progressive increase in latitude errors toward the north of Europe. It was as if the original source map had been drawn in a way that took account of the curvature of the earth. We were able to eliminate the possibility that this was merely the result of an error in scale. We found that the length of the degree of latitude on the *original source map* had been longer than the length of the degree of longitude. We were not able to determine whether the greater average length of the degree of latitude was the result of an arbitrary ratio between the degrees of latitude and longitude based on the ratio at a given parallel of latitude, or whether it was *progressive*, that is, whether the parallels grew gradually farther apart with distance from the equator. In the first case, we would have the method ascribed to Ptolemy by Nordenskiöld;[1] in the second, we would have something like the Mercator projection, a projection based on spherical trigonometry (Note 5).

Another part of the Piri Re'is Map pointed to the use of spherical map

---

[1] See Note 9.

projections in remote times. This was the Caribbean sector, which suggested that the original cartography may have been done on something like the azimuthal equidistant polar projection. Furthermore, the accuracy of the latitudes and longitudes in this sector strongly suggested that whoever converted one projection into the other knew what he was doing and had correct longitudes and latitudes for the Caribbean.

As our studies extended from map to map we accumulated more and more evidence of the ancient existence, in an era long before Greece, of spherical trigonometry and its application to mapmaking. The Zeno Map of the North illustrated the conversion of an originally spherical sort of projection to the flat portolano one.

The antiquity of the knowledge of spherical trigonometry, and of its application to mapmaking in sophisticated map projections became quite clear in the case of the Oronteus Finaeus Map of Antarctica. Here no one had tried any conversion of the ancient map to the portolan projection. Here our long and intensive examination slowly revealed that the original map must have been drawn on a projection like the Cordiform projection known and used in the Renaissance. It was a projection with curved meridians. Our cartographic friends in the U.S. Air Force had suggested such a projection,[2] but it is hardly possible to specify its precise character, excepting that it is unthinkable that it could have been drawn without spherical trigonometry. It is a fascinating fact that this map, according to geological clues, appears to represent an antiquity greater than that of most of the other maps—an antiquity that may exceed several times over the antiquity of the oldest written records hitherto known to us.

The portolan charts of the Mediterranean and the Black Sea, dating apparently from the 14th and 15th Centuries, were found by us to have an accuracy in the ratio of latitude to longitude, that, according to Strachan (Note 7), implied the use of spherical trigonometry.[3] The De Canerio Map of 1502 threw much additional light on this problem. In the first place, it incorporated a map of the Mediterranean and Black Seas that was essentially the same as the common original of the 14th and 15th Century portolan charts—the map Nordenskiöld referred to as the "normal portolano." We sought to solve the projection of this map first by plane trigonometry, finding, as with the Piri Re'is Map, extraordinary accuracy of longitude, but progressively increasing errors of latitude with distance from the equator. We tried different ways of resolving this problem. We applied the Mercator projection in spreading the parallels, while leaving the meridians alone,

---

[2] See Note 22.

[3] That is, a sophisticated map projection was constructed by spherical trigonometry and then the geographical points were found on it, their locations in latitude and longitude having been ascertained by the use of instruments capable of considerable precision.

a very artificial test. This did not work. Finally, we applied spherical trigonometry, which did work. It is interesting that the use of spherical trigonometry appears to have produced an oblong grid like the grids discovered empirically on the Piri Re'is Map and the Chinese map of Yü Chi Thu.

2. The Twelve-Wind System.

It appears that there may be a connecting link between the hypothetical body of ancient maps we have discussed and the remote civilization from which they may have been derived; that link is the so-called "twelve-wind system." This system appears to stem from the farthest antiquity.

To begin with, scholars have long been aware of the fact that the type of portolan design known as the "eight-wind system" and used in the portolan charts of the Middle Ages and the Renaissance was preceded, in antiquity, by another type, the twelve-wind system, which we have discussed. No portolano based on the twelve-wind system was known until we discovered that system in the Venetian Chart of 1484.[4] The presence of the twelve-wind system in this chart, plus the fact that it proved to have been constructed by trigonometry, is good evidence of its origin in antiquity.

Now, from the standpoint of the history of science, the twelve-wind system is of very special importance. This system involved, as we have previously pointed out, the division of the circle into twelve arcs of 30 degrees each, or six arcs of 60 degrees. (See Fig. 69.) It involved the division of the circle into 360 degrees. This fact relates the system, in most interesting fashion, to Babylonian science. The Babylonians had a numbering system based on sixty, and on decimals. They are supposed to have invented the 360-degree circle, and the divisions of time we still use today.

The Babylonians also had a zodiac, and this was divided into twelve signs of 30 degrees each. The constellations of the zodiac did not precisely coincide with the twelve signs, as was natural enough, since the latter were mathematical divisions.

Now the stars were used in ancient times, in navigation, as E. G. R. Taylor points out (199:40), and so the zodiac and the other constellations of the northern and southern hemispheres (Note 21) were a sort of map written in the sky.[5] The relationships of the Babylonians and Phoenicians in ancient times were very close, and we can easily imagine that the Phoenicians might have applied these basic elements of Babylonian science to mapmaking. The result of any such effort would have been the twelve-wind system.

We must not allow ourselves to be confused by the fact that the 360-degree

---

[4] But see Note 16.

[5] This idea was beautifully developed by a little-known 19th Century writer.
(32)

circle is used in modern navigation. This method of dividing the circle is not modern; it is the oldest way of dividing the circle known to man. Furthermore, since it involves counting by tens, it alone can explain how the ancient source map of the Antarctic, probably drawn ages before either Phoenicians or Babylonians existed, had on it the circle that Oronteus Finaeus took for the Antarctic Circle, but which we have shown may have been the 80th parallel. The implication from this is that the 360-degree circle and the twelve-wind system were ancient before the rise of Babylonia and long before Tyre and Sidon were built by the Phoenicians. Babylonian science was thus, perhaps, a heritage from a much older culture.

There are curious connections and comparisons that can be made between the ancient sciences of Greece, Egypt, Babylonia, and China, not to neglect either India or Central America. I have assembled some passages referring to these connections, showing particularly that both the Babylonians and the Chinese had numbering systems that could fit in very well with decimals of the twelve-wind system (see Note 17).

3. The Age of the Cartographic Tradition: Geological Problems.

It is necessary now to attempt to interpret such evidence as there is of the antiquity of the cartographic tradition represented by our maps. The evidence is geological in nature. We have, first, some indications of changes in shorelines since the maps were made, but these are uncertain. It is possible to dismiss them simply as bad mapmaking. Perhaps the most impressive example is the large bay shown on the Ibn Ben Zara Map of Spain at the present location of the delta of the Guadalquivir River. Here the indication is that a delta thirty miles wide and fifty miles long has been built since the original map was drawn. This would, of course, involve a great length of time. There are suggestions (also in the Ibn Ben Zara Map) of a lowered sea level. Despite the extraordinary general accuracy of the map many islands are shown in the Aegean Sea that do not now exist, and a number of islands are shown larger than they are now. This again might be bad mapmaking, but there is no necessity to adopt that conclusion. The case is the same with the indications of remnants of ice age glaciers in Sweden, Germany, England, and Ireland on the Benincasa and Ibn Ben Zara Maps and on the Ptolemaic Map of the North. This evidence hangs together and points in the same direction—towards a great antiquity for the beginnings of the cartographic tradition.

The most important evidence for the age of the maps, however, is to be found in those showing the Antarctic, especially in the maps of Mercator, Piri Re'is, and Oronteus Finaeus. All of these maps appear to show the continent at a time when there was a temperate climate there. Some geological evidence, in the form of three sedimentary cores from the bottom of the Ross Sea, has been presented to suggest that such a warm period may indeed have existed there down to about 6,000 years ago.

Since this matter is now primarily a geological one, it may be advisable to introduce the discussion with a brief description of the present conditions in the Antarctic, a summary of the present geological ideas regarding the climatic history of the continent, and the causes of ice ages generally.

A glance at any good up-to-date map of the Antarctic will reveal that the continent is entirely covered by ice. It was formerly thought that the central areas were a very high plateau, covered by an ice cap about a mile thick. It has now been discovered that there is no high central plateau, but instead the rock surface of the continent is no farther elevated above sea level than the surfaces of North America and Europe. The ice cap, therefore, is much thicker than formerly supposed, and in places approaches two miles in depth. In some areas the rock surface dips below sea level so that, if the ice were melted, very important inland lakes or seas, and numerous bays, would appear.

In discussing the history of this great ice cap, geologists first of all say that it has existed for millions of years—since Miocene or even Pliocene times. If their views are correct it is extremely unlikely that man could have mapped Antarctica when its coasts were free of ice. Man as we know him probably did not exist then.[6]

Present geological opinions, however, may not be correct. There is, in the first place, quite a mystery about the ancient climate of Antarctica. One might suppose that since the presence of the ice seems to be due to the fact that the continent is located at the South Pole, the climate there must have been glacial since the beginning of the geological record, that is, for about two billion years. But nothing is farther from the truth. A few years ago I summarized the known evidence on this matter as follows:

. . . Those who may be inclined to disbelieve that Antarctica could have possessed a warm climate 10,000 years ago[7] must be reminded of the evidence that Antarctica has many times possessed such a climate.

So far as we know at present, the very first evidence of an ice age in Antarctica comes from the Eocene Epoch. That was about 60,000,000 years ago. Before that, for one and a half billion years, there is no suggestion of polar conditions, though very much earlier ice ages existed in other parts of the earth (85:59).

I quoted from Thomas R. Henry, author of *The White Continent*, a summary of evidence showing that in the Edsel Ford Mountains in Antarctica there are folded layers of sedimentary rock 15,000 feet thick that must have been deposited by flowing rivers when the continent was ice-free:

---

[6] So far, creatures of a human type, who may have had more intelligence than they are credited with, have been traced back to about 1,700,000 years ago. Very little, however, is known with certainty about the origins of man.

[7] As suggested by the Hough cores.

The greater part of the erosion probably took place when Antarctica was essentially free of ice since the structure of the rocks indicates strongly that the original sediments from which they were formed were carried by water.[8] Such an accumulation calls for an immensely long period of tepid peace in the life of the rampaging planet. (85:59)

Another writer on ancient climates has described the evidence of warm coralline seas stretching right across Antarctica several hundred million years ago. (85:245)

A reasonable deduction from this evidence presented in my earlier book, *Earth's Shifting Crust* (1958), is that Antarctica may not have been at the South Pole in the periods when the continent was warm. There are at least three ways of accounting for this possibility. One is to suppose a change in the position of the earth's axis of rotation. A second is to suppose that the continents are not fixed rigidly to the body of the earth, but rest on semi-molten non-crystalline rock over which they may gradually wander, and the third suggestion is that the whole crust of the earth may be displaced at times, moving over the soft inner body, much as the skin of an orange, if it were loose, might shift over the inner part of the orange, all in one piece.

When I was writing *Earth's Shifting Crust* I considered these alternative ideas. The objections to the theory of changes of axis were formidable indeed. It seemed that no force on the earth could account for such a thing, and any force acting on the earth from interstellar space (such as collision with another body) would probably destroy all terrestrial life.

The theory of continental drift was first proposed by Alfred Wegener in 1912. He had long been perplexed by finding fossil evidence of plants and animals in places far removed from the climatic zones where similar species exist today. He supposed that the continents had originally constituted one land block which was broken up into pieces that gradually drifted apart. His suggestion of the force that might have done this and sent the continents drifting was not acceptable to physicists, however, and as a result his theory was accepted only by a minority of scientists, most of whom were biologists. In the last few years it has been given a new lease on life by the suggestion of a new "mechanism" to account for the drift. This is the contribution of the Canadian physicist, H. Tuzo Wilson, and is based on the supposed effects of sub-crustal convection currents of molten rock moving in the depths of the earth (218).

Only one aspect of the revived continental drift theory may concern us here: the rate at which the sub-crustal currents are thought to move. Wilson estimates it at about one centimeter a year, or one meter a century. This would be one kilometer (about two thirds of a mile) in 100,000 years. Since, according to

---

[8] Naturally there are no flowing rivers now in Antarctica, for everything is frozen, except for a little melting in a few places during the Antarctic summer.

Wilson, it is the sub-crustal currents that move the continents, continental drift must be very slow.

Wegener advanced the continental drift theory to explain not only the distribution of fossil species but also the occurrence of ice ages, and it is clear that the revived theory may well explain glacial periods that occurred hundreds of millions of years ago (such as those of the Permo-Carboniferous Period) because a continent can drift a long way (a thousand miles, perhaps) in a hundred million years. But this does not help us with recent ice ages. There have been four of these in North America within the last million years. The last one, which ended only about 8,000 years ago, apparently reached its apogee (or maximum) only 10,000 years before it disappeared.[9] Hence, the vast climatic change that melted a continental glacier covering four million square miles of North America probably took place in no more than 10,000 years. From a geological standpoint this is incredibly rapid. It has not been accounted for by any hypothesis accepted by geologists. The puzzle of the ice ages remains one of the unsolved problems of geology.[10]

In *Earth's Shifting Crust*, I attempted, with the help of my collaborator, James Campbell, to state the case for the third alternative—the displacement of the earth's whole crust as one piece over the inner layers—and to suggest a mechanism to account for it. The book was written to explain, among other things, the most recent ice age in North America. I felt that if we could find an acceptable explanation for this ice age, which is the ice age we know most about, we would be more likely to be on the right track regarding the ice ages of long ago, about which we know very little. This procedure accorded with the sound principle of working from the known to the unknown.

Campbell and I started with the assumption that the earth's crust rests upon a very weak layer below, a layer that is virtually liquid. Then, adopting a suggestion made by the engineer Hugh Auchincloss Brown we considered that a force sufficient to displace the whole crust over this weak layer might be provided by the centrifugal effects of the ice caps themselves. In Antarctica, for example, the center of mass of the great ice cap is, because of the shape of the continent, about 300 miles from the pole. As the earth rotates, the eccentricity creates a centrifugal effect that works horizontally on the crust, tending to displace it toward the equator. Campbell calculated the mathematical effects being produced theoretically by the Antarctic ice cap at the present time, and found that they were approximately of the right magnitude to lead to failure and fracturing of the crust. Such failure could, in turn, lead to a shifting of the entire crust of the earth. The hypothesis of such a mechanism to account for crust displacement is attractive

---

[9] See Chapter II of "Earth's Shifting Crust."

[10] A recent theory proposed by Professors Ewing and Donn, of Columbia University and City College respectively, has called attention to a possible contributory factor in the recent ice ages, but has serious weaknesses.

because the mechanism would provide a continuing force that could displace the crust a great distance and would also explain why the displacements would stop short of moving glaciated continents all the way to the equator. The ice caps, as they were moved farther from the pole, would displace the crust over a great distance because their centrifugal effects would increase. The movement of the crust, however, would eventually stop because the ice caps would melt away in the temperate zones.[11]

Applying this theory to the last ice age in North America, Campbell and I supposed that at the peak of the glacial period the ice accumulation was sufficient to start a crustal displacement. The effect would be to pull North America southward toward the equator, and the movement would continue until Hudson Bay or the Province of Quebec, which were at the center of the ice cap, and which had been, according to our theory, at the North Pole, reached their present latitudes. At this point the ice cap would be sufficiently reduced by melting to bring the movement to a stop. The crust would have been displaced 2,000 miles along the 90th meridian of West Longitude.

But now, if North America had been brought about 2,000 miles southward, what about the rest of the Western Hemisphere? Since the whole crust would have moved as one unit, obviously the whole hemisphere would have been displaced by the same amount. Thus South America would also move southward. East Asia, on the other side of the earth, would move, naturally, in the opposite direction— northward. A good part of *Earth's Shifting Crust* is devoted to presenting evidence showing that the climate grew drastically colder in Siberia at this particular time.

We come to Antarctica. It is clear, of course, that if the Western Hemisphere shifted 2,000 miles southward along the 90th meridian, Antarctica must have moved correspondingly. Hence, it must have been 2,000 miles to the north before the displacement. This would have put it outside the Antarctic Circle, in a temperate or cold-temperate climate. During the movement it would gradually have grown colder, an ice cap forming and developing slowly until it reached its present dimensions.

The Oronteus Finaeus Map, as we interpret it, shows most of the coasts without ice. But the inland waterways, connecting the Ross, Bellingshausen and Weddell Seas, are not shown. This may mean that glaciation was well advanced in some parts of Antarctica when the original maps were made. On the other hand, it seems that rivers may still have been flowing from the mountains near the coasts, and that the estuaries along the shores of the Ross Sea (despite its present closeness to the Pole) were not yet filled with ice. Antarctica apparently had not yet really frozen up. If we accept the Ross Sea cores as valid evidence, we may conclude that its shores were covered by the ice cap by about 6,000 years ago. We can hardly

---

[11] For the best brief summary of the theory see the Foreword written by Albert Einstein for "Earth's Shifting Crust" (Note 18).

imagine that the sailors and mapmakers of our ancient hypothetical world-encircling lost civilization were visiting the Ross Sea in the stormy and frigid period when the continent was entering the ice age. Hence, it does not seem unreasonable to allow at least 2,000 years for the final transition from the picture shown by the Oronteus Finaeus Map to the present conditions. Indeed, it would perhaps more likely have involved a period twice or three times as long.

It seems, then, that the cartographic tradition of the ancient maps may well extend back in time at least to the period when the glaciers were withdrawing in the Northern Hemisphere. For this reason we should not lightly dismiss the suggestions of late glacial conditions in the maps of Ptolemy, Benincasa, and Ibn Ben Zara.

CHAP
TER VIII

A
CIVILIZA
TION
THAT
VANISHED

The evidence presented by the ancient maps appears to suggest the existence in remote times, before the rise of any of the known cultures, of a true civilization, of a comparatively advanced sort, which either was localized in one area but had worldwide commerce, or was, in a real sense, a *worldwide* culture. This culture, at least in some respects, may well have been more advanced than the civilizations of Egypt, Babylonia, Greece, and Rome. In astronomy, nautical science, mapmaking and possibly ship-building, it was perhaps more advanced than any state of culture before the 18th Century of the Christian Era. It was in the 18th Century that we first developed a practical means of finding longitude. It was in the 18th Century that we first accurately measured the circumference of the earth. Not until the 19th Century did we begin to send out ships for purposes of whaling or exploration into the Arctic or Antarctic Seas. The maps indicate that some ancient people may have done all these things.

Mapping on such a scale as this suggests both economic motivations and economic resources. Organized government is indicated. The mapping of a continent like Antarctica implies much organization, many exploring expeditions, many stages in the compilation of local observations and local maps into a general map, all under a central direction. Furthermore, it is unlikely that navigation and mapmaking were the only sciences developed by this people, or that the application of mathematics to cartography was the only practical application they made of their mathematical knowledge.

Whatever its attainments may have been, however, this civilization disappeared, perhaps suddenly, more likely by gradual stages. Its disappearance has implications we ought to consider seriously. If I may be permitted a little philosophizing, I would like to suggest that there are four principal conclusions to which we are led.

1. The idea of the simple linear development of society from the culture of the paleolithic (Old Stone Age) through the successive stages of the neolithic (New Stone Age), Bronze, and Iron Ages must be given up. Today we find primitive

cultures co-existing with advanced modern society on all the continents—the Bushmen of Australia, the Bushmen of South Africa, truly primitive peoples in South America, and in New Guinea; some tribal peoples in the United States. We shall now assume that, some 20,000 or more years ago, while paleolithic peoples held out in Europe, more advanced cultures existed elsewhere on the earth, and that we have inherited a part of what they once possessed, passed down from people to people.

2. Every culture contains the seeds of its own disintegration. At every moment forces of progress and of decay co-exist, building up or tearing down. All too evidently the destructive forces have often gained the upper hand; witness such known cases as the extinctions of the high cultures of ancient Crete, Troy, Babylon, Greece, and Rome, to which it would be easy to add twenty others. And, it is worth noting that Crete and Troy were long considered myths.

3. Every civilization seems eventually to develop a technology sufficient for its own destruction, and hitherto has made use of the same. There is nothing magical about this. As soon as men learned to build walls for defense, other men learned how to tear them down. The vaster the achievements of a civilization, the farther it spreads, the greater must be the engines of destruction; and so today, to counter the modern worldwide spread of civilization, we have atomic means to destroy all life on earth. Simple. Logical.

4. The more advanced the culture, the more easily it will be destroyed, and the less evidence will remain. Take New York. Suppose it was destroyed by a hydrogen bomb. After some 2,000 years, how much of its life could anthropologists reconstruct? Even if quite a few books survived, it would be quite impossible to reconstruct the mental life of New York.

When I was a youth I had a plain simple faith in progress. It seemed to me impossible that once man had passed a milestone of progress in one way that he could ever pass the same milestone again the *other* way. Once the telephone was invented it would stay invented. If past civilizations had faded away it was just because they had not learned the secret of progress. But Science meant *permanent* progress, with no going back, and each generation was pressing on further and further, rolling back the frontiers of the unknown. This process would go on forever.

Most people still feel this way, even in spite of two world wars, and the threat of universal annihilation in a third. The two world wars shook the faith of many in progress, but even without the very sad story of the century we live in, there never was any good basis for the belief that progress is an automatic process. Progress or decline in civilization is just a balance sheet between what the human race creates in a given period and what it destroys. Sometimes for a while our race creates more than it destroys, and there is "progress"; then for a while it destroys more effectively—more scientifically, let us say—than it creates, and we have decline. Compare, for example, the time it took for saturation bombing by the American and British Air Forces in World War II to destroy most of the cities of Germany, including golden Dresden, and its priceless heritage of medieval architecture, with the time it took to build those cities. Think of the destruction, in one instant, by American bombers, of the oldest monastery in the West, the Abbey of Monte Cassino.

But the sad story of destruction, whereby man destroys almost as much as he creates (even in the best of times), does not begin with the 20th Century. Consider the question of libraries. There is something particularly upsetting about the burning of a library. Somehow it symbolizes the whole process. The ancient world of Rome and Greece had many libraries. The most famous of these was the Library of Alexandria, founded in Egypt by Alexander the Great three centuries before the Christian Era. Five hundred years later it is said to have contained about one million volumes, and into it was gathered the entire knowledge of the ancient western world—the technology, the science, the literature, and the historical records.

This library, the heritage of untold ages, was burned. The details are not very well known, but we think there were at least three burnings. The first happened when Julius Caesar captured Alexandria. The citizens resisted him, and in the battle about a third of the Library was destroyed. Caesar is said to have called a public meeting of the citizens and lectured them, sadistically accusing them of being guilty of the destruction—because they had resisted him! In his view Rome had a perfect right to conquer Egypt, and so the Alexandrians were guilty of misconduct in resisting him. This is the way people still think today.

There is evidence that most of the library—restored and enormously enlarged after the time of Julius Caesar—was destroyed by a Christian mob, inflamed by the preaching of a fanatical bishop, who pointed out to them—rightly, of course—that the library was no more than a repository of heathen teachings, and therefore a veritable timebomb, ticking away, preparing an explosion that could wreck the Christian world. But how can we afford to point the finger at the ignorant mob? We have had our book-burnings in the 20th Century. And I don't refer only to Hitler's infamous Burning of the Books. The libraries of America are combed relentlessly by gimlet-eyed agents of various self-appointed saviors of morality and religion. The books just disappear off the shelves! Thousands of them, every year! And, of course, American libraries have recently been the particular objects of anti-American mobs in several countries. (59, 100, 159, 189, 205)

The final chapter in the destruction of the Library of Alexandria was a burning carried out by the Arabs after their conquest of Egypt in the 7th Century. There are two stories. According to one, the conquering Caliph said, on being asked what to do with the library, that anything in it contrary to Islamic teaching should be destroyed, and everything else was in the Koran already. The library was therefore entirely destroyed (100:95–97). The other version is that the hot, dusty, dirty Arab legions, just out of the desert, found the enormous Roman baths of the capital city ready for use, but out of fuel for heating the water, and that the parchments from the library furnished the fuel. Sad as this reason for the destruction was, it was at least morally more justifiable than the others.

The Romans were guilty of another destruction of a library, which is important for our story. In the year 146 B.C. they burned the great city of Carthage, their ancient enemy and their incalculable superior in everything relating to science. The library of Carthage is said to have contained about 500,000 volumes, and these no doubt dealt with the history and the sciences of Phoenicia as a whole.

If the reader asks, how much of the total of ancient knowledge was lost by these and innumerable other acts of destruction, we will say 90 per cent or more. A few facts may give him a general idea. The most famous scientist of ancient times was Aristotle; his thought dominated the world for fifteen hundred years. He wrote many works, and it might be thought that these works, at least, would have been preserved from destruction. Not so. Only one work of his survives, the *Constitution of Athens*. All his other so-called "works" are only edited and re-edited versions of his students' notes. As I think of the kind of notes most of my students take in lectures, I shudder through and through, and wonder how much of Aristotle's real thought really does survive. Furthermore, Aristotle wrote many literary works that were considered marvels of style. All of these are lost.

Plato is an equally famous figure in the history of civilization. His dialogues, including his great *Republic*, have survived. But how many know that these were only his *popular* works? Every one of those he regarded as his serious scientific and technical works has been lost. With the great Greek tragedians, Aeschylus, Euripides, Sophocles, the story is the same. We possess only a handful—about 10 per cent—of the plays they wrote.

What we have, then, of ancient cultural products is only a sample and not necessarily a representative sample either. On the contrary, whole aspects of ancient culture have been consigned to oblivion. What fragments we have come from books considered of value to the people who dominated the Church and State in the centuries after the dissolution of the ancient civilization. The churchmen were interested in moral questions; the educated laymen—mostly aristocrats—continued to devote themselves to the great classics of arts and literature. Science, however, was neglected.

But if it is true that we have lost so much, still we have preserved much more than some people suppose. When I began this work I was aware of no definite evidence for the existence of an ancient advanced world civilization, though I was aware that others believed it had existed. Now that I have found, in the maps, evidence I accept as decisive in answering this question in the affirmative, I see additional evidence on every hand.

The reader will quite naturally wonder how, if once a great civilization existed over most of the earth, it could disappear leaving no traces except these maps?

For an answer to this we must cite one of the best known principles of human psychology: *We find what we look for*. I do not mean by this that we never find anything by accident. But rather, we usually overlook, neglect, and pass by facts unless we have a motive to notice them. It was Darwin who said that to make new discoveries one had to have a theory (not a fixed dogmatic theory, of course, but an *experimental* hypothesis). With the theory of evolution people began to look in new directions, and they found new facts, by the thousands, which supported and verified the theory. The same thing had happened a half-century before with the geological theory of Sir Charles Lyell. It happened in the beginning of modern astronomy, when Copernicus proposed a new theory of the solar system. Hitherto people have not seriously believed that an advanced civilization could have preceded the civilizations now known to us. The evidences have been, therefore, neglected.

But if we take a glance at the history of archaeological research in the 19th Century we see that it consists mainly of the rediscovery of lost civilizations. Jaquetta Hawkes, in her fascinating anthology of the writings of some of the principal archaeologists of all periods (86), devotes a section to "Lost Civilizations."

The story begins in Mesopotamia, about 1811, when Claudius Rich began the rediscovery of Babylon. It continued with Paul Emile Botta, Henry Layard, and Henry Rawlinson who brought Assyria back into history. Egypt came back into history after Champollion solved the problem of Egyptian hieroglyphics, and in the fourth quarter of the century, Schliemann brought Troy out of the mists of legend, and Sir Arthur Evans gave substance to the myths of Crete. More recently still an advanced culture, with strangely modern luxuries, that flourished on the banks of the Indus River 5,000 years ago has joined the ranks of lost civilizations rediscovered.

But is this all? Is the process at an end? Are no more lost civilizations waiting to be discovered? It would be contrary to history itself if this were the case. Unimaginative people made fun of all these discoveries in turn and often hounded the discoverers. The same sort of person today accepts all that has been discovered in the past, but denies there is anything more to be discovered.

Let us start our review of the evidences with Egypt. Scholars are in disagreement about the particular achievements of the Egyptians in science, but they are in good agreement about some particular aspects of them. Egyptian knowledge of astronomy and geometry as early as the Fourth Dynasty has been shown to be remarkable. The Egyptians had a double calendar which has been described as "the most scientific combination of calendars that has yet been used by man" (77:7). This calendar system may have been in use as early as 4241 B.C. One historian of science writes:

It may be, as some indeed suspect, that the science we see at the dawn of recorded history was not science at its dawn but represents the remnants of the science of some great and as yet untraced civilization. (77:12)

Some of the scientific knowledge possessed by ancient peoples can hardly be accounted for in view of the crudeness of the scientific instruments they are supposed to have possessed. The Mayans, for example, are supposed to have measured the length of the tropical year with incredible precision. Their figure was 365.2420 days, as against our figure of 365.2423 days. They are also supposed to have measured the length of a lunation, with an error of less than .0004 of a day (10:150). How did they achieve these results?

George Rawlinson, in a discussion of Babylonian science, made the statement: "The exact length of the Chaldean year is said to have been 365 days, 6 hours, and 11 minutes, which is an excess of two seconds only over the true length of the sidereal year" (173:II,576). He also remarked, "There is said to be distinct evidence that they [the Chaldeans] observed the four satellites of Jupiter and strong reason to believe that they were acquainted likewise with the seven satellites of Saturn. . . ." (173:II,577)

This knowledge may, of course, have been derived by the Mayans, the Babylonians, the Egyptians by the use of instruments or methods of which we know nothing. But it is at least possible that such knowledge came to them as a heritage from the same ancient unknown people who made our maps.

The fact that vast areas of ancient science have remained unknown to us has recently been revealed in startling fashion by the discovery of a computer designed and built in ancient times. It was found by divers in 1901 in the wreck of a Greek galley that had been sunk off the Greek island of Antikythera in the 1st Century B.C. Transported to the National Museum at Athens, and carefully cleaned over a long period of time, it was finally examined by Professor Derek de Solla Price of Yale. He found it to be a planetarium, a machine to show the risings and settings of the known planets, and therefore very complicated. But what was particularly astonishing about it was the sophistication of the gearing system, which, Dr. Price said, was essentially modern.

It is obvious, of course, that if this great tradition of technical and mechanical knowledge was lost to history, the same could well have happened to geographical and cartographical knowledge possessed by the Greeks, whether discovered by them or inherited from older peoples.

Perhaps it should be noted here in passing that the loss of ancient scientific knowledge was not confined to the period of the fall of ancient civilization. The Arabs preserved much of it, and much of it was undoubtedly passed on to medieval Europe. Perhaps we hear echoes of some of it in the remarkable mechanical ideas of the medieval monk Roger Bacon, or even in some of the ideas of Leonardo da Vinci. A considerable loss seems to have occurred in the Renaissance itself. This was partly because of the invention of printing. The printing presses in the 15th and 16th Centuries were monopolized by two classes of books: religious tracts (Catholic and Protestant), and humanist books dealing with arts and letters. Science was of very little interest at the time, and scientific manuscripts just lay about and were allowed to rot away. Lord Francis Bacon is supposed to have drawn attention to this deplorable neglect of scientific documents.

I am aware of a good many other indications of this kind, scattered all over the world, suggesting the ancient tradition of an advanced culture, but as yet their investigation is so incomplete that there is no point in mentioning them.[1] There is one matter, however, which I cannot forbear to mention, despite its rather controversial character, because I did investigate it myself.

Just outside Mexico City there is a round step pyramid, which, long ago, was swamped by lava from a volcano not far off. This is the pyramid of Cuicuilco. The pyramid is not a mere mound, but a complex stone structure reflecting a comparatively advanced society. The lava flow swirled around three sides of the pyramid and covered about sixty square miles of territory to a depth of from five to thirty feet. The layer of volcanic rock thus formed is called the Pedrigal.

Geologists who examined the Pedrigal and tried to estimate, by the condition of its surface and the amount of loose sediment accumulated over it, how long ago it was formed, came up with a figure of about 7,000 years. This would have meant that the Mexican pyramid was older by far than the pyramids of Egypt, the oldest of which date back about 5,000 years. Archaeologists could not accept this, and generally took the view that the pyramid probably dated no earlier than the 7th or 8th Centuries A.D. The development of the new technique of radiocarbon dating after World War II threw new light on this question.

Radiocarbon dating was developed by the nuclear scientist Willard F. Libby, of the University of Chicago. It was based on the discovery that a very small percentage of the carbon contained in the carbon dioxide of the atmosphere is radioactive, and, like all radioactive substances, loses mass at a measurable rate. Radioactive carbon (Carbon 14) radiates away half its mass in about 5,000 years. All living things taking carbon dioxide from the air will, during their lifetimes, contain the same percentage of radiocarbon as the atmosphere, but after their death any new supply from the atmosphere is cut off, while the amount already absorbed continues to decay. After a time the percentage of radiocarbon in the body of the plant or animal will be less than that in the atmosphere, and by accurately measuring the difference it becomes possible to determine the lapse of time since the death of the plant or animal. This gives us a method of "absolute dating" for archaeological and geological materials. Despite many complexities, it is regarded as generally dependable, within a certain margin of error, for the period of the last forty thousand years.

---

[1] Two recent developments of great interest have provided new evidence of scientific achievements in what we refer to as the Stone Ages. One consists of evidence of the use of an advanced lunar calendar as far back as 35,000 years ago (133), and the other is the discovery, by the use of a computer, that the builders of Stonehenge were really good astronomers. (87–88)

The first radiocarbon date for the Cuicuilco Pyramid was found by Dr. Libby (124). He used a sample of charcoal found under the Pedrigal in direct association with pottery fragments similar in style to the pottery of the known "Archaic Period" of the Indian civilization of Mexico. The result was a finding of an age of 2,422 years, with a margin of error of 250 years either way. It appeared from this that the carbon came from a tree that died or was destroyed some time between 209 B.C. and 709 B.C. It was not certain, however, that this dated the lava flow, for the charcoal was not directly associated with the lava. The wood might have been burned by humans (perhaps for cooking) sometime before the lava flow. But the position of the charcoal directly under the lava suggested that no great period of time may have elapsed between the burning of the wood and the lava flow.

Additional radiocarbon dates subsequently amplified our information on Cuicuilco. Between 1957 and 1962 a number of samples of charcoal, collected from different depths beneath the Pedrigal, were dated in the radiocarbon laboratory of the University of Southern California (UCLA).[2] One of these samples was directly associated with the lava, and gave an age of about 414 A.D., but was considered by the archaeologists, in the light of other evidence, to be probably about 200 years older. The consensus of specialists was that the flow probably occurred about 200 A.D.

This would appear at first to demolish the claim that the pyramid was very old. It would appear that it might have been built by the same people who built the other pyramids near Mexico City. There is, however, another aspect of the matter which would appear to have been overlooked. It seems that the archaeologists who have discussed the date of Cuicuilco have not, in some cases, attentively read the text of the report made by the man who excavated the pyramid for the Government of Mexico in 1920. He was Byron S. Cummings, an American archaeologist.[3]

Cummings dug down through the Pedrigal, below which he found a stratum of earth with fragments of pottery and figurines of the Archaic culture. He then dug further. At the bottom of the Archaic layer he found a deposit of volcanic ash. He extended his excavation down through the ash, and below it found evidences of an entirely different culture, one that must have preceded the Archaic. He considered that the evidence of the pottery and figurines here showed a level of culture higher than the Archaic, but unconnected with it. As he sank his trenches

---

[2] "Radiocarbon," Supplement of the American Journal of Science, Vol. 5, pp. 12–13, and Vol. 6, pp. 332–334.

[3] Cummings was assisted in his excavation of the pyramid by Dr. Manuel Gamio and by Jose Ortiz of the "Direccion de Anthropologia," the anthropology office of the Mexican Government. Funds were provided by the National Geographic Society of Washington, D.C. Cummings' report was published in 1933 by the University of Arizona Press, Tucson (56).

deeper, he came to the bottom of this layer, and to another layer of volcanic ash. He dug through this, and came upon another layer of artifacts—fragments of pottery and figurines. These resembled those in the second layer, but they were cruder. Finally, at a depth of eighteen feet, Cummings came upon a pavement that had surrounded the Pyramid of Cuicuilco and which had evidently been built when the pyramid was built.

Cummings made an estimate of the time required to accumulate the eighteen feet of sediment between the underside of the Pedrigal and the temple pavement. He estimated, first, the age of the Pedrigal lava flow at 2,000 years, and here came very close to the truth. Then he measured the thickness of the sediments that have accumulated on the top of the Pedrigal since it was formed, and used this as a measuring stick to estimate the time required to accumulate the sediments below. He came to an estimate of 6,500 years for the time required to accumulate these eighteen feet of sediments.

In answer to the argument that the rate of accumulation of the sediments may have been different and more rapid in the period before the eruption of the volcano, Cummings pointed out that a great lapse of time was clearly indicated by the nature of the sediments themselves. The three culture layers are separated by two layers of volcanic ash, and over each layer of ash is a thick layer of sterile soil, with no indication of vegetation. In each case the development of a new layer of humus-rich top soil over the sterile layer probably took time on the order of centuries, and only after this process was completed did a new layer of artifacts appear. The evidence, according to Cummings, suggested that, first, the pyramid was abandoned, for some reason, by the people who built it; then, much later, a crude people with crude pots and tools occupied the region around the pyramid. After a lapse of time, an eruption of one or more of the neighboring volcanoes eliminated the occupation, depositing a layer of volcanic ash. A further considerable period elapsed, new top soil was formed, and the area was again occupied, this time by an advanced people whose artifacts suggested they were the descendants of the people preceding them. A process of cultural development would appear to have taken place in some other region perhaps nearby. Again, after a considerable time, another eruption of the volcanoes seems to have eliminated this advanced culture, and this time resulted in a complete culture break, for the third people to occupy the region, those of the Archaic culture, appear to have had no connection with their predecessors. Only after all these things had taken place was the Pedrigal formed.

A check on Cummings' estimate of 6,500 years, for the time required to accumulate all the sediments, is provided by the radiocarbon samples referred to above. They were taken at various depths below the Pedrigal, though at a distance of about 1,000 feet from the pyramid. They all consisted of charcoal. Arranged in the order of depth below the lava, their approximate dates were as follows:

Table A: Cuicuilco Radiocarbon Dates

Sample Numbers	Depth (Approx.)	Age	Margin (± Yrs.)
UCLA-228, Cuicuilco A-2	Associated with lava	414 A.D.	65
UCLA-205, Cuicuilco B-1	4 ft. 6 in.	160 A.D.	75
UCLA-206, Cuicuilco B-2	7 ft. 6 in.	15 A.D.	80
UCLA-602, Cuicuilco B-17	7 ft. 6 in.	240 B.C.*	80
UCLA-208, Cuicuilco B-4	7 ft. 8 in.	150 B.C.	150
UCLA-603, Cuicuilco B-18	7 ft. 11 in.	280 B.C.	80
UCLA-207, Cuicuilco B-3	8 ft. 1 in.	650 B.C.*	70
UCLA-209, Cuicuilco B-5	8 ft. 8 in.	350 B.C.	70
UCLA-594, Cuicuilco B-9	14 ft. 3 in.	610 B.C.	80
UCLA-210, Cuicuilco B-6	15 ft. 0 in.	2030 B.C.*	60
UCLA-595, Cuicuilco B-10	15 ft. 0 in.	540 B.C.	100
UCLA-596, Cuicuilco B-11	15 ft. 4 in.	610 B.C.	100
UCLA-597, Cuicuilco B-12	16 ft. 8 in.	1870 B.C.	100
UCLA-598, Cuicuilco B-13	16 ft. 8 in.	1870 B.C.	100
UCLA-211, Cuicuilco B-7	17 ft. 6 in.	4765 B.C.*	90
UCLA-212, Cuicuilco B-8	19 ft. 0 in.	2100 B.C.	75
UCLA-600, Cuicuilco B-15	20 ft. 8 in.	1980 B.C.	100
UCLA-599, Cuicuilco B-14	21 ft. 6 in.	1900 B.C.	200
UCLA-601, Cuicuilco B-16	21 ft. 6 in.	2160 B.C.	120

* Samples UCLA-602, 207, 210, and 211 anomalous.

If we disregard the samples out of chronological order (UCLA-602, 207, 210, 211), which suggest disturbances in the sediments through digging operations (or other causes) in ancient times, and compare the accumulation of sediments with the lapse of time between each pair of consecutive samples, we find there are very wide variations in the rate of accumulation.

Table B: Rate of Sedimentation

Sample Numbers	Accumulation	Time	Rate (Approx.)
UCLA-228, 205	4 ft. 6 in.	254 yrs.	1':56 yrs.
UCLA-205, 206	3 ft. 0 in.	145 yrs.	1':48 yrs.
UCLA-206, 208	0 ft. 2 in.	165 yrs.	1':990 yrs.
UCLA-208, 603	0 ft. 3 in.	130 yrs.	1':520 yrs.
UCLA-603, 209	0 ft. 9 in.	70 yrs.	1':93 yrs.
UCLA-209, 594	5 ft. 7 in.	260 yrs.	1':48 yrs.
UCLA-594, 596	1 ft. 1 in.	00	0:00
UCLA-596, 597	1 ft. 4 in.	1,260 yrs.	1':948 yrs.
UCLA-597, 212	2 ft. 4 in.	230 yrs.	1':100 yrs.
UCLA-212, 601	2 ft. 6 in.	60 yrs.	1':25 yrs.

If we accept the dates of 414 A.D. and 2160 B.C. for the top and bottom of our column of sediments (Table A), we can suppose that 21½ feet of sediment accumulated in 2,574 years before the eruption of the Pedrigal, at an average rate of a foot in 119 years. The variations in the rate may mean simply that the sediments were much disturbed in ancient times, or they may reflect changes in the rate of accumulation related to periods of volcanic eruption, when the rate would have been rapid, and to periods following eruptions when there was no human occupation and very little vegetation, when it would be very slow. The samples were all taken from a human occupation site, that is, from mounds under the Pedrigal containing the ruins of buildings, where the rate of accumulation of sediment would naturally have been faster. The essential point is that while the radiocarbon samples taken near the pyramid give us approximate dates for various phases of the Archaic or Pre-Classical cultures in the area, they have not, so far, dated the pyramid. No excavation appears to have been made below the pavement mentioned by Cummings as surrounding the pyramid. It appears from the evidence that the structures near the pyramid under the Pedrigal, that have now been dated, were probably the work of people who occupied the region after the abandonment of the pyramid.

If this is the case, we have the date of 2160 B.C. as a minimum date for the abandonment of the pyramid. This does not date its construction. Cummings gives reasons to believe (see Note 19) that the structure was in use for a long period of time. Since its scale and advanced construction imply an advanced people possibly flourishing in Mexico four or five thousand years ago, we may have here a relic of the people who navigated the whole earth, and possessed the advanced sciences necessary to make our ancient maps.

A word of caution. I am not expecting that these remarks regarding the Pyramid of Cuicuilco will be regarded as final in a scientific sense. I mean to suggest only a possibility. I would suggest that there should now be a re-examination of that pyramid, and of several other sites in Mexico and in South America, to determine whether, in fact, they may not be related to the ancient civilization which the maps so strongly indicate must once have existed, and which must have been worldwide, at least so far as exploration and mapmaking were concerned. Repeatedly, during the last hundred years, discoveries have been made, which were claimed by the discoverers to indicate the existence of an ancient advanced civilization. These alleged discoveries were disregarded or discredited by archaeologists as the products of sheer imagination or fakery. The task of disinterring and re-examining these old and perhaps mistakenly rejected discoveries will be a long one; that of finding new evidence in the field has not yet been begun. The research project is one for many hands, many years, and much money.

Outside the archaeological field there are two areas in which there is worthwhile evidence of an ancient world civilization. There is, first, the problem of the origin of the principal families of speech and the various groups of languages. Some scholars have claimed that most languages betray evidences of an original common language, ancestral to all the groups of language (such as the Indo-European, etc.). One of these was Arnold D. Wadler, who spent a lifetime on the problem. I do not know whether his conclusions are valid, but his book (214) shows, it seems to me, a scientific approach. It is interesting that a tradition of a universal language seems to be common in ancient literature. In Genesis we read, of course, "And the whole earth was of one language and one speech." Lincoln Barnett, in his *Treasure of Our Tongue*, remarks, "The notion that at one time all men spoke a single language is by no means unique to Genesis. It found expression in ancient Egypt, in early Hindu and Buddhist writings and was seriously explored by several European philosophers during the 16th Century. . . ." (24:46)

The other line of research is comparative mythology. For some years, with my anthropology classes, I have been pursuing research in mythology, and one concept that has emerged from our studies, and with great clarity I may say, is the virtual identity of the great systems of mythology throughout the world. The same pattern, the same principal deities, appear everywhere—in Europe, in Asia, in North and South America, in Oceania. Table C below lists the Gods of the Four Elements —Air, Earth, Fire, Water—as they are found in mythologies all over the world.

There have been many theories of mythology. One of them attributed the similarities in the myths to a common origin in Egypt. This has been generally rejected, because the diffusion of Egyptian myths to America, India, China, and Oceania cannot be proved. If there was diffusion, the point of origin must lie farther back, in a culture earlier than Egypt. Another theory attributes the similarities to instinct. Its proponents argue that the myths derive from instincts that are the same in all men. This theory is weak because, in the first place, modern psychologists tend to doubt the existence of such instincts, and, secondly, insofar

Table C

Gods of the Four Elements in Various Pantheons*

	FIRE	AIR	EARTH	WATER
EGYPT	Re	Shu	Geb, Gea	Nu, Nunu†
BABYLONIA	Girru	Anu	Enlil	Ea
HEBREW	Gabriel	Raphael	Raashiel	Rediyas
PHOENICIA	Ouranos	Aura	Gea	Ashera
PERSIA‡	Atar	Ahura Mazda	Ameretet	Anahita
INDIA	Agni	Yayu	Prithivi	Varuna
CHINA	Mu-King	How-Chu	Yen-Lo-Wang	Mo-Hi-Hai
JAPAN	Ama-Terashu	Amida	Ohonamochi	Susa-No O
IRAN‡	Asha; Atar Oeshma	Vohu Manah Oka Manah	Spenta Armati Bushyasta	Hauvatet Apaosha
NORSE	Thor	Tyr	Odin	Njord
INCA	Manco-Capac	Supay	Pachacamac	Viracocha
AZTEC	Ometecutli	Tezcatlipoca	Omeciuatl	Tlaloc
MAYAN	Kulkulcan	Bacabs	Voltan	Itzamna
SLAV	Swa	Byelun	Raj	Peroun
FINNS	Fire-Girl	Ukko	Ilmatar	Kul Uasa

* Prepared by the anthropology class at Keene State College.

† The gods of the four elements in Egypt were different in different periods.

‡ Persian and Iranian mythologies were not the same; in Iranian mythology the four gods of the elements have their opposites, representing the good and evil aspects.

as they may exist they can apply only to the most general themes, such as love, hate, mystical feeling, etc. The resemblances between the myths, as the table shows, are really too specific to be attributed to general instincts.

We have, then, a general conclusion. The evidence for an ancient worldwide civilization, or a civilization that for a considerable time must have dominated much of the world in a very remote period, is rather plentiful—at least potentially. We have manifold leads, which further research can hardly fail to develop.

Acknowledgments

I have always felt that science is a social process. As a student of the history of science I have learned that every advance of science is the product of many minds working together, even if it may seem that only one individual is responsible. For this reason I wish to give credit here to the many people who have contributed to this work.

For the initiation of the project I am deeply indebted to Mrs. Ruth Verrill of Chiefland, Florida, who first drew my attention to the work of Captain Arlington H. Mallery.

From the beginning, my students took an active part in the investigation and many of them made important contributions. Among them were Ernest Adams, Ronald Bailey, Ruth Baraw, George Batchelder, Richard Cotter, Don Dougal, Clayton Dow, James D. Enderson, Leo Estes, Sidney B. Gove, William Greer, Gary Howard, Alfred Isroe, Warren Lee, Marcia Leslie, Loren Livengood, John F. Malsbenden, John Poor, Frank Ryan, Alan Schuerger, Robert Simenson, Lee Spencer, Margaret Waugh, Robert Jan Woitkowski, and my sons, Frederick and William. Many of them were mentioned in the book.

From the third year of the investigation we were fortunate to receive the co-operation of Captain Lorenzo W. Burroughs, chief of the Cartographic Unit of the 8th Reconnaissance Technical Squadron, at Westover Air Force Base, Westover, Massachusetts. Captain Burroughs and his commanding officer, Colonel Harold Z. Ohlmeyer, examined our findings regarding the Piri Re'is and Oronteus Finaeus Maps, and endorsed them in letters. Captain Burroughs, however, did much more than check our conclusions. His many suggestions were invaluable to us. In addition, he encouraged the members of his cartographic staff to contribute their own time to assisting us. The officers and men of the Air Force who contributed their work were: Captain Richard E. Covault, Captain B. Farmer, Chief Warrant Officer Howard D. Minor, Master Sergeant Clifton M. Dover, Master Sergeant David C. Carter, Technical Sergeant James H. Hood, Staff Sergeant James L. Carroll, Airman First Class Don R. Vance, and Airman Second Class R. Lefever.

Numerous individuals throughout the country took an interest in our work and contributed valuable information and advice. These include William Briesemeister, chief cartographer of the American Geographical Society, now retired, who read the manuscript of this book and contributed suggestions for the preparation of the various maps in it; David C. Ericson of the Lamont Geological Observatory, the late Archibald T. Robertson of Boston, Robert L. Merritt, of Cleveland, Ohio, who read the manuscript and contributed much scholarly information; Miss Elizabeth Kendall, of Washington, D.C., who very kindly provided me with translations of articles from the Soviet press relating to Piri Re'is; and to my aunt, Mrs. Norman Hapgood, for many suggestions. I am indebted also to Niels West, Dino Fabris, Mrs. Margaret Allen, Paul R. Swann; to Dr. Arch Gerlach, Walter Ristow, and Richard W. Stephenson, of the Library of Congress; to Miss Nordis Felland, librarian of the American Geographical Society, to E. Pognon and M. Hervé of the Bibliothèque Nationale in Paris; to I. Hatsukade of the National Diet Library in Tokyo, and to the staff of the Hispanic Society of America, in New York. In addition, I am grateful for the co-operation extended to me, in my search for ancient maps of the East, by the former Sultan of Zanzibar and by Emperor Haile Selassie of Ethiopia.

I am indebted to the United States Department of State for copies of the letters relating to the discovery of the Piri Re'is Map. Also, I wish to thank the American Geographical Society for making arrangements for me with the American Embassy in Ankara, Turkey, for special photographs of the Piri Re'is Map.

I must not forget a deep personal debt to Mr. and Mrs. Philip Martin of Keene, New Hampshire, my most excellent photographers, who bore with me through the innumerable rephotographings of countless maps, for which I was always in a hurry, and to my typist, Miss Eileen Sullivan, who worked under equally bad conditions without ever losing her patience.

I am most deeply indebted to Dr. J. K. Wright, former director of the American Geographical Society, who, in addition to writing the Foreword to this book, read the entire manuscript and contributed many very valuable suggestions for improvements in both content and style, and to Charles W. Halgren, of Caru Studios in New York, who, in addition to preparing our drawings for reproduction, gave us much valuable advice on technical matters.

Notes

Note 1: A Biography of Piri Re'is[1]

Piri Muḥyi 'l-Din **Re'is,** Ottoman navigator and cartographer, was probably
of Christian (Greek) origin and is described as nephew of the famous corsair
Kemāl Re'is (on the latter see the Bonn dissertation by Hans-Albrecht von Burski,
Kemal Re'īs, ein Beitrag zur Geschichte der türkischen Flotte, Bonn 1928 and
especially J. H. Mordtmann, *Zur Lebensgeschichte des Kemāl Re'is,* in M.S.O.S.,
xxxii., part 2, Berlin 1929, p. 39–49 and p. 231 sq.), who was probably a renegade.
His father is said to have been a certain Ḥādjdjī Meḥmed, while he himself in the
preface to his sailing-book calls himself the son of Ḥādjdjī Ḥakīrī, which is perhaps
only to be taken as a name chosen to rhyme with Pīrī (cf. Sinān b. 'Abd al-Mannān
or Dāwūd b. 'Abd al-Wudūd and similar rhyming names of fathers of renegades
usually formed with 'Abd). As Hakīrī cannot be an 'alam but at most a *makhlaṣ,*
the pure Turkish descent of Pīrī is more than doubtful, if he is not called simply
Ḥakīrī Meḥmed, i.e. bore a name for which there is evidence, for a later period
it is true, in the *Sidjilli-i 'othmānī,* ii. 239. The same source (ii. 44) says that the
corsair's full name was Pīrī Muḥyi 'l-Din Re'is. In any case it may safely be assumed
that Pīrī is to be taken as a *takhalluṣ,* while the real name ('*alam*) was probably
Meḥmed—the combination Pīrī Meḥmed was quite customary in the xvi[th] century—i.e. an 'alam to which Muḥyi 'l-Din corresponded as *khiṭāb* (cf. *Isl.,* xi.,
1921, p. 20, note 3). Of the life of Pīrī Re'is, who made many voyages under his
uncle Kemāl Re'is (d. 16th Shawwāl 916 = Jan. 16, 1511) and later distinguished
himself under Khair al-Din Barbarossa (q.v.; July 4, 1546) we only know that on
these raids he had acquired an unrivalled knowledge of the lands of the Mediterranean. He afterwards held the office of *ḳapudan* of Egypt and in this capacity
sailed from Suez on voyages to the Persian Gulf and the Arabian Sea. In 945
(1547) he occupied 'Aden (cf. *Die osmanische Chronik des Rustem Pascha,* ed. by
Ludwig Forrer [*Turk. Bibl.,* xxi., Leipzig 1923], p. 174 sqq. with full commentary).
In 959 (1551) he lost on the coast of Arabia several of his 30 ships, took the port
of Masḳaṭ and carried off a number of its inhabitants as slaves. He then laid siege
to Hormuz but raised the siege and returned to Baṣra, having accepted bribes to
do so, it is said (according to Pečewī, 'Ali, Ḥādjdjī Khalifa, *Tuḥfat al-Kibār,* first
edition, fol. 28 according to J. v. Hammer, G.O.R., iii. 415). A report that an
enemy fleet was approaching decided him to return hurriedly home with only 3
galleys but with all the treasure he had collected. He was wrecked on the island
of Bahrain, but succeeded with two ships in reaching Suez, then Cairo. Ḳobād

---

[1] **From Encyclopedia of Islam.**

Pa<u>sh</u>a, the governor of Ba<u>s</u>ra, had in the meanwhile reported to the Porte that the expedition had been a failure, which resulted in an order for the execution of Pīrī Re'is being sent to Cairo. He was beheaded there, in 962 (1554–1555), it is said, but probably rather in 959 or 960 and his estate sent to Stambul. After his death envoys are said to have arrived from Hormuz representing the plundered inhabitants to demand the return of the treasure he had carried off; they were naturally not successful. The post of *kapudan* of Egypt was given to another noted corsair, Murād, the dismissed sandjakbey of Ḳaṭif (probably the same as survives in the proverb, according to H. F. v. Diez, *Denkwurdigenkeiten von Asien*, part i., Berlin 1811, p. 55, as *Murād kaptan*).

Pīrī Re'is is generally known as the author of a sailing-book of the Aegean and Mediterranean known as *Baḥrīye* in which he describes all the coasts he had voyaged along with an account of the currents, shallows, landing-places, bays, straits and harbours. Pīrī Re'is had already begun the work in the reign of Selīm I (d. Sept. 1520) although he says in the preface that he did not begin it until 927 (end of 1520), in order to make the dedication to Sulaimān the Magnificent be more impressive. He presented the completed atlas to the latter in 930 (1523). Paul Kahle has published an edition with text and translation based on the known manuscripts, entitled *Pīrī Re'is, Baḥrīye. Das türkische Segelhandbuch für das Mittellandische Meer vom Fahre 1521* of which so far (middle of 1935) vol. i., text, part I and vol. ii., part L, section I—28 have been published, Leipzig and Berlin 1926. Separate sections had been previously published, e.g. H. F. v. Diez, op. cit.; E. Sachau, *Sizilien, in Centenario delle Nascita di Michele Amari*, ii., Palermo 1910, p. I sqq.; R. Herzog, *Ein türkisches Werk über das Agäsische Meer aus dem Fahr 1520*, in *Mitteilungen des Kaiserl. Deutschen Archälog. Instituts, Athenische Abteilung*, xxvii., 1902, p. 417 sqq.; E. Oberhummer, section Zypern, in: *Die Insel Zypern*, Munich 1903, pp. 427–434. Other sections in Carlier de Pinon, ed. E. Blochet (with pictures) and K. Foy, in M.S.O.S., part ii., xi., 1908, p. 234 sqq. Cf. thereon F. Taeschner in Z.D.M.G., lxxvii. (1923), p. 42 with other references.

The so-called "Columbus map," found in October 1929 by <u>Kh</u>alīl Edhem Bey in the Seray Library in Stambul, according to his signature on it of the year 1513, seems also to go back to Pīrī Re'is; it is in Turkish in bright colours on parchment, 85 by 60 cm., and represents the western part of a map of the world. It comprises the Atlantic Ocean with America and the Western strip of the Old World. The other parts of the world are lost. It has been supposed that this is the same map as Pīrī, according to a statement in his *Baḥrīye*, presented to Sultan Selīm in 1517 which would explain its preservation in the Imperial Library. On it cf. Paul Kahle, *Impronte Colombiane in una Carta Turea del 1513*, in *La Cultura*, year x., vol. I, part 10, Milan-Rome 1931; do., *Una mapa de America hecho por el turco Piri Re'is, en el ano 1513, basandose en una mapa de Celon y en mapas portugueses*, in *Investigacion y Progreso*, v., 12, Madrid 1931, p. 169 sqq.; "C" in *The Illustrated London News*, clxxx., No. 4845 on Febr. 27, 1932, p. 307: *A Columbus Controversy—and two Atlantic charts* (with reproduction); P. Kahle, *Die verschollene Columbus-Karte von 1498 in einer türkischen Weltkarte von 1513* (with 9 maps, 52 pp., Berlin and Leipzig 1933); also Eugen Oberhummer, *Eine türkische Karte zur Entdeckung Amerikas*, in *Anzeiger der Akademie der Wissenschaften in Wien*, phil.-hist. Kl., 1931, pp. 99–112; do., *Eine Karte des Columbus in türkischer Überliejerung*, in *Mitteilungen der Geographischen Ges. in Wien*, lxxvii., 1934, p. 115 sqq. and lastly P. Kahle in *Geographical Review*, 1933, pp. 621–638.

Bibliography: Ḥādjdjī Khalifa, *Djihānnumā*, Stambul 1145, p. 11; do., *Tuḥfat al-Kibār fī Esfar āl-Biḥār*, Stambul 1142, p. 28a; do., *Kashf al-Zunūn*, ed. G. Flügel, ii. 22 sqq. (No. 1689); Mehmed Thureiyā, *Sidjilli-i ʿothmānī*, ii. 44: Kahle, op. cit., Introduction: Hans v. Mžik, *Pīrī Reʾis und seine Baḥrīje*, in *Beiträge zur historische Geographie* etc., ed. by Hans v. Mžik, Leipzig and Vienna 1929, pp. 60–76.

(FRANZ BABINGER)

Note 2: State Department Correspondence

No. 102.   ISTANBUL, July 26, 1932

SUBJECT: *Photograph of Map for Library of Congress.*

The Honorable
THE SECRETARY OF STATE,
Washington

SIR:

Today there arrived the Department's unsigned instruction, No. 13 of July 15, file No. 103.7/2409. Also today there came to lunch, Yusuf Akcora Bey, who is not only a Deputy of the Grand National Assembly, but also President of the new Turkish Historical Society and presiding officer at their recent congress held in Ankara. After lunch he produced a copy of the July 23rd LONDON ILLUSTRATED NEWS and took great pride in turning to pages 142–143 containing the first translation into English of any of his writings. It was an account of the very map, now in the possession of the Turkish Government, of which your No. 13 requested a photograph! He told me that although the map was supposed to be in the Palace Library at Istanbul, it was really for the present in his possession in Ankara, and that he would gladly provide me with a photograph thereof the next time he went up there, which will be in the course of the next week or two. He was as much surprised as I at the amazing coincidence of my having received a request from you for this map, a few hours before his showing me the copy of the ILLUSTRATED NEWS containing the photograph now in his possession. I suggest this information be communicated to the Library of Congress, so that by consulting that copy of the English magazine they will be able to see a reproduction of the map and read an article about it while waiting for the photograph which will shortly follow. It purports to be a map by a Turkish Admiral prepared upon information received from Christopher Columbus.

Respectfully yours,

CHARLES H. SHERRILL

103.7
CHS:FM

No. 111   ISTANBUL, August 4, 1932

SUBJECT: *Photograph of Map for Library of Congress.*

The Honorable
     THE SECRETARY OF STATE,
          Washington

SIR:

Referring to your instruction No. 13, File No. 103.7/2409, of July 15, you will please find enclosed two copies of reproductions of a portion of the map referred to in this instruction. In my No. 102 of July 26, I reported that the map is now in the personal possession of Yusuf Akcora Bey, President of the History Congress, and is actually in his house at Ankara, and that he promised me reproductions thereof. Upon receipt of the two enclosed copies I discovered that they reproduce only a portion of the map, for use with his article of July 23 in the LONDON ILLUS-TRATED NEWS. I have written asking him to permit me to have a photograph made (as per your No. 13), but this request, although he will surely grant it, will be delayed for several weeks as he will not return to Ankara before the end of September or beginning of October. I had tea with him July 29 at his country place at Cooz-tepe out beyond Scutari and had quite a talk with him upon the interest aroused about the Occident by the discovery of this map, and congratulated him on the timeliness of his article in the London magazine. This morning had another talk with him there by telephone. He was working last night with the Gazi till 4 A.M. at the Dolma Bagtche Palace upon the new Turkish History, in which this map is to figure, but promises I shall have fuller photographs when he next visits Ankara.

Respectfully yours,

CHARLES H. SHERRILL

Enclosures:
1,2,: Two copies of reproductions of portion of map.

103.7/2409
CHS:on

SUBJECT: *Photographs of Map for Library of Congress.*

The Honorable
    THE SECRETARY OF STATE,
        Washington

SIR:

Referring to the Department's instructions No. 13, File No. 103.7/2409 of July 15, 1932, and to my No. 102 of July 26, I beg further to report that yesterday there came to luncheon with me at the Embassy, Yusuf Akcora Bey, Deputy and President of the Historical Congress held during the first fortnight in July at Ankara. He tells me that the 1513 map of the Turkish Admiral has now been returned to the Library of the old Seraglio at Seraglio Point in Stamboul, and he promises to secure me permission to inspect it. He corrected the impression which I previously held that two copies of the map which I sent with my No. 102 represented only a portion of a larger map. This is not true, for it is the copy of the entire map desired by the Library of Congress. He brought with him and presented me with seven more photographs explaining them to me as follows:

"A" shows two pages of the book in which the Turkish Admiral describes how he made these maps as the result of what he personally saw on his own voyages and having completed the maps and the book, presented them to Sultan Selim I in Egypt, who was then carrying out the invasion of that country.

"B" shows the Straits of Gibraltar with portions of the Spanish and African coasts. The small town depicted amid the four hills at the top of the photograph is labelled Granada. Note the larger of the two inscriptions which follow the N.E. compass line. He there places the title "Spain or Rumeli" which is probably a reference to the fact that the Turks call the European, as contradistinguished from the Asiatic shore of the Bosphorus, "Rumeli."

"C" shows the French Riviera shore from Nice running easterly toward Italy.

"D" shows the Italian shore from Pisa (up in the left hand corner) down through to Civitavecchia and the mouth of the Tiber.

"E" shows the Island of Rhodes.

"F" gives a group of islands, but my Turkish friend did not seem to be able to locate them.

"G" is another map from the book, but as it bears no Turkish titles discloses nothing of its whereabouts.

Yusuf Akcora Bey was insistent upon calling my attention to the fact that all the writing on these maps is in Turkish and not Arabic, although written in Arabic script. This is equally true of the writing upon the larger map sent you in my No. 102. He translated to me the long inscription which states that this map was not the result of the Turkish Admiral's own investigations, but is copied from Christopher Columbus' map, which unfortunately is now lost. It carried the name of Columbus no less than eight times, and is chiefly interesting for the contemporary statement that Columbus was of Genoese birth, and not of Spanish as certain Spanish writers have recently claimed. It tells that Columbus, the Genoese, incited thereto by an ancient book, applied to the Genoese authorities for ships and money to enable him to cross the seas to Japan and China. This recalls the fact (not generally known) that up to the day of his death, Columbus thought the western islands which he discovered were part of Zipango or Japan. When the Genoese authorities refused him, he next applied to the "Bey of Spain," who finally gave him two ships. With these he sailed westerly across the great sea and finally reached land. There came down to meet him the entirely naked inhabitants who shot at the invaders with arrows pointed with fish bones. Columbus finally pacified them to the extent of trading beads for fish. Presently he remarked a golden bracelet on a woman's arm, and this led to his trading beads for gold, which came from mines up in the hills. Later he found that the natives also had real pearls, and he then traded beads for pearls. He returned to deliver his booty to the "Bey of Spain," who sent him back on a second voyage, of which we learn little except that he carried with him onions and barley, which were soon cultivated by the natives. I hope to get further information about this map later on, and will promptly forward the same.

Respectfully yours,

CHARLES H. SHERRILL

Enclosures:
Nos. 1–7   Photographic copies of 1513 maps of a Turkish Admiral.

103.7
CHS:on

No. 272   ANKARA, December 23, 1932

SUBJECT: *Photograph of Map for Library of Congress.*

The Honorable
    THE SECRETARY OF STATE,
        Washington

SIR:

Referring to your No. 58 of November 29, I beg to report that immediately upon its receipt, December 13, I wrote to Yusuf Akcura [*sic*] Bey (President of last July's Historical Congress) asking for the information requested by you for the Library of Congress. I have had two long talks with him yesterday and today upon the subject of your No. 58, and he tells me that, upon receiving my letter, he instituted systematic and widespread inquiry to ascertain "whether there is any chance whatever that the original Spanish or Italian chart of Columbus is still preserved in the Turkish archives or among the papers of the family of Piri Reis." Piri Reis was finally beheaded, so it is more than doubtful if anything will be found, because beheading was generally followed by complete confiscation of the deceased's effects.

There is enclosed herewith a copy of his letter mentioned in the LONDON ILLUSTRATED NEWS, volume 181, 1932, pages 142 and 143, which you requested.

In regard to the "literal transliterations and translations of the place names and other legends in the western half of the map," please see my letter to Mr. J. Brent Clarke, Chief of the Department's Mail Room, dated December 19, which explains the possible sending on of the map and translations to me after my departure from Washington to my house in New York. On December 19, I also wrote my house to send any such mailing tube back to Mr. Wallace Murray with a line saying that it was being returned in compliance with the Department's instruction, No. 58 of November 29, 1932. In regard to these translations by Ali Nur Bey, he has consulted a number of Turks versed in maritime matters, but even those specialists were, in some instances, puzzled by the nautical words used by that distinguished Turkish navigator back in 1513. Although the language used is Turkish, the characters are Arabic, and this means that there is frequent omission of vowels from the words written on the map, which of course complicates the transliteration into English or Spanish or Italian of the geographical place names. Notwithstanding this difficulty, I think that Ali Nur Bey has been surprisingly successful in his efforts.

Respectfully yours,

CHARLES H. SHERRILL

Enclosure:
1. Copy of letter.

103/7/892.3
CHS:er

Enclosure No. [1] to Despatch No. 272 of December [23, 1932] from the Embassy at Istanbul.

COPY:
TÜRK TARIHI TETKIK CEMIYETI
        Reisligi

To the EDITOR of the "Illustrated London News,"
        London, England

DEAR SIR,

We had the pleasure of reading the explanation given in the issue of your valuable magazine dated February 25, 1932 regarding the map drawn by Piri Reis, one of the Turkish geographers.

We kindly request the publication in your magazine [of] the following lines containing the results of the studies we have made on the original of this map now in our possession, and aiming to complete your exposes as well as to correct certain mistakes in it.

We enclose, as documents to prove our statement, photographic copies of two pages from the book "Bahriye" of Piri Reis, and five small maps taken from the same book.

We shall be glad to see them, with this explanatory article, published in your worthy magazine.

The map in question is drawn on a gazelle skin by Piri Reis who had made a name for himself among the Western and Eastern Scholars[1] through his detailed geographical book on the Mediterranean Sea entitled "On the Sea" and which testifies to his capacity and knowledge in his profession. Piri Reis is the son of the brother of the famous Kemal Reis who was the Turkish admiral in the Mediterranean Sea at the last quarter of the fifteenth century. History records Piri Reis Bey's last official post as admiral of the Fleets in the Red Sea and the Indian Ocean. Piri Reis wrote and completed the above-mentioned map in the city of Gelibolu (Gallipoli) in the year 1513, and four years after this date, i.e. in the year 1517, he presented personally to Selim I, the conqueror of Egypt, during the presence of the latter there.[2]

As the same thing will be noticed in the maps of ancient and mediaeval times, the map of Piri Reis contain [sic] important marginal notes regarding the history and the geographical conditions of some of the coasts and islands.[3] *All these marginal notes* with hundreds of lines of explanation *were written in Turkish*. Three lines only, which from the title and head lines of the map, were written in Arabic; and this is done to comply with the usual traditional way which is noticed on all the Ottoman Turkish monuments up [to] the very latest centuries. These three lines in Arabic testify that the author is the nephew of Kemal Reis, and that the work [was] written and compiled of [sic] Gelibolu in the year 1513.

---

[1] There are 207 fine charts drawn by Piri Reis in this book of his wherein correct and scientific informations [sic] profusely presented. (See enclosed copies of the maps of Malta, Sicily, Corsica, environs of Gibraltar, Venice.) This book "Bahriye" (= On the Sea) was prepared by Piri Reis, at Gelibolu, as a tentative work eight years after the above-mentioned map was made, and seven years later, improved and rewritten by the author it was presented to Kanuni Süleyman (Soliman the Magnificent) in Istanbul. Several copies of this work are found in the libraries of Stanbul (in the Museum of Topkapi, in the Naval Museum and in the libraries of Aya-Sofya (St. Sofia) and Nuri Osmaniye as well as in the Dresden Library, in the University Library at Bologne and in the Bibliothèque Nationale in Paris. Kahle, a Bonne University professor, in addition to the articles he had published on this book, has taken the initiative in recent years in reproducing it.

[2] In the introduction of his book Piri Reis gives some explanations about maps, at the same time, he has the following lines concerning his own map: "Previously I prepared a more detailed map and presented it to Selim (Sultan Selim) in Egypt and he congratulated me. In this map I recorded the latest information about the Indian and Chinese seas which was set forth on the maps prepared at that time; but which was not known here." (The photographic reproduction of the original of this statement is enclosed under No. 7.)

[3] See marginal notes on the west edge of the map.

The map in our possession is a fragment and it was out of from [sic] a world chart on large scale. When the photographic copy of the map is carefully examined, it will be noticed that the lines of the marginal noted [sic] on the eastern edges have been cut half away.[4]

In one of these marginal notes the author states in detail the maps he had seen and studied in preparing his map. In the marginal note describing the Antilles Islands, he states that he has used Christopher Columbus' chart for the coasts and islands. He sets forth the narratives of the voyages made, by a Spaniard a slave in the hands of Kemal Reis, Piri Reis' uncle, who under Christopher Columbus made three voyages to America. He also states, in his marginal notes regarding the South American coast that he saw the charts of four Portuguese discoverers. That he has made use of Christopher Columbus' chart is made clear in the following lines of his:

In order that these islands and their coasts might be known Columbus gave them these names and set it down on his chart. The coasts (the names of the coasts) and the islands are taken from the chart of Columbus.

The work essentially was a world map. Therefore Piri Reis had made a study of some of the charts which represented the world, and according to his personal statement, he has studied and examined the maps prepared at the time of Alexander (the Great), the "Mappa Mundis" and the eight maps in fragments prepared by the Muslims.[5]

Piri Reis himself plainly explains, in one of the marginal notes in his map, how his map was prepared:—

This section explains the way the map was prepared. Such a map is not owned by any body at this time. I, personally, drawn [sic] and prepared this map. In preparing this map, I made use of about twenty old charts and eight Mappa Mundis, i.e. of the charts called "Jaferiye" by the Arabs and prepared at the time of Alexander the Great and in which the whole inhabited world was shown; of the chart of [the] West Indies; and of the new maps made by four Portugueses [sic] containing the Indian and Chinese countries geometrically represented on them. I also studied the chart that Christopher Columbus drew for the West. Putting all these material [sic] together in a common scale I produced the present map. My map is as correct and dependable for the seven seas as are the charts that represent the seas of our countries.

Piri Reis, in a special chapter in his book "Bahriye" mentions the fact that in drawing his map he has taken note of the cartographical traditions considered international at that time. The cities and citadels are indicated in red lines, the deserted places in black lines, the rugged and rocky places in black dots, the shores and sandy places in red dots and the hidden rocks by crosses. In short:

---

[4] See marginal note on the South-East corner of the map.

[5] It has been rightly said by the latest Oriental Scholars that it is a mistake to talk of an Arabic Civilization. They should talk of an Islamic Civilization. Because those who have created this civilization, were not all Arabs, although they used the Arabic language there were among them more Persians and Turks than Arabs.

1. The picture of the map printed in your magazine is the Turkish work of a Turkish sailor of the name Piri Reis. The work is not in Arabic. Except the three lines as an introduction to the map, the rest are written entirely in Turkish. Even the places on the Atlantic coast of Africa bear the Turkish geographical names like Babadagi (= Father Mount), Akburun (= White Cap), Yesilburun (= Green Cap), Kizilburun (= Red Cap), Kozluburun (= Walnut Cap), Altin Irmak (= Gold River) and Güzel Körfez (= Handsome Gulf).[6]

2. The map was finished and presented to the Padishah in 919 . . . (1513), and certainly not in 929 (1523).[7]

3. Piri Reis states that in preparing this work he has made use of the charts prepared in the Islamic World, the Portuguese charts and even the chart of Christopher Columbus. But this is not a copy; it is an original work.

4. The map in our possession is a fragment. If the other fragment were not lost, there would have been in our possession a Turkish chart drawn in 1513 representing the old and new worlds together. As Christopher Columbus' voyages took place in the latter part of the fifteenth century, it can be said that such a map which was made a little after the new discoveries, is one of the early charts to have contained in it all the continents of the world.

Sincerely yours,

(s) . . . .*

Department of State
WASHINGTON
March 20, 1933

In reply refer to
HA 103.7/2537

LIBRARIAN OF CONGRESS,
    Washington, D.C.

SIR:

The Department refers to previous correspondence regarding a map which is said to have been prepared by a Turkish Admiral upon information received from Columbus, and encloses a copy of a further despatch on the subject, No. 374 of February 21, 1933, from the American Ambassador at Istanbul.

Very truly yours,

For the Secretary of State:
E. WILDER SPAULDING,
Assistant to the Historical Adviser

Enclosure:
    From Embassy, Istanbul,
    No. 374, February 21, 1933.

---

[6] See West coast of Africa at the chart.

[7] The date is clearly read on the title lines of the chart: "Tis'a ashara va tis'a mi'a" (= nine hundred and nineteen).

* Source of the preceding letters is The National Archives, Washington, D.C.

**Enclosure No. 2 to Despatch No. 374
of February 21, 1933, from the Embassy at Istanbul**

TRANSLATION

PIRI REIS MAP

> SOURCE: LETTER FROM YUSUF
> AKCURA BEY, Deputy
> of Istanbul at the
> GRAND NATIONAL ASSEM-
> BLY OF TURKEY, dated
> February 19, 1933

AKCURAOCIU YUSUF
Deputy of Istanbul

ANKARA, Keciören, February 19, 1933

MR. AMBASSADOR:

Upon receipt of your esteemed favour of December 17, 1932, I immediately wrote a circular letter to the directors of our various museums and to some friends who are engaged in the study of questions dealing with the maritime history of Turkey, requesting them to assist me in my search for the original map of Christopher Columbus and of the descendants of Piri Reis.

I have received up to the present time nine replies. All of my correspondents maintain that none of the Turkish museums or archives contain the testament or will of our famous armiral [sic]. The Direction of the Evkaf likewise does not know of any certificate relating to a pious fund signed by him or by his heirs.

Piri Reis having been beheaded in Egypt in 1564, it is very likely that he had no opportunity to convey his property and his precious documents to his heirs. Besides there is no reliable information as to his having had heirs.

However, Djevdet Bey, one of our historians, informed me that Commander Saffet Bey who died in 1912 told him of having found in the archives of the Ministry of Marine a document mentioning the granting of a pension to the descendants of Piri Reis by Sultan Selim III (1789–1807).

On the basis of this information I will continue my investigations. I shall not fail to communicate to you the results which they may yield.

Please accept, Mr. Ambassador, the assurance of my sincere and most cordial regards.

(signed) AKCURA YUSUF

General CHARLES H. SHERRILL,
Ambassadour [sic] of the United States to Turkey

Note 3: The Legends on the Piri Re'is Map*

1. There is a kind of red dye called vakami, that you do not observe at first, because it is at a distance . . . the mountains contain rich ores. . . . There some of the sheep have silken wool.

II. This country is inhabited. The entire population goes naked.

III. This region is known as the vilayet of Antilia. It is on the side where the sun sets. They say that there are four kinds of parrots, white, red, green and black. The people eat the flesh of parrots and their headdress is made entirely of parrots' feathers. There is a stone here. It resembles black touchstone. The people use it instead of the ax. That it is very hard . . . [illegible]. JPe saw that stone.

[NOTE: Piri Reis writes in the "Bahriye": "In the enemy ships which we captured in the Mediterranean, we found a headdress made of these parrot feathers, and also a stone resembling touchstone."]

IV. This map was drawn by Piri Ibn Haji Mehmed, known as the nephew of Kemal Reis, in Gallipoli, in the month of muharrem of the year 919 (that is, between the 9th of March and the 7th of April of the year 1513).

V. This section tells how these shores and also these islands were found.

These coasts are named the shores of Antilia. They were discovered in the year 896 of the Arab calendar. But it is reported thus, that a Genoese infidel, his name was Colombo, he it was who discovered these places. For instance, a book fell into the hands of the said Colombo, and he found it said in this book that at the end of the Western Sea [Atlantic] that is, on its western side, there were coasts and islands and all kinds of metals and also precious stones. The above-mentioned, having studied this book thoroughly, explained these matters one by one to the great of Genoa and said: "Come, give me two ships, let me go and find these places." They said: "O unprofitable man, can an end or a limit be found to the Western Sea? Its vapour is full of darkness." The above-mentioned Colombo saw that no help was forthcoming from the Genoese, he sped forth, went to the Bey of Spain [king], and told his tale in detail. They too answered like the Genoese. In brief Colombo petitioned these people for a long time, finally the Bey of Spain gave him two ships, saw that they were well equipped, and said:

"O Colombo, if it happens as you say, let us make you kapudan [admiral] to that country." Having said which he sent the said Colombo to the Western Sea. The late Gazi Kemal had a Spanish slave. The above-mentioned slave said to Kemal Reis, he had been three times to that land with Colombo. He said: "First we reached the Strait of Gibraltar, then from there straight south and west between the two . . . [illegible]. Having advanced straight four thousand miles, we saw an island facing us, but gradually the waves of the sea became foamless, that is, the sea was becalmed and the North Star—the seamen on their compasses still say star—little by little was veiled and became invisible, and he also said that the

---

*From "The Oldest Map of America," by Professor Dr. Afet Inan. Ankara, 1954, pp. 28–34. The Roman numerals refer to the key map.

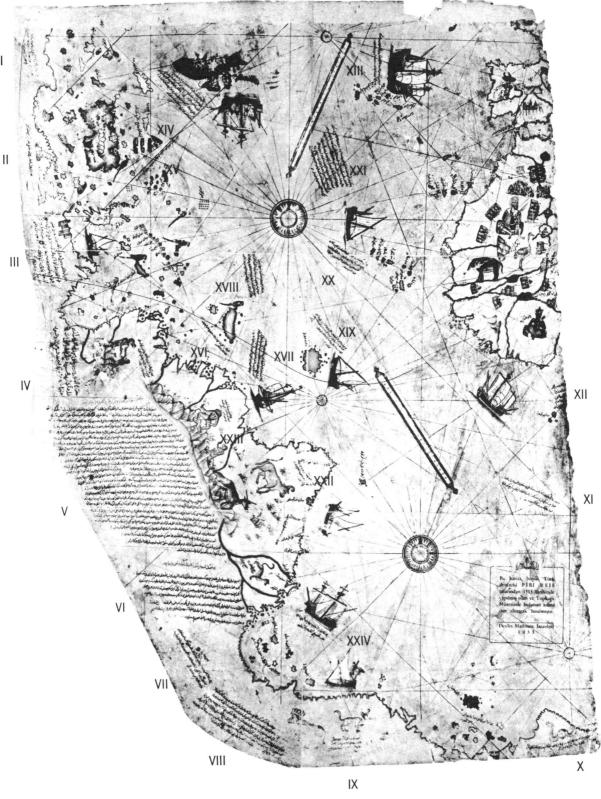

Figure 100. The Piri Re'is Map, with Roman numerals corresponding to the translated legends. After A. Afet Inan.

stars in that region are not arranged as here. They are seen in a different arrangement. They anchored at the island which they had seen earlier across the way, the population of the island came, shot arrows at them and did not allow them to land and ask for information. The males and the females shot hand arrows. The tips of these arrows were made of fishbones, and the whole population went naked and also very . . . [illegible]. Seeing that they could not land on that island; they crossed to the other side of the island, they saw a boat. On seeing them; the boat fled and they [the people in the boat] dashed out on land. They [the Spaniards] took the boat. They saw that inside of it there was human flesh. It happened that these people were of that nation which went from island to island hunting men and eating them. They said Colombo saw yet another island, they neared it, they saw that on that island there were great snakes. They avoided landing on this island and remained there seventeen days. The people of this island saw that no harm came to them from this boat, they caught fish and brought it to them in their small ship's boat [filika]. These [Spaniards] were pleased and gave them glass beads. It appears that he [Columbus] had read—in the book that in that region glass beads were valued. Seeing the beads they brought still more fish. These [Spaniards] always gave them glass beads. One day they saw gold around the arm of a woman, they took the gold and gave her beads. They said to them, to bring more gold, we will give you more beads, [they said]. They went and brought them much gold. It appears that in their mountains there were gold mines. One day, also, they saw pearls in the hands of one person. They saw that when; they gave beads, many more pearls were brought to them. Pearls were found on the shore of this island, in a spot one or two fathoms deep. And also loading their ship with many logwood trees and taking two natives along, they carried them within that year to the Bey of Spain. But the said Colombo, not knowing the language of these people, they traded by signs, and after this trip the Bey of Spain sent priests and barley, taught the natives how to sow and reap and converted them to his own religion. They had no religion of any sort. They walked naked and lay there like animals. Now these regions have been opened to all and have become famous. The names which mark the places on the said islands and coasts were given by Colombo, that these places may be known by them. And also Colombo was a great astronomer. The coasts and island on this map are taken from Colombo's map.

VI. This section shows in what way this map was drawn. In this century there is no map like this map in anyone's possession. The—hand of this poor man has drawn it and now it is constructed. From about twenty charts and Mappae Mundi—these are charts drawn in the days of Alexander, Lord of the Two Horns, which show the inhabited quarter of the world; the Arabs name these charts Jaferiye—from eight Jaferiyes of that kind and one Arabic map of Hind, and from the maps just drawn by four Portuguese which show the countries of Hind, Sind and China geometrically drawn, and also from a map drawn by Colombo in the western region I have extracted it. By reducing all these maps to one scale this final form was arrived at. So that the present map is as correct and reliable for the Seven Seas as the map of these our countries is considered correct and reliable by seamen.

VII. It is related by the Portuguese infidel that in this spot night and day are at their shortest of two hours, at their longest of twenty two hours. But the day is very warm and in the night there is much dew.

VIII. On the way to the vilayet of Hind a Portuguese ship encountered a contrary wind [blowing] from the shore. The wind from the shore . . . [illegible] it [the ship]. After being driven by a storm in a southern direction they saw a shore opposite them they advanced towards it [illegible]. They saw that these places are good anchorages. They threw anchor and went to the shore in boats. They saw people walking, all of them naked. But they shot arrows, their tips made of fish-bone. They stayed there eight days. They traded with these people by signs. That barge saw these lands and wrote about them which. . . . The said barge without going to Hind, returned to Portugal, where, upon arrival it gave information. . . . They described these shores in detail. . . . They have discovered them.

IX. And in this country it seems that there are white-haired monsters in this shape, and also six-horned oxen. The Portuguese infidels have written it in their maps. . . .

X. This country is a waste. Everything is in ruin and it is said that large snakes are found here. For this reason the Portuguese infidels did not land on these shores and these are also said to be very hot.

XI. And these four ships are Portuguese ships. Their shape is written down. They travelled from the western land to the point of Abyssinia [Habesh] in order to reach India. They said towards Shuluk. The distance across this gulf is 4200 miles.

XII. .... on this shore a tower
    .... is however
    .... in this climate gold
    .... taking a rope
    .... is said they measured

[NOTE: The fact that half of each of these lines is missing is the clearest proof of the map's having been torn in two.]

XIII. And a Genoese kuke [a type of ship] coming from Flanders was caught in a storm. Impelled by the storm it came upon these islands, and in this manner these islands became known.

XIV. It is said that in ancient times a priest by the name of Sanvolrandan (Santo Brandan) travelled on the Seven Seas, so they say. The above-mentioned landed on this fish. They thought it dry land and lit a fire upon this fish, when the fish's back began to burn it plunged into the sea, they reembarked in their boats and fled to the ship. This event is not mentioned by the Portuguese infidels. It is taken from the ancient Mappae Mundi.

XV. To these small islands they have given the name of Undizi Vergine. That is to say the Eleven Virgins.

XVI. And this island they call the Island of Antilia. There are many monsters and parrots and much logwood. It is not inhabited.

XVII. This barge was driven upon these shores by a storm and remained where it fell. . . . Its name was Nicola di Giuvan. On his map it is written that these rivers which can be seen have for the most part gold [in their beds]. When the water had gone they collected much gold [dust] from the sand. On their map. . . .

XVIII. This is the barge from Portugal which encountered a storm and came to this land. The details are written on the edge of this map. [NOTE: see VIII.]

XIX. The Portuguese infidels do not go west of here. All that side belongs entirely to Spain. They have made an agreement that [a line] two thousand mile.

to the western side of the Strait of Gibraltar should be taken as a boundary. The Portuguese do not cross to that side but the Hind side and the southern side belong to the Portuguese.

XX. And this caravel having encountered a storm was driven upon this island. Its name was Nicola Giuvan. And on this island there are many oxen with one horn. For this reason they call this island Isle de Vacca, which means, Ox Island.

XXI. The admiral of this caravel is named Messir Anton the Genoese, but he grew up in Portugal. One day the above-mentioned caravel encountered a storm, it was driven upon this island. He found much ginger here and has written about these islands.

XXII. This sea is called the Western Sea, but the Frank sailors call it the Mare d'Espagna. Which means the Sea of Spain. Up to now it was known by these names, but Colombo, who opened up this sea and made these islands known, and also the Portuguese, infidels who have opened up the region of Hind have agreed together to give this sea a new name. They have given it the name of Ovo Sano [Oceano] that is to say, sound egg. Before this it was thought that the sea had no end or limit, that at its other end was darkness. Now they have seen that this sea is girded by a coast, because it is like a lake, they have called it Ovo Sano.

XXIII. In this spot there are oxen with one horn, and also monsters in this shape.

XXIV. These monsters are seven spans long. Between their eyes there is a distance of one span. But they are harmless souls.

Note 4: Blundeville's Directions for Constructing the Portolan Design *

"Of the Mariners Carde and of the marking thereof."

"First drawe with a pair of compasses a secrete circle which may be put out, so great as you shall think meet for your carde, which circle shall signifie the Horizon, then divide that circle into foure equall quarters, by drawing two Diameters crossing one another, in the center of the foresaide circle with right angles, whereof the perpendicular line is the line of North and South, and the other crossing the same is the line of East and West, at the foure ends of which crosse Diameters you must set downe the foure principall windes, that is, East, West, North, and South, making the North parte with a flower deluce in the toppe, and the East parte with a crosse, as you may see in the figure following. Then divide everie quarter of the saide circle with your compasses into two equall partes, setting downe pricks in the middest of everie quarter, through which pricks, and also through the centre of the circle drawe two other crosse lines, which must extende

---

* Heathecote (89) quotes the directions given by Blundeville in "Blundeville his Exercises," 1594, for constructing the portolano design.

somewhat beyond the circumference of the Horizon, which two crosse lines together with the first two crosse lines shall divide the circle into 8 partes, and thereby you shall have the eight principall windes. That done, divide everie eight part of the saide Horizon into two equall partes by drawing other two crosse lines through the centre and extending somewhat beyond the circumference of the Horizon as before, whereby the whole circle shall be divided into 16 partes, which shall suffice without making anie more divisions, which woulde cause a confusion of lines, and at the end of every one of these 16 lines you must drawe a little circle, whose center must stande upon the circumference of the Horizon, everie one whereof must bee also divided into 16 partes by the helpe of 16 lines, diversely drawne from the center of one little circle to another, in such order as the figure here placed more plainely sheweth to the eie, than I can expresse by mouth. And these little circles do signifie 16 little Mariners Compasses, the lines whereof signifying the winds, do shew how one place beareth from another, and by what winde the shippe hath to saile. But besides these little circles there is woont to be drawne also another circle somewhat greater than the rest upon the verie center of the Horizon, which circle by reason of the 16 lines that were drawne passing through the same, is divided into 16 parts, and the Mariners doe call this circle the mother compas."

Note 5: Strachan on Map Projections

Richard W. Strachan discusses here the features of a number of projections that appear to have been involved with one or more of the ancient maps. The reader may be reminded that map projections are mechanical and mathematical devices for transferring points from the round earth to flat paper, and they are therefore artificial and complex. The earth is virtually a sphere, and only a globe can correctly represent all of it in correct proportion. Various projections may represent parts of the earth on flat paper with sufficient accuracy for practical purposes, but they all have their faults, and they all distort the earth very badly in one way or another. The mapmaker tries to select the particular projection that is best for mapping the area he wants to map, the projection that has the most advantages and the fewest disadvantages. Strachan here defines five projections, but there are many more. For more detailed discussions the reader may consult Deetz and Adams (60).

AZIMUTHAL: An azimuthal projection is one in which the earth is projected onto a flat plane held tangent to it at one point. The tangent point may be at a pole, on the equator, or anywhere else desired. The class of azimuthal projections includes several of interest, namely the stereographic, gnomonic and azimuthal equidistant, which are described below.

STEREOGRAPHIC: A stereographic projection results if the earth is projected onto a plane from a point on the earth opposite to the point of tangency.

The advantages of this projection are that it can show a whole hemisphere without great distortion, and that great circles* through the tangent point plot as straight lines.

GNOMONIC: A gnomonic projection is obtained when the projection of the earth onto a tangent plane is done from the center of the earth. The advantage of this projection is that any great circle plots as a straight line.

AZIMUTHAL EQUIDISTANT: This projection is one in which the distance scale along any great circle through a tangent point is constant. A polar azimuthal equidistant projection shows the meridians of longitude as straight radial lines, and the parallels as equally spaced concentric circles. It cannot be visualized as being projected, but rather as being constructed, by setting up a scale and transferring the features of the earth onto this scale point by point. It has the great advantage that the whole earth may be shown. Also it has a constant distance scale from the point of tangency (or "pole"); and, like all other azimuthal projections, all angles measured from the point of tangency are true.

CORDIFORM: No recent literature describes the cordiform projection, and to get some sort of definitive statement on it, it was necessary to refer to Nordenskiöld. He lists three cordiform projections (147:86–92). The first he ascribes to the cartographer Sylvanus, noting its similarity to Ptolemy's "homeother" projection. The second, also similar to Ptolemy's homeother projection, was used by the geographer Apianus in a map he drew in 1520. The third was described by Johannes Werner in 1514 as an invention of his own. The second of these projections was the one used by Oronteus Finaeus, and later by Mercator. It is often referred to as "Werner's Second Projection" since he discussed all three. The details of this projection are as follows:

(a) The pole is the center for the parallels of latitude, which are concentric circles or portions of concentric circles.

(b) The size (diameter, circumference, spacing) of the parallels is adjusted to give the true proportion between the length of the degree of longitude at the equator and at other latitudes. That is, the sizes of the parallels are changed to give the right length of a degree of longitude at any latitude.

(c) At the equator, the length of the degree of latitude is equal to the length of the degree of longitude.

MERCATOR PROJECTION: This projection is of the cylindrical type, in which a cylinder, placed around the earth and touching it along a circle, has projected onto it (from inside outwards) a representation of the earth's surface. The cylinder is then cut lengthwise, and flattened out to form a map. If the cylinder is tangent to the earth at the equator, the meridians become vertical, parallel, equidistant, straight lines; and the equator becomes a horizontal straight line across the center of the map.

---

* A "great circle" is a circle about the earth that equals the full circumference. For example, the equator is a great circle, but it is the only parallel of latitude which is a great circle. All other parallels are shorter, and are therefore not great circles. A great circle actually describes a plane that cuts the earth in half, and may be drawn about the earth in any direction. (C.H.)

The Mercator Projection is of the cylindrical type, but is not constructed by geometric projection. The parallels of latitude are derived mathematically. Its main feature is that the parallels of latitude are spaced ever farther apart with increasing distances from the equator, so as to maintain, at every point, the correct ratio between the degree of latitude and the degree of longitude; which, of course, actually grow shorter toward the poles. The advantage of this projection is that it is "conformal"—that is to say angles may be measured correctly at any point, and distances may be measured directly over small changes in latitude. For purposes of navigation, course lines are straight lines, whose directions may be directly measured from the chart.

Note 6: Plane v. Spherical Trigonometry in the Piri Re'is Map

Strachan a number of times calculated the positions of the five projection points on the Piri Re'is Map both by plane and by spherical trigonometry. Each time it seemed that the calculations by spherical trigonometry were at variance with the geography of the map. The following comparison will illustrate the point. It was made in 1960, on the assumptions that the center of the map was at Syene, on the Tropic of Cancer in Longitude 32½° East, that the radius of the circle was drawn from this point to the North Pole, and that the base line for latitude was the Piri Re'is Equator.

Plane Trigonometry	Spherical Trigonometry
1. 50.9 N; 30.5 W	1. 38.1 N; 46.1 W
2. 25.7 N; 36.4 W	2. 18.7 N; 42.0 W
3. 0.0 N; 34.4 W	3. 0.0 N; 34.0 W
4. 22.1 S; 18.5 W	4. 18.2 S; 23.1 W
5. 37.0 S; 2.6 E	5. 32.9 S; 07.1 W

We see that by plane trigonometry the maximum spread of latitude is 87.9 degrees, and that of longitude is 39 degrees. By spherical trigonometry, on the other hand, the spread of latitude is only 71 degrees while the spread of longitude is 53.2 degrees. Thus the effect of the use of spherical trigonometry would be to compress latitude and exaggerate longitude in a way that apparently does not fit the map at all.

Note 7: Strachan on the Necessity of the Use of Trigonometry for Mapping Large Areas

Massachusetts Institute of Technology,
Cambridge, Massachusetts,
April 18, 1965

Mr. Charles Hapgood
Keene State College
Keene, N.H.

Dear Charlie,

In reply to your question as to the possibility of mapping very large areas of the earth's surface without the use of mathematics, I must say that I am convinced that the source map or maps from which your ancient maps are derived must have had some mathematical foundation, for one or more of the following reasons:

(1) The determination of place locations relative to one another on a continent requires at least geometric triangulation methods. Over large distances (of the order of 1,000 miles) corrections must be made for the curvature of the earth, which require some understanding of spherical trigonometry.

(2) The location of continents with respect to one another requires an understanding of the earth's sphericity, and the use of spherical trigonometry.

(3) Cultures with this knowledge, plus the precision instruments to make the required measurements to determine location would most certainly use their mathematical technology in creating maps and charts. The application of the portolan grid system to these old maps must surely be after the fact; it would not be at all useful in the construction of such maps.

(4) Your discovery of the apparent use of the mercator type of projection in the Piri Re'is, De Canerio and Chinese maps came as a very great surprise.* Yet, in view of the technology required to make maps of such accuracy this discovery seems less startling, because, for one thing, of the great utility of the mercator type projection in navigation. This is, of course, more complex than the simple geometric projection. With this (which involves latitude expansion with increasing latitude) a trigonometric coordinate transformation method must be used.

I hope that this answers your question.

Best regards,

Dick [Strachan]

---

* This was later found to be mistaken (see p. 35).

Note 8: Strachan on the Construction of the Piri Re'is Grid

19 AGASSIZ ST., APT. 24
CAMBRIDGE 40, MASS.
2 August 1960

Mr. Charles Hapgood
296 Court St.
Keene, New Hampshire

DEAR CHARLIE,

I just received your letter (of 7-30-60), and thought that I'd better write and clear up some misunderstandings on your part. You seem to feel that the grid which Frank* drew lines up pretty closely with the locations of the points which I figured out, *which is true*, but that the grid was figured pretty independently of the latitudes and longitudes of the points, *which is false*. The fact is, that the grid was *derived* from the latitude and longitude of the points. They are directly related, and had the Piri Re'is chart [the parchment on which the map is drawn] not stretched or shrunk some, they would coincide. And quite probably, some of the discrepancies are due to the relative inaccuracy of my math; that is, I worked only to slide rule accuracy which is not extremely precise. The mathematicians have a name for what was done to finally draw the grid; it is a conversion from polar (or circular) to rectangular (or grid-like) coordinates. The actual conversion was in changing the locations of our five points as expressed on the circumference of a circle to their locations on a rectilinear grid, as is latitude and longitude by plane sailing. Let me go through the steps leading to the formation of the grid so that you will see what I mean.

1. We are given the locations of the five points with respect to another point (Syene). These positions are given in terms of distance (radius of circle) and in terms of angles. This relationship in terms of angles and distances is called *polar* (Syene being the pole here). We also know the location of Syene in latitude and longitude—the rectangular system. This is the key to our finding the latitude and longitude locations of the five points; it is the starting point.

2. As seen in the figure, we are able to find the latitude and longitude of each point in turn using our known distance and angle from Syene and applying a little trigonometry. If you don't follow the math or my notation system, just take my word for it. This gives us then the latitude and longitude of the five scattered points.

3. If we consider these points to be on a chart, as they are, we may draw a network of latitude and longitude lines through the points to the top and bottom

---

* Frank Ryan.

and to the sides of the chart to our (now) blank latitude and longitude scales. Thus, we have five latitudes located on our latitude scale and five longitudes on our longitude scale. Now we are in business. We next measure with a ruler the distance on the chart between any two points on the latitude scale. Now we set up a simple proportion to find the chart length of one degree of latitude:

$$\text{Length of one degree} = \frac{\text{length between measured points}}{\text{degrees between measured points}}$$

And so we find the length of a degree of latitude on our latitude scale. We do the exact thing over again with the longitude scale to find the length of one degree of longitude on the chart.

4. Now, knowing the length of a degree of latitude and longitude on the chart scale, and knowing the latitudes and longitudes of five points, we can start at any point and draw in our latitude and longitude grid with spacing to suit our purposes. The latitudes and longitudes of our original five points should be the same as they actually are when we use the grid which we have just constructed to find them.

Why do we find some discrepancies on the Piri Re'is chart between the actual (calculated) positions of the points and the positions of these points as found using the grid which Frank has drawn, after I have just proven that they must be the same? Well, if the assumptions which we used to find the positions of the five points are accurate, and if the points were originally drawn on the chart correctly, then either the chart has physically distorted (shrunk, warped, etc.) through the centuries, or the precision of the mathematics is not good enough. Actually, any of these errors may be suspect:

a) original assumptions inaccurate
b) chart not drawn accurately
c) chart distorted through age
d) insufficient mathematical precision

We can only hope to find where the error lies by comparing the positions of known points as given on the chart with the actual positions of these points. In my opinion, we will be able to find the cause of the error *if enough such positions are compared.*

After working with these minor problems and eventually solving them we hope, the BIG QUESTION still remains; who drew the original chart(s)? I wonder if we ever will know.

I received the package which you sent me, and I am working on it. There is certainly much food for thought there, before I can reply with anything definite.

I'll see you later,

DICK

Note 9: Nordenskiöld's Comment on the Projection of Marinus of Tyre

Referring to the projection of Marinus of Tyre, which he identifies with Ptolemy's equidistant-rectangular, or equidistant-cylindrical projection, Nordenskiöld remarks:

For the sake of brevity I shall name this projection after Marinus of Tyre, who, according to Ptolemy, used it for his charts, but I suppose that it had already been employed by earlier, unknown cartographers. The meridians and parallels are equidistant straight lines, forming right angles to each other, and so drawn that the proper ratio between the degrees of latitude and longitude are maintained on the map's mean or main parallel. When the equator is selected for this purpose the net of graduation becomes quadratic. The 26 special maps in all older manuscripts of Ptolemy are drawn on this projection. . . . (147:85)

There may well be some connection between the oblong grids found by us on the Piri Re'is and Chinese maps, and in the Spanish sector of the Ben Zara Map, and the projection described here by Nordenskiöld. In this connection our solution of the De Canerio Map raises a problem, for in its case a solution by spherical trigonometry yielded a similar oblong grid.

Note 10: Route of the Norwegian-British-Swedish Expedition Across Queen Maud Land

The expedition, with equipment for taking depth soundings through the ice cap, left Maudheim, on the coastal ice shelf, at 71° S. Lat., and 11° W. Long. (just northeast of Cape Norvegia). They crossed some shelf ice in an east-southeasterly direction, and reached the 500-meter contour line of the continental ice cap at 71.5° S., and 7° W. Directly east of here, at distances of 4 and 5 degrees were the Witte Peaks and Stein's Nunataks. Slightly to the northeast, at a distance of 3 degrees of longitude were the Passat and Boreas Nunataks. These were all comparatively low features. Here the expedition was 150 kilometers from Maudheim. The profile showed that for another fifty kilometers the surface of the continent was below sea level. At one point (A, in Fig. 48) the ice extended 1,000 meters below sea level.

Just before the expedition reached the 1,000-meter contour, they passed over a subglacial "island" rising a couple of hundred meters above sea level. This island would be at 6° W. and 71° 40′ S. and would measure about 30 km. across. Beyond this, to the 1,000-meter contour, the surface was again below sea level, for about 40 km. (B on Fig. 48).

Just after the 1,000-meter contour, the subglacial surface rose steeply above sea level, reaching an altitude under the ice of about 750 meters, or about half a mile. The expedition was now (when the surface came above sea level) about 225 km. from Maudheim. At 280 km. from Maudheim the surface again dipped below sea level (C on Fig. 48). The subglacial mountain, or mountain range, indicated here lay approximately in Lat. 72° S., and Long. 5–6° W., and was about 55 km. across from northwest to southeast. A relatively slight change in sea level (about 200 meters) from depression of the land or rise of the sea could have divided this land mass into two islands.

About 120 km. farther on, the group began to pass over a higher submerged mountain range. They established an advance base at a point where a mountain peak just reached the surface of the ice cap, still at an altitude of about 1,000 meters. This was in 72.3° S. Lat., and 3.5° W. Long. About 20 miles beyond this point the Regula Mountains were reached, with peaks rising to elevations of about two miles. The submerged mountains were certainly a part of the Regula Mountains covered by the ice. Three times more, after this, the surface dipped below sea level, indicating subglacial "islands." Naturally, the party traveled on the surface of the ice cap, and did not climb the peaks that showed above the surface of the ice eastward and westward of their line of march. At one point they were forced to detour to the west to avoid a high mountain, Shubert Peak (2,710 meters). To the west of them they passed Mount Ropke (2,280 meters), and farther on they passed Speiss Peak (2,420 meters) about 20 miles to the west. After this they reached Penck Trough, where again the surface dipped below sea level, and then reached the Neumeyer Escarpment, at an elevation of 2,500 meters, the beginning of the interior plateau. They had crossed New Schwabenland. The route ended in 74.3° S. and 0.5° E. at a point where the ice surface was about 2,700 meters above sea level.

If the southern part of the Piri Re'is Map represents this coast, then it shows terrain a considerable distance in both directions from this line. It shows the sea advanced to the base of the Neumeyer Escarpment, and the various mountains as islands. Toward the east a number of inland mountain ranges are shown, while to the west a peninsula may represent what is now Cape Norvegia or Maudheim (A in Fig. 48). If the inland mountain ranges are the Muhlig–Hofmann and Wholthat Ranges, then the Piri Re'is Map shows the Antarctic coast from about 10° W. to 15° E. longitude.

Note 11: Strachan on the Oronteus Finaeus Projection

19 AGASSIZ STREET
CAMBRIDGE, MASS.
29 September, 1962

Professor Charles Hapgood
Keene Teachers College
Keene, New Hampshire

DEAR CHARLIE,

I was very interested to have the opportunity to see for the first time a picture of the Oronteus Finaeus Map of 1531. The chart projection, as Nordenskiold clearly states, and again as noted by Captain Burroughs, is the Cordiform Projection. This projection appears to be some modified form of the simple conic projection; the parallels of latitude are circles centered at the pole and equally spaced as in the conic; however, the meridians are curved (with the exception of the 90th) in the Finaeus while being straight in the simple conic. The meridional curvature is designed to minimize the area distortion of the projection. Nordenskiold credits Ptolemy's "homeother" projection as the basis for the cordiform. In any event the Cordiform Projection is probably derived mathematically, requiring the use of geometry and trigonometry to transcribe accurately the features of the earth (i.e., a sphere) onto it.

Coming to the main question at hand, we ask how did Oronteus Finaeus draw his map? It is apparent that Oronteus used some earlier source map. This we must concede. We must presume that the source map was equipped with a grid of some sort; it must have had one to have been drawn originally and there would seem to be no reason to have removed it. The simplest method of transferring a map or figure from one grid to another is by point to point transference. That is to say, a point located at a given latitude and longitude on the source map must be relocated to the same latitude and longitude on the secondary grid, regardless of the relative shapes of the grids. This process does not require knowledge of cartography or mathematics; any child can do it. This appears to me to be a logical method of transcription, and is the way that I would do it. Notice that this neither involves nor introduces in any way any new knowledge. It is merely a way of transplanting data from one grid to another and does not depend upon the shapes of the grids, or upon any method by which the original intelligence was placed upon the primary grid. By this theory, Oronteus Finaeus could not have correctly located the South Pole as nearly as he did by his own knowledge. He merely placed it as shown on his source map. And he need not have known a bit of mathematics to have accomplished this job.

Does this help?

Sincerely yours,
(DICK)
RICHARD W. STRACHAN

Note 12: A Comment by Nordenskiöld on the First World Map of Oronteus Finaeus

Several years before the publication of the cordiform map [his map of 1538] Oronteus Finaeus had constructed another map also on Werner's second projection, but modified in such a manner that the map of the world here is divided into two parts, the one embracing the northern hemisphere, with the North Pole for a centre of the parallel circles, and the other, the southern hemisphere, with the South Pole as a centre. It is of this map that a facsimile is given on Pl. XLI [our Fig. 49]. It is dated 1531, but is generally found inserted in *Novus Orbis Regionem ac Insularum veteribus incognitarum, Parisiis, 1532*. It was afterwards reprinted from the same block, but with a new title legend from which the name of Oronteus was omitted, in the edition of the Geography of Glareanus printed Brisgae 1536, and in an edition of *Pomponius Mela, Parisiis apud Christianum Wechelum 1540*.

"The map of Oronteus Finaeus finally had the honour of being copied, although with some modifications, by Gerard Mercator, for one of his first maps, of which I give a facsimile on Plate XLIII. . . . (147)

Note 13: A Possible Connection Between the Oronteus Finaeus Map of 1538 and the Hadji Ahmed Map of 1559

Nordenskiöld suggests that the 1538 map of Oronteus Finaeus may have served as a model for the Hadji Ahmed Map of 1559:

It is evident that the original map of Finaeus also served as a model for the large Turkish Cordiform Map engraved on wood at Venice by Hhaggy Ahhmed from Tunis, and dated year 967 of the Mohammedan chronology, which corresponds with our year 1559. The blocks for this map, which, for some reason or other, had been sequestrated, probably before the issue of the print, were discovered in 1795 in the depositaries of the Venetian Council of Ten, and are now preserved at the Biblioteca Marciana. This discovery seems to have produced a certain sensation. It gave rise to a whole literature introduced by a paper of Abbe Simon Assimani, and to various fables concerning the manner in which the old blocks came into the possession of the Venetian Government. (147:89)

Despite Nordenskiöld's view expressed here, a close comparison of the two maps appears to me to reveal many more points of difference than of agreement.

Note 14: Hough's Interpretation of the Ross Sea Cores

The log of core N-5 shows glacial marine sediment from the present to 6,000 years ago. From 6,000 to 15,000 years ago the sediment is fine-grained with the exception of one granule at about 12,000 years ago. This suggests an absence of ice from the area during that period, except perhaps for a stray iceberg 12,000 years ago. Glacial marine sediment occurs from 15,000 to 29,500 years ago; then there is a zone of fine-grained sediment from 30,000 to 40,000 years ago, again suggesting an absence of ice from the sea. From 40,000 to 133,500 years ago there is glacial marine material, divided into two zones of coarse- and two zones of medium-grained texture.

The period 133,000–173,000 years ago is represented by fine-grained sediment, approximately half of which is finely laminated. Isolated pebbles occur at 140,000, 147,000, and 156,000 years. This zone is interpreted as recording a time during which the sea at this station was ice free, except for a few stray bergs, when the three pebbles were deposited. The laminated sediment may represent seasonal outwash from glacial ice on the Antarctic continent.

Glacial marine sediment is present from 173,000 to 350,000 years ago, with some variation in the texture. Laminated fine-grained sediments from 350,000 to 420,000 years ago may again represent rhythmic deposition of outwash from Antarctica in an ice-free sea. The bottom part of the core contains glacial marine sediment dated from 420,000 to 460,000 years by extrapolation of the time scale from the younger part of the core (96:257–59).

Note 15: Gerard Mercator

Gerard Kramer (1512–1594), who took the name of Gerardus Mercatorius, and is known as Mercator, was the leading cartographer of the 16th Century and the founder of scientific cartography. He deserves this title because of his invention of the famous "Mercator Projection," still the most widely used of all map projections, especially for purposes of navigation.

According to Asimov (16:58–59) Mercator was at first under the influence of Ptolemy. However, he seems to have abandoned the Ptolemaic ideas sometime after he established a center for geographical studies at Louvain in 1534. From a comparison of his maps with the portolan charts, it is my impression that he probably abandoned Ptolemy because he realized the superiority of the portolanos. It seems evident, not only from his maps of the Antarctic, but also from his maps of South America, that he made use of the ancient maps. It would seem, in the latter case, that he used an ancient map in 1538, but abandoned it in favor of the explorers' accounts in his 1569 Atlas.

Mercator's maps were unique in their artistry. They incorporated, however, the 16th Century misconception of the size of the earth, as we have seen. In consequence, his distances are in fact less accurate than those of the ancient maps, which were based on a comparatively accurate estimate of the circumference. Mercator, then, while he laid the basis for modern scientific cartography, did not attain the technical level of the ancients.

It would be a matter of great interest to discover his source maps. In the hope of doing so, my student, Alfred Isroe, transferred from Keene State College to the University of Amsterdam, and spent a considerable time during the academic year 1964–1965 in a search for them. Despite the excellent cooperation of the Dutch authorities, the search was fruitless, and it appears that the source maps may have perished. If so, this would be only one more instance of the careless treatment of ancient manuscripts of practical or scientific value in the Renaissance. The humanists devoted themselves to collecting and restoring the manuscripts of ancient classical literature, but their interest rarely extended to physical science. Between the religious frenzy of the Reformation and the aristocratic bias of the humanists, the printing presses, as already mentioned, were largely monopolized by non-scientific material. Thus, the Renaissance was not only an age of recovery of ancient learning, but also possibly a period in which a large part of the scientific heritage of the past was lost.

Note 16: Attempts to Adapt the Twelve-Wind System to the Compass

N. H. de Vaudrey Heathecote, B.S., of University College, London, in an essay on *Early Nautical Charts* (89) found references to the use of the twelve-wind system on the compass. After discussing the usual 32-point compass card, he says:

Another system appears to have been in use in which the "wind" was divided, not into four quarter-winds, but into six "sixth-winds," so that there were *two* points on the compass card between, for example, N.W. and N.N.W., instead of the single point corresponding to N.W. by N. I have myself seen only one chart marked according to this division of the compass; it is one contained in a collection of charts by various Venetian cartographers of the second half of the fifteenth century (British Museum MS., Egerton 73, fol. 36). Breusing says (Breusing, "La Toleta de Marteloio," Zeitsch. f. wiss. Geog. 2, 129) that this was the French system. It strikes one as less convenient than the Mediterranean system, and certainly does not appear to have been in very wide use.

Note 17: Clagett on Science in Antiquity

On Egyptian Geometry:

The most advanced of the Egyptian mathematical achievements were in geometry. . . .

The Egyptians knew how to determine the areas and volumes of a number of figures; they could find the area of a triangle and a trapezium, the volume of a cylindrical granary and of the frustrum of a square pyramid, and perhaps even the area of the surface of a hemisphere (although the last is doubtful). Their proficiency in geometry was certainly fostered by their high development in architectural engineering and surveying. . . . (52:26)

On Babylonian Mathematics:

When we turn to Mesopotamia, we find from at least 1800 B.C. a Babylonian mathematics more highly developed than the Egyptian. Although it too had strong empirical roots that are clearly present in most of the tablets that have been published, it certainly seems to have tended to a more theoretical expression. The key to the advances made by the Babylonians in mathematics appears to have been their remarkably facile number system, which demands brief characterization.

(1) Although it had certain features of both the decimal and sexagesimal systems, it was primarily a sexagesimal system. That is to say, it was based on sixty and powers of sixty.

(2) It was a system highly general and abbreviatory in character. All numbers could be made with only two symbols.   ▽ = 1 and   ◁ =10. Using these symbols the numbers from 1 to 59 can be represented thus: ▽▽ = 2; ◁◁▽=21; etc. Numerous tricks were used to save writing all the symbols out in a string. Not only could these symbols be used to represent numbers from 1 to 59, but they could also be used to write the numbers 1 to 59 times any power of 60. Thus, unless one knew what order of magnitude was being considered from the details of the problem being worked, he could not know whether the two symbols   ◁◁   by themselves on a tablet without supporting text equalled 20, or 20×60, or 20×60^2, or 20×60^{-1}, etc.

What is more, these same symbols changed their value as their position changed; that is to say, this system was a place-value system, as is our own decimal system. In our system as the symbol changes position in the following numbers it changes its value in that it stands for a higher power of ten: 00.1, 1, 10, 100, etc. So in the number ▽◁▽ the symbol ▽ has the value of 1 in the last position and sixty the first position, the whole number being 71 (if the   ▽   in the last position represents 60, then the   ▽   in the first position represents 60^2 and the whole number is 4260:60^2 plus 11×60).

It was not until very late that the Babylonian system developed what in the decimal system is called zero—i.e., a sign for the absence of any units of a given power of ten indicated by position; thus in our system 101 means of course one hundred, no tens, and one unit. Instead of using that sign we could agree that we would simply leave a space, writing 1 1. This was done until the very last stages of the Babylonian system, when a zero sign or its equivalent was developed. . . .

Freed from the drudgery of calculation by this really remarkable system of calculation . . . the Babylonians made extraordinary advances in algebra. . . . (52:28–30)

B. L. van der Waerden sums up the characteristics of Babylonian astronomy thus:

. . . The fundamental ideas of Babylonian Astronomy are: the idea of the periodical return of celestial phenomena, the artificial division of the Zodiac into 12 signs of 30 degrees each, the use of longitude and latitude as coordinates of stars and planets, and the approximation of empirical functions by linear, quadratic and cubic functions, computed by means of arithmetical progressions of first, second, and third order. (216:50)

Needham notes the early appearance in China of a number system with affinities to the Babylonian:

. . . Decimal place-value and blank space for the zero had begun in the land of the Yellow River earlier than anywhere else, and decimal metrology had gone along with it. By the first century B.C. Chinese artisans were checking their work with sliding calipers decimally graduated. Chinese mathematical thought was always profoundly algebraic; not geometrical. . . . (145:118–119)

We see here suggestions that Babylonian and Chinese science were linked, either through contemporary contacts or through inheritance from a common source.

Taking Babylonian, Chinese, Egyptian, and Greek science together, we may note that there was a very considerable development of geometry in Egypt, but apparently no algebra. There was a remarkable development of algebra in Babylonia and in China, but no special development of geometry. Remarkable likenesses existed in the number systems of Babylonia and China, but this number system had no similarity to that of Egypt.

We have seen that the science reflected in the maps implies, however, the possession of all of these elements by *one culture*. Geometry is present in the portolan design; the Babylonian division of the Zodiac is present in the twelve-wind system; so are the units of sixty (six units of sixty in the circle). The included decimal system for counting the 360 degrees of the circle are present in the Oronteus Finaeus Map (for, as we have seen, the 80th parallel must have been drawn on that map by the people who mapped Antarctica).

Is it possible that what we have here is evidence that all these different scientific achievements were once the possesions of the unknown people who originally drew these maps, and that, in the dissolution of their culture, various remnants survived, some in one place, some in another? Let us suppose that a "carrier people" —an intermediary people (like the seafaring Phoenicians)—were the ones to inherit all these aspects of science from the ancient source. Let us further suppose that the "carrier people" brought this science, by trade contacts, separately to our known civilizations of antiquity; the Babylonians and the Chinese took some elements of this ancient heritage, the Egyptians others, and American Indian peoples perhaps still others. We must not, of course, omit India. It is here, apparently, that our symbol for zero appeared, and it is in India more than in any other country on earth that the traditions of an ancient great world civilization are still preserved.

Note 18: Einstein on the Theory of Earth's Shifting Crust *

I frequently receive communications from people who wish to consult me concerning their unpublished ideas. It goes without saying that these ideas are very seldom possessed of scientific validity. The very first communication, however, that I received from Mr. Hapgood electrified me. His idea is original, of great simplicity, and—if it continues to prove itself—of great importance to everything that is related to the history of the earth's surface.

A great many empirical data indicate that at each point on the earth's surface that has been carefully studied, many climatic changes have taken place, apparently quite suddenly. This, according to Hapgood, is explicable if the virtually rigid outer crust of the earth undergoes, from time to time, extensive displacement over the viscous, plastic, possibly fluid inner layers. Such displacements may take place as the consequence of comparatively slight forces exerted on the crust, derived from the earth's momentum of rotation, which in turn will tend to alter the axis of rotation of the earth's crust.

In a polar region there is continual deposition of ice, which is not symmetrically distributed about the pole. The earth's rotation acts on these unsymmetrically deposited masses, and produces centrifugal momentum that is transmitted to the rigid crust of the earth. The constantly increasing centrifugal momentum produced in this way will, when it has reached a certain point, produce a movement of the earth's crust over the rest of the earth's body, and this will displace the polar regions toward the equator.

Without a doubt the earth's crust is strong enough not to give way proportionately as the ice is deposited. The only doubtful assumption is that the earth's crust can be moved easily enough over the inner layers.

The author has not confined himself to a simple presentation of this idea. He has also set forth, cautiously and comprehensively, the extraordinarily rich material that supports his displacement theory. I think that this rather astonishing, even fascinating, idea deserves the serious attention of anyone who concerns himself with the theory of the earth's development.

To close with an observation that has occurred to me while writing these lines: If the earth's crust is really so easily displaced over its substratum as this theory requires, then the rigid masses near the earth's surface must be distributed in such a way that they give rise to no other considerable centrifugal momentum, which would tend to displace the crust by centrifugal effect. I think that this deduction might be capable of verification, at least approximately. This centrifugal momentum should in any case be smaller than that produced by the masses of deposited ice.

---

* The Foreword Dr. Albert Einstein wrote for "Earth's Shifting Crust," by Charles H. Hapgood. New York: Pantheon, 1958.

Note 19: Cummings on the Pyramid of Cuicuilco

In his pamphlet (56:40) Cummings, after describing the excavation of successive pavements about the Pyramid of Cuicuilco, writes:

. . . These six pavements, with their six corresponding shrines, all lying at different levels below the carbonized stratum that indicates the time of the eruption of Xitli and the coming of the Pedrigal, speak in a language that is clear and convincing. The lowest pavement lies more than 18 feet below the surface. Eighteen feet of gradual fill on top and 12 to 20 feet of debris overlying the base, all accumulated probably before the Christian era, and the composite condition of the structure itself, bespeak a lapse of time that pushes its builders back into the dim beginnings of things in the Valley of Mexico.

Cummings' further comments on the probable age of the pyramid should be quoted in full.

THE GREAT AGE OF THE TEMPLE

Cuicuilco tells its own story quite clearly. Its crude cyclopean masonry without mortar of any kind, its massive conical form, and its great elevated causeways for approach instead of staircases, all demonstrate that the structure was the work of primitive men, and that its builders hardly knew the rudiments of architecture. Its base lies buried beneath from 15 to 20 feet of accumulated debris, which in turn was covered with three lava flows that have crowded around its slopes and piled up upon each other in rapid succession to the depth of 10 to 20 feet. The old temple had been so completely covered with rock and soil, volcanic ash and pumice in successive strata that the noses of the lava streams as they pushed around the mound and crawled up its slopes were nowhere able to touch the walls of the ancient structure. Centuries must have elapsed and several eruptions of old Ajusco must have buried its platforms and slopes under successive mantles of ash and pumice before Xitli poured fourth its baptisms of fire. Two and a half to 3 feet of surface soil have accumulated above this scorched and blackened stratum that marks the footprints of Xitli's consuming blasts. Careful measurements were taken in several places of this accumulation since the eruption of Xitli, and of the accumulation directly beneath between the pavement surrounding the temple and the blackened stratum just underlying the lava, and the story was always the same. If it has taken 2,000 years for the deposit to form since the eruption of Xitli, then by the same yardstick it took some 6,500 years for the debris beneath the lava to have accumulated, and so Cuicuilco fell into ruins some 8,500 years ago. But some will say, "What evidence have we that the deposit overlying the structure beneath the lava did not form rapidly on account of successive showers of volcanic ash?" True, volcanic deposits played their part, but the presence of three strata of rock and organic soil containing stone implements, pottery and figurines of three quite different types of workmanship, and separated by two thick barren strata of volcanic ash and pumice mingled with streaks of sand, indicates the passing of centuries rather than months or years. The two massive enlargements of the temple, and the repeated reinforcement of the great causeways leading to the top of the structure, represent no brief period of the active use of this ancient center of religious ceremonials. The presence of 18½ feet of deposit on top of the original platform, and the six successive pavements with their corresponding platforms or altars, all buried therein before Xitli erupted, demonstrates further the long use of this lofty pile as a sacred gathering place. The late Mrs. Nuthall, a noted Aztec scholar, found that the word Cuicuilco, signified a place for singing and dancing. Everything about the structure bears out that interpretation. Here men and women met to pay tribute to the great spirits who, they thought, controlled their lives and their destinies. Here they danced and sang in honor of their gods and for the benefit of their fellow men seemingly through many centuries of time. Cuicuilco stands out as a monument to the religious zeal and to the organized power and perseverance of the earliest inhabitants of the Valley of Mexico. It is a great temple that records devotion to their gods and subservience to the will of great leaders. It shows the beginning of that architecture that developed into the pyramids and altars of Teotihuacan. It certainly gives evidence of being the oldest temple yet uncovered on the American continent. Everything about it so far revealed bears out its great antiquity, and it should serve as a strong incentive to the further investigation of the archaic culture of Mexico.

Note 20: Brogger on an Ancient "Golden Age" of Navigation

Vilhjalmur Stefansson (192) mentioned a reference by a Norwegian historian to a great age of navigation in ancient times:

To those of us brought up in the pedagogic tradition of forty and more years ago, where navigation of the high seas was supposed to have started with the Phoenicians, it is more than a little against the grain to believe that man swarmed over at least three of the oceans, the Atlantic, Indian, and Pacific, during remote periods. In fact, about the only group of scholars to whom that type of thinking appears to be natural or ingrained is the archaeologists, particularly those who devote themselves to the late Stone and various Bronze ages. . . .

Professor A. W. Brogger created no great stir, or at least aroused no storm of protest, when at an international congress of archaeologists at Oslo in 1936 he lectured, as president of the congress, about a golden age of deep-sea navigation which he thinks may have been at its height as much as three thousand years before Christ and which was on the decline after 1500 B.C., so that the very period which we used to select as the beginning of real seamanship, the Phoenician, is shown as having been (by that theory) at the bottom of a curve, which thereafter rose slowly until it attained a new high in the navigational cycle of the Viking Age which started less than fifteen hundred years ago.

That man of the Old World discovered the Americas, from Brazil to Greenland, during Brogger's golden age of navigation five thousand years ago, and perhaps earlier, rests merely on possibilities and probabilities. As yet we cannot prove it certain, though we can prove it likely.

Note 21: Representations of the Constellations as Denoters of Latitudes on the Piri Re'is Map

In the light of the apparent connection of the twelve-wind system with the signs of the ancient Babylonian zodiac, and because of various evidences that sailors in ancient times used the constellations in navigation, it would not be particularly strange to find that the representations of the constellations were used on ancient maps. Some of these representations seem to be suggested by the ships and animals on the Piri Re'is Map.

This suggestion was first made to me by the late Archibald T. Robertson, of Boston, a scholar in the esoteric lore of ancient navigation and astronomy. He suggested that Piri Re'is might have copied (and misinterpreted) some of his animals and ships from the ancient source maps he used, adding others to suit his fancy.

Robertson suggested that the great snake shown in what we take to be Queen Maud Land, Antarctica, may have originally been intended to represent the constellation Hydra (the Snake), a constellation which is visible in the southern sky (during the spring equinox of the Northern Hemisphere) only in Latitude 70°–72° South, the correct latitude of the Queen Maud Land coast. The ship lying off the coast of what looks like Argentina, he suggested, might represent the constellation Argo (the Ship), visible at that season in Latitude 55° South, as would be correct. Following this line of reasoning, we might suggest that other ancient constellations might be represented by the bull in the center of Brazil (Taurus) and the wolf-like creature in the south (Lupus). The bull is shown in the equatorial region of Brazil, which would be correct, since Taurus is an animal of the Zodiac.*

It would be natural for Piri Re'is, or any other Arab, Medieval, or Renaissance mapmaker, to misunderstand these figures on the ancient source maps, and to take them for references to historical events or local fauna. It would be natural for them to add others of their own, which, as we have seen, Piri Re'is did. For the unfamiliar ships on the source maps, he might have quite naturally substituted those he knew, the ships of the 16th Century, connecting them with known or presumed historical events, such as the voyages of St. Brendan and Diaz. He seems also to have drawn upon the medieval bestiaries for fabulous animals, in accordance with the habits of mapmakers of his day.

---

* **The Zodiac was confined to the equatorial constellations—the path of the sun.**

Note 22: Ohlmeyer and Burroughs Correspondence

8 RECONNAISSANCE TECHNICAL SQUADRON (SAC)
UNITED STATES AIR FORCE
WESTOVER AIR FORCE BASE
MASSACHUSETTS

Reply to
Attn of: RTC   6 July 1960

SUBJECT: Admiral Piri Reis World Map

To: Professor Charles H. Hapgood
      Keene Teachers College
      Keene, New Hampshire

DEAR PROFESSOR HAPGOOD:

Your request for evaluation of certain unusual features of the Piri Reis World Map of 1513 by this organization has been reviewed.

The claim that the lower part of the map portrays the Princess Martha Coast of Queen Maud Land Antarctica, and the Palmer Peninsula is reasonable. We find this is the most logical and in all probability the correct interpretation of the map.

The geographical detail shown in the lower part of the map agrees very remarkably with the results of the Seismic profile made across the top of the ice cap by the Swedish-British-Norwegian Antarctic Expedition of 1949.

This indicates the coastline had been mapped before it was covered by the ice cap.

The ice cap in this region is now about a mile thick. We have no idea how the data on this map can be reconciled with the supposed state of geographical knowledge in 1513.

HAROLD Z. OHLMEYER
Lt. Colonel, USAF
Commander

8TH RECONNAISSANCE TECHNICAL SQUADRON (SAC)
UNITED STATES AIR FORCE
WESTOVER AIR FORCE BASE, MASS.

14 Aug 61

Mr. Charles H. Hapgood
Keene Teachers College
Keene, N.H.

DEAR PROFESSOR HAPGOOD:

It is not very often that we have an opportunity to evaluate maps of ancient origin. The Piri Reis (1513) and Oronteus Fineaus [sic] (1531) maps sent to us by you, presented a delightful challenge, for it was not readily conceivable that they could be so accurate without being forged. With added enthusiasm we accepted this challenge and have expended many off duty hours evaluating your manuscript and the above maps. I am sure you will be pleased to know we have concluded that both of these maps were compiled from accurate original source maps, irrespective of dates. The following is a brief summary of our findings:

  a. The solution of the portolano projection used by Admiral Piri Reis, developed by your class in Anthropology, must be very nearly correct; for when known geographical locations are checked in relationship to the grid computed by Mr. Richard W. Strachan (MIT), there is remarkably close agreement. Piri Reis' use of the portolano projection (centered on Syene, Egypt) was an excellent choice, for it is a developable surface that would permit the relative size and shape of the earth (at that latitude) to be retained. It is our opinion that those who compiled the original map had an excellent knowledge of the continents covered by this map.

  b. As stated by Colonel Harold Z. Ohlmeyer in his letter (July 6, 1960) to you, the Princess Martha Coast of Queen Maud Land, Antarctica, appears to be truly represented on the southern sector of the Piri Reis Map. The agreement of the Piri Reis Map with the seismic profile of this area made by the Norwegian-British-Swedish Expedition of 1949, supported by your solution of the grid, places beyond a reasonable doubt the conclusion that the original source maps must have been made before the present Antarctic ice cap covered the Queen Maud Land coasts.

  c. It is our opinion that the accuracy of the cartographic features shown in the Oronteus Fineaus [sic] Map (1531) suggests, beyond a doubt, that it also was compiled from accurate source maps of Antarctica, but in this case of the entire continent. Close examination has proved the original source maps must have been compiled at a time when the land mass and inland waterways of the continent were relatively free of ice. This conclusion is further supported by a comparison of the Oronteus Fineaus [sic] Map with the results obtained by International Geophysical Year teams in their measurements of the subglacial topography. The comparison also suggests that the original source maps (compiled in remote antiquity) were prepared when Antarctica was presumably free of ice. The Cordiform Projection used by Oronteus Fineaus [sic] suggests the use of advanced mathematics. Further, the shape given to the Antarctic continent suggests the possibility, if not the probability, that the original source maps were compiled on a stereographic or gnomonic type of projection (involving the use of spherical trigonometry).

d. We are convinced that the findings made by you and your associates are valid, and that they raise extremely important questions affecting geology and ancient history, questions which certainly require further investigation.

We thank you for extending us the opportunity to have participated in the study of these maps. The following officers and airmen volunteered their time to assist Captain Lorenzo W. Burroughs in this evaluation: Captain Richard E. Covault, CWO Howard D. Minor, MSgt Clifton M. Dover, MSgt David C. Carter, TSgt James H. Hood, SSgt James L. Carroll, and A1C Don R. Vance.

LORENZO W. BURROUGHS
Captain, USAF
Chief, Cartographic Section
8th Reconnaissance Technical Sqdn (SAC)
Westover Air Force Base, Massachusetts

Note 23: The Vinland Map and the Tartar Relation *

There has just been published, as this book goes to press, an account of the discovery of a world map apparently drawn about 1450, showing a part of North America. The story of the discovery of the map, as given in the press, is most interesting, and the map itself is certainly evidence of medieval (probably Norse) visits to America before Columbus.

The reader of this book will naturally want to know whether this new map has any connection with the maps I have discussed. The answer is, no. The Vinland Map has no apparent connection with the ancient sea charts. Despite the presence of America on it, it is a typical example of unscientific medieval cartography. Land shapes, except for Greenland, are exceedingly inaccurate, even in the Mediterranean. The map is further evidence for the fact we have repeatedly emphasized—that geographers of the Middle Ages were simply incapable of constructing the highly scientific portolan charts.

The fact that the representation of Greenland is reasonably good may be owing to the influence of the far superior Zeno Map of the North, which the Zeno brothers are supposed to have drawn about sixty years earlier. I have presented evidence to support the view that the Zeno map must have been a copy of an ancient map originally drawn with the use of trigonometry, and must have been based on a true knowledge of the size of the earth. The comparative accuracy of the Vinland Map, then, as to Greenland, may be the result of the persisting influence of the ancient cartography.

---

* **The Vinland Map and the Tartar Relation, Thomas E. Manston and George D. Painter. New Haven: The Yale University Press, 1965.**

Note 24: Communication from the Turkish Embassy in Washington Regarding Piri Re'is

TURKISH EMBASSY
WASHINGTON, D.C.

January 28, 1965

Mr. Robert L. Merritt
Attorney at Law
Hippodrome Building
Cleveland 14, Ohio

DEAR MR. MERRITT,

In reply to your letter of October 16, 1964, I am pleased to enclose a photocopy of the Piri Reis Map received recently from Ankara.

To the best of our knowledge the original map has been drawn on gazelle hide.

A short biographical information on Piri Reis is given below which we think might be of interest to you.

"He was born at the town of Karaman, near Konya, Turkey. The exact date of his birth is unknown. In his early youth he joined his uncle Kemal Reis, a well known pirate. He distinguished himself during the operations of his uncle's small fleet on French and Venetian coasts. When Kemal Reis had abandoned piracy and joined the Imperial Ottoman Fleet during the reign of Beyazit II (1481–1512) Piri Reis followed suit and was appointed captain. The battles of Modon and Inebahti (Lepanto) made him famous. According to historian Von Hammer, "he gained an awesome fame" for his deeds in these expeditions.

"Piri Reis, whose real name was Ahmet Muhiddin, stayed with the Ottoman Fleet during the reigns of Yavuz Selim (1512–1520) and Suleiman the Magnificent (1520–1566). He served as an aide to Barbaros Hayrettin Pasha, Great Admiral of the Imperial Ottoman Fleet. In 1551 he was elevated to the rank of Commander in Chief of the Fleet of Egypt, then a dependency of the Ottoman Empire. In an expedition launched the same year with 31 vessels he seized the port of Masqat on the Arap peninsula and laid siege to the islands of Hurmuz in the Persian Gulf. The islanders offered him treasures, which he accepted as spoils of war and lifted the siege. On his way back, news reached him that a powerful Portuguese fleet had blockaded the entrance to the Persian Gulf. He loaded up all the treasures he had gotten from the islanders on three ships and leaving the remaining 28 in Basra he sailed to Istanbul. While passing through the Portuguese blockade he lost one of his vessels but managed to return safely to Egypt with the other two. The Governor of Egypt, one of his political opponents, misrepresented the facts to the Emperor in Istanbul reporting that "Piri Reis had returned with only 2 ships though he sailed at the beginning with 31," without mentioning the treasures he had brought with him. Emperor Suleiman flew into a rage and in a fit of anger ordered his execution, thus committing one of the very few fateful mistakes of his 46 years rule. Piri Reis was executed in Egypt in 1554.

"Kitabi Bahriye—The Navy's Book," which is the most famous of his works, is considered as an excellent geography book of his times. He also prepared a map of the world which has been reproduced in recent years. He wrote many poems too."

Sincerely yours,
Ali Suat Çakir
Second Secretary

Note 25: History of Navigation and Ship Building

In the course of research on the maps, much attention was paid to ancient navigation and ship building, particularly to the influence that the Phoenicians may have exerted on the Greeks in these areas. For reasons of space it seemed best to eliminate a detailed discussion of them in this volume, but some of the references have been retained in the Bibliography (Nos. 34, 38, 41, 49, 86, 99, 102, 114, 190, 211, 212, 213).

Appendix

The Piri Reis Map: Mathematical Considerations by Richard W. Strachan

Trigonometric projection based on the equator

I

The Piri Reis Map of 1513 is a typical portolano in appearance. The usual portolan chart is characterized by groups of eight, sixteen, or thirty-two lines radiating from one or more centers on the chart, like the spokes of a wheel. These lines, or rhumbs, are equally spaced at angles of 45, 22½ or 11¼ degrees apart. It has hitherto been supposed that this system of radial lines originated as actual course lines between various ports, that is, compass courses. It has not been supposed that any mathematical system underlay these portolan charts.

It is this assumption that has now been destroyed by the discoveries made by Professor Hapgood and his students. They have proved that, in the cases of several of these maps, the portolan design is based on geometry and may be translated by plane or spherical trigonometry into the terms of modern latitudes and longitudes.

II

In the case of the Piri Re'is Map, Hapgood found that the five minor projection centers shown on the surviving fragment (its western section) evidently had been placed on the perimeter of a circle with a center somewhere to the eastward of the torn edge. Rhumbs from these points, when projected to the east, proved to meet in Egypt at the intersection of the Tropic of Cancer with the Meridian of Alexandria. It appeared from this that the complete map (which included Asia) may have had sixteen of these minor projection centers (22½ degrees apart) on the perimeter of the circle.

A mathematician would consider this graphically as a polar (or circular) type of construction. The problem was to convert this polar projection into the rectangular coordinate system which is used today. Reference points used in the rectangular coordinate system are located by intersection of lines of latitude and longitude, the lines which form the type of grid with which we are all familiar. It is not difficult to convert from polar to rectangular coordinates provided several points, angles, distances, or a combination of these factors are known. Hapgood made the following assumptions:

1. The center of the portolan grid was located at the intersection of the Tropic of Cancer and the Meridian of Alexandria, that is, at 23° 30′ N, 30° E.

2. The radius from this center to the perimeter of the circle on which the minor projection points are located is 69.5°, or 3° longer than the distance from the Tropic of Cancer to the North Pole. The drawing of the projection involved, then, an overestimate of the circumference of the earth amounting to about 4.5°.

3. Projection Point III on the map was presumed to lie precisely on the Equator.

With these assumptions, we have enough information to solve for the geographical positions (latitude and longitude) of all five projection points.*

Sketching in the knowns and unknowns on a triangle as shown below, we have:

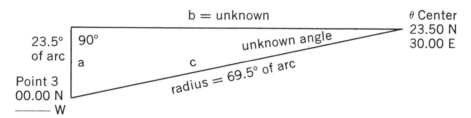

As seen in the figure, we know the length of the two sides of the triangle (which is a right triangle by construction). We may solve for the unknown angle, θ, first, and knowing θ and the length of one side, we can find the length of side X, which you can see is the latitude difference between the center and Point III. We may solve as follows:

$$\text{Sin } \theta = \frac{a}{c} = \frac{23.50}{69.50} = .3381 \qquad \theta = 19.75$$

$$\text{Then cos } \theta = \frac{b}{c} \qquad b = c \text{ cos } \theta = (69.5)(\cos 19.75)$$

$$b = (69.5)(.94108) = 65.41° \text{ of arc}$$

Subtracting the longitude of the center (since it is east Longitude) from the length of side b, we find the longitude of Point 3.

So:
$$\begin{array}{r} 65.41 \\ -30.00 \text{ E} \\ \hline 35.41 \text{ W} \end{array}$$

Thus we have the position of Point 3 to be:

$$\begin{array}{ll} 00.00 \text{ N} & \text{(given)} \\ 35.41 \text{ W} & \text{(calculated)} \end{array}$$

---

* It appeared, after a trial, that as between plane and spherical trigonometry, plane trigonometry gave better results in terms of a grid fitting the geography. Therefore, in this case, we have used plane trigonometry. (See Note 6)

The calculations for the other points are slightly different. For these calculations we know the length of one side of a triangle (which we construct), and the angle θ, which we find by addition or subtraction knowing the original θ, and knowing that the angles between the rhumb lines are $22\frac{1}{2}°$. The solution for Point II is illustrated:

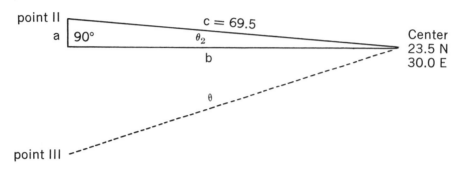

$$\theta_2 = 22.5 - \theta = 22.50 - 19.75 = 2.75$$

Then

$$a = c \sin \theta_2 = (69.50)(.04798) = 3.335° \text{ of arc}$$

and latitude of Point 2 equals:

$$23.50 + 3.34 = 26.84° \text{ N}$$

also

$$b = c \cos \theta_2 = (69.5)(.99885) = 69.42° \text{ of arc}$$

and longitude of Point 2 equals:

$$69.42 - 30.00 = 39.42° \text{ W}$$

The positions of the remaining points are:

Point	Latitude	Longitude
1	53.15 N	32.86 W
2	26.84 N	39.42 W
3	00.0	35.41 W
4	23.09 S	21.45 W
5	39.36 S	0.35 E

III

Knowing the positions of the five points, it is necessary to know only one more thing to be able to construct the rectangular coordinate grid. This is the direction of True North on the Piri Reis Map. Once Hapgood and his co-workers had established the North direction, it was a simple matter to draw in a grid, merely using the lines indicated on the map itself, which intersect the five projection points, running North-South or East-West.

Supplement to Mathematical Considerations: In calculating the positions of the five Projection Points on the Piri Re'is Map, based on the pole [see Fig. 17] we would proceed as follows for the new grid:

(1) Assume center position 23.5 N, 30.0 E.
(2) Assume length of radius, 69.5 degrees.
(3) Assume (decide from geometry I should say) point II to be on same latitude as tropic and center, i.e., rhumb from pt II goes through center, must be same latitude.
(4) Find longitude of *point II*.

radius length = 69.5

pt II ———————————————————————— 23.5 N
   23.5 N   30. E
   ? longitude

longitude (λ) = 69.5° − 30° E =   39.5° W

L = 23.5° N from map

(5) Do other points similar to last time:

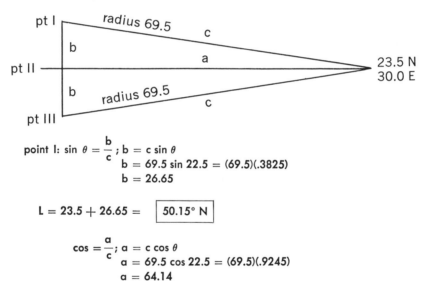

point I: $\sin \theta = \dfrac{b}{c}$; $b = c \sin \theta$
         b = 69.5 sin 22.5 = (69.5)(.3825)
         b = 26.65

L = 23.5 + 26.65 =   50.15° N

    $\cos = \dfrac{a}{c}$; $a = c \cos \theta$
         a = 69.5 cos 22.5 = (69.5)(.9245)
         a = 64.14

λ = 64.14 − 30.00 =   34.14° W

point III: same calculation for a & b; they are the same (triangle for pt III is upside down from pt I)

thus b = 26.65
    L = 26.65 − 23.50 =   3.15° S

    a = 64.14

    λ = 64.14 − 30.00 =   34.14° W

point IV:

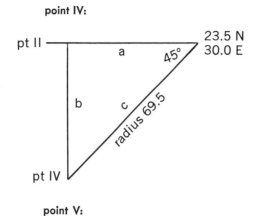

$b = c \sin \theta$
$b = 69.5 \sin 45° = (69.5)(.707)$
$b = 49.10$

$L = 49.10 - 23.50 =$ | 25.60° S |

by geometry $a = b = 49.10$

$\lambda = 49.10 - 30.00 =$ | 19.10° W |

point V:

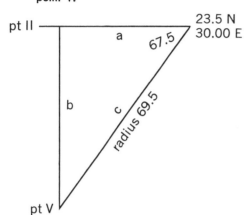

$b = c \sin \theta$
$b = (69.5)(\sin 67.5) = (69.5)(.924)$
$b = 64.30$

$L = 64.30 - 23.50 =$ | 40.80° S |

$a = c \cos \theta$
$a = (69.5)(\cos 67.5) = (69.5)(.382)$
$a = 26.55$

$\lambda = 30.0 - 26.55 =$ | 3.35° E |

These calculations are more accurate than those of my last letter. Note some small changes.

Pt	L	
I	50.15° N	34.14° W
II	23.50° N	39.50° W
III	3.15° S	34.14° W
IV	25.60° S	19.10° W
V	40.80° S	3.35° E

These figures, of course, had to be adjusted to take account of the Eratosthenian error of 4½ per cent in the circumference of the earth. This was done by adding 4½ per cent to latitudes from the North Pole and to longitudes from the meridian of Alexandria.

For the De Canerio Map

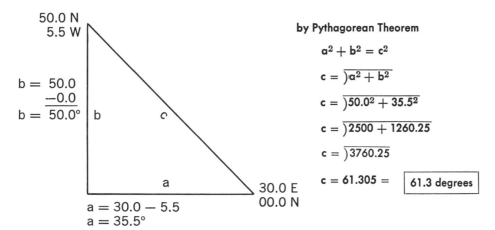

50.0 N
5.5 W

b = 50.0
−0.0
b = 50.0°

b

c

a

30.0 E
00.0 N

a = 30.0 − 5.5
a = 35.5°

by Pythagorean Theorem

$$a^2 + b^2 = c^2$$

$$c = \sqrt{a^2 + b^2}$$

$$c = \sqrt{50.0^2 + 35.5^2}$$

$$c = \sqrt{2500 + 1260.25}$$

$$c = \sqrt{3760.25}$$

$$c = 61.305 = \boxed{61.3 \text{ degrees}}$$

Spherical Trigonometry of the De Canerio Map*

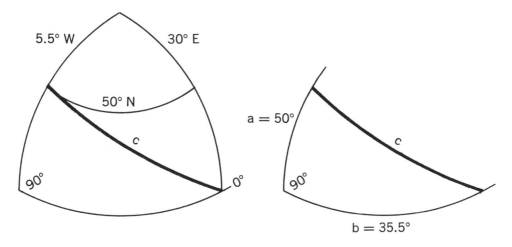

5.5° W   30° E

50° N

c

90°   0°

a = 50°

c

90°

b = 35.5°

From spherical trigonometry:

$$\cos c = \cos a \cos b$$

log cos a = 9.80807 − 10
log cos b = 9.91069 − 10
─────────────────────────
log cos c = 58   26′40″
(or about 58.4)†

* Calculation by E. A. Wixson.
† An independent calculation by Dr. John M. Frankland, of the Bureau of Standards, gave a closely similar result of 58° 27′.

Geographical Tables

Table 1: Piri Re'is Map of 1513

Locality	True Position	Piri Re'is Map	Errors
(a) AFRICA			
1. Annobon Islands	2.0 S	2.0 S	0.0
	6.0 E	0.0	6.0 W
2. Cavally River			
3. Cape Palmas	4.0 N	5.0 N	1.0 N
	8.0 W	2.5 W	5.5 E
4. St. Paul River	7.0 N	6.0 N	1.0 S
	11.0 W	7.5 W	3.5 E
5. Mano River	7.5 N	6.0 N	1.5 S
	12.0 W	9.0 W	3.0 E
6. Freetown	8.5 N	7.5 N	1.0 S
	15.5 W	12.0 W	3.5 E
7. Bijagos Islands	11.0 N	10.0 N	1.0 S
	16.0 W	15.0 W	1.0 E
8. Gambia River	13.5 N	13.0 N	0.5 S
	16.5 W	16.0 W	0.5 E
9. Dakar	15.0 N	14.0 N	1.0 S
	17.0 W	17.5 W	0.5 W
10. Senegal River	16.0 N	15.0 N	1.0 S
	16.5 W	16.0 W	0.5 E
11. Cape Blanc	21.0 N	21.0 N	0.0
	17.0 W	18.0 W	1.0 W
12. Cape Juby	28.0 N	25.5 N	2.5 S
	13.0 W	15.0 W	2.0 W
13. Sebu River	34.3 N	32.0 N	2.3 S
	9.0 W	8.0 W	1.0 E

Locality	True Position	Piri Re'is Map		Errors
(b) EUROPE				
14. Gibraltar	36.0 N	35.0 N		1.0 S
	5.5 W	7.0 W		1.5 W
15. Guadalquivir River	37.0 N	37.0 N		0.0
	6.3 W	7.0 W		0.7 W
16. Cape St. Vincent	37.0 N	36.0 N		1.0 S
	9.0 W	11.0 W		2.0 W
17. Tagus River	38.5 N	38.0 N		0.0
	9.0 W	11.0 W		2.0 W
18. Cape Finisterre	43.0 N	43.0 N		0.0
	9.0 W	12.0 W		3.0 W
19. Gironde River				
20. Brest	48.0 N	48.0 N		0.0
	5.0 W	8.0 W		3.0 W
(c) NORTH ATLANTIC ISLANDS				
21. Cape Verde Islands	15–17 N	14–19 N		0.0
	22–25 W	23–29 W		0.0
22. The Canary Islands	27–29 N	26–28 N		1.0 S
	13–17 W	14–20 W		1.0 W
23. Madeira Islands	36.6 N	31.0 N		5.6 N
	17.0 W	17.0 W		0.0
24. The Azores	37–39 N	36–40 N		0.0
	25–31 W	25–32 W		0.0

(d) THE CARIBBEAN ON GRID B

The northward shifting of the geography of the main grid by 4.4°, to agree with the line through Point III as the equator, pushed the geography of the Caribbean about 4° west. This error is taken account of in the following table.

25. Cuba

	True Position	Piri Re'is Map		Errors
(a) Gulf of Guacanayabo	20.5 N	18.0 N		2.5 S
	77.5 W	88.0 W	−4.0	7.5 W
(b) Guantanamo Bay	20.0 N	18.0 N		2.0 S
	75.0 W	86.0 W	−4.0	7.0 W
(c) Bahia de Nipe	21.0 N	21.5 N		0.5 N
	77.5 W	85.0 W	−4.0	3.5 W

Locality	True Position	Piri Re'is Map		Errors
(d) Bahia de la Gloria	22.0 N	22.0 N		0.0
	77.5 W	88.0 W	−4.0	6.5 W
(e) Camaguey Mountains	21.0 N	20.0 N		1.0 S
	77–79 W	85–89 W	−4.0	4.0 W
(f) Sierra Maestra Mountains	20.0 N	18.0 N		2.0 S
	76–77 W	84–86 W	−4.0	4.0 W
26. Andros Island	23–25 N	26.0 N		2.0 N
	76–77 W	92–96 W	−4.0	12.0 W
27. San Salvador (Watling)	24.0 N	26.5 N		2.5 N
	74.5 W	84.5 W	−4.0	6.0 W
28. Isle of Pines	22.0 N	16.0 N		6.0 S
	83.0 W	91.0 W	−4.0	4.0 W
29. Jamaica	18.0 N	15–16 N		2.5 S
	77.0 W	86.0 W	−4.0	5.0 W
30. Hispaniola (Santo Domingo, Haiti)	18–20 N	16–19 N		1.0 S
	68–74 W	74–76 W	−4.0	0.0
31. Puerto Rico	18–19 N	21.0 N		1.5 N
	66–67 W	74.0 W	−4.0	3.5 W

(e) SOUTH AND CENTRAL AMERICAN COASTS ON GRID B

Locality	True Position	Piri Re'is Map		Errors
32. Rio Moroni?	6.0 N	11.0 N		5.0 N
	54.0 W	59.0 W	−4.0	1.0 W
33. Corantijn River?	6.0 N	10.0 N		4.0 N
	57.0 W	60.0 W	−4.0	1.0 E
34. Essequibo River?	7.0 N	10.0 N		3.0 N
	58.0 W	61.5 W	−4.0	1.0 E
35. Orinoco River	9–10 N	14.0 N		4.5 N
	61–63 W	67.0 W	−4.0	1.0 W
36. Gulf of Venezuela	11–12 N	13.0 N		1.5 N
	71.0 W	76.0 W	−4.0	1.0 W
37. Pt. Gallinas	12.5 N	14.5 N		2.0 N
	72.5 W	77.5 W	−4.0	1.0 W
38. Magdalena River?	11.0 N	12.5 N		1.5 N
	75.0 W	79.5 W	−4.0	0.5 W
39. Gulf of Uraba	8.0 N	10.5 N		2.5 N
	77.0 W	79.0 W	−4.0	2.0 E
40 Honduras (Cape Gracias a Dios)	15.0 N	13.0 N		2.0 S
	83.0 W	86.0 W	−4.0	1.0 E
41. Yucatán	21.0 N	15.0 N		6.0 S
	88.0 W	96.0 W	−4.0	4.0 W

Locality	True Position	Piri Re'is Map	Errors

(f) SOUTH AND CENTRAL AMERICAN COASTS (on the main grid of the map and on the inset grid, Grid C)

Locality	True Position	Piri Re'is Map	Errors
42. Cape Frio	23.0 S 42.0 W	23.0 S 38.0 W	0.0 4.0 E
43. Salvador	13.0 S 39.0 W	13.5 S 38.0 W	0.5 S 1.0 E
44. San Francisco River	11.5 S 36.5 W	10.5 S 37.0 W	1.0 N 0.5 W
45. Recife (Pernambuco)	8.0 S 35.0 W	7.0 S 34.5 W	1.0 N 0.5 E
46. Cape São Rocque	5.0 S 36.0 W	6.5 S 36.0 W	1.5 S 0.0
47. Rio Parahyba	3.0 S 42.0 W	4.0 S 40.0 W	1.0 S 2.0 E
48. Bahia São Marcos	2.5 S 44.0 W	4.0 S 45.0 W	1.5 S 1.0 W
49. Serras de Gurupi, de Desordam, de Negro			
50. The Amazon (No. 1) Pará River	0.5 S 48.0 W	0.5 S 48.0 W	0.0 0.0
51. The Amazon (No. 2) Pará River	0.5 S 48.0 W	2.0 N 50.0 W	2.5 N 2.0 W
52. The Amazon (No. 2) western mouth	0.0 50.0 W	4.0 N 53.0 W	4.0 N 3.0 W
53. Island of Marajó			
54. Essequibo River	7.0 N 58.5 W	5.0 N 60.0 W	2.0 S 1.5 W
55. Mouths of the Orinoco	9–10 N 61–63 W	6.5–7.5 N 62.5 W	2.5 S 0.0
56. Peninsula of Paria	10.5 N 62–63 W	7–9 N 61.0 W	2.5 S 1.5 E

(g) CARIBBEAN ISLANDS ON THE MAIN GRID

Locality	True Position	Piri Re'is Map	Errors
57. Martinique	14.5 N 61.0 W	14.5 N 59.0 W	0.0 2.0 E
58. Guadaloupe	16.3 N 61.5 W	17.0 N 60.0 W	0.7 N 1.5 E

Locality	True Position	Piri Re'is Map		Errors
59. Antigua	17.0 N 62.0 W	19.0 N 59.0 W		2.0 N 3.0 E
60. Leeward Islands	17–18 N 61–63 W	17–21 N 60–63 W		0.0 0.0
61. Virgin Islands	18.5 N 64.5 W	26–28 N 62–65 W		c. 9.0 N 0.0

(h) CENTRAL AMERICAN COAST

Between the Peninsula of Paria and the Gulf of Venezuela there is a break in the map, and the coast between these points has been omitted, resulting in a loss of about 4½° of west longitude. This has been compensated for by adding 4½° to each finding of west longitude.

Locality	True Position	Piri Re'is Map		Errors
62. Gulf of Venezuela	11–12 N 71.0 W	10–11 N 65.0 W	 +4.5	0.0 1.5 E
63. Magdalena River	11.0 N 75.0 W	10.0 N 68.0 W	 +4.5	1.0 S 2.5 E
64. Atrato River	8.0 N 77.0 W	8.0 N 73.0 W	 +4.5	0.0 0.5 W
65. Honduras (Cape Gracias a Dios)	15.0 N 82.5 W	17.0 N 72.0 W	 +4.5	2.0 N 6.0 E
66. Yucatán	21.0 N 88.0 W	24.0 N 77.0 W	 +4.5	3.0 N 6.5 E

(i) THE LOWER EAST COAST OF SOUTH AMERICA AND SOME ATLANTIC ISLANDS (on the main grid of the map)

Here a mistake of compilation has resulted in the omission of all of the coast between Cape Frio and Bahia Blanca. About 900 miles of coastline is missing, with a net loss of 16° of latitude going south, and 20° of longitude going west. The table includes adjustment for these errors.

Locality	True Position	Piri Re'is Map		Errors
67. Bahia Blanca	39.0 S 62.0 W	22.0 S +16 40.0 W +20		1.0 N 2.0 E
68. Rio Colorado	40.0 S 62.0 W	22.5 S +16 41.0 W +20		1.5 N 1.0 E
69. Gulf of San Mathias	42.5 S 64.0 W	25.0 S +16 42.5 W +20		1.5 N 1.5 E
70. Rio Negro (Argentina)	41.0 S 63.0 W	25.0 S +16 43.0 W +20		0.0 0.0
71. Rio Chubua	44.0 S 65.0 W	27.0 S +16 47.5 W +20		1.0 N 2.5 W

Locality	True Position	Piri Re'is Map		Errors
72. Gulf of San Gorge	47.0 S	27.5 S	+16	3.5 N
	66.0 W	45.0 W	+20	1.0 E
73. Bahia Grande	50–52 S	30.0 S	+16	5.0 N
	69.0 W	47.0 W	+20	2.0 E
74. Cape San Diego (near the Horn)	55.0 S	35.0 S	+16	4.0 N
	65.0 W	46.5 W	+20	1.5 W
75. Falkland Islands	52.0 S	30–32 S	+16	5.0 N
	60.0 W	43–45 W	+20	4.0 W

At this point there appears to be another break in the map, with the omission of Drake Passage. This involves a further loss of about 9° of latitude. The total latitude adjustment now amounts to 25°.

76. The South Shetlands	61.0 S	33–34 S	+25	2.5 N
	60.0 W	40–43 W	+20	0.0
77. South Georgia	Anomalous. See below.			

(j) ANTARCTICA

78. The Palmer Peninsula	65.0 S	36.0 S	+25	4.0 N
	60.0 W	40.0 W	+20	0.0
79. The Weddell Sea	67–75 S	37.0 S	+25	c. 8.0 N
	20–60 W	30–40 W		

At this point the deficiency of west longitude is compensated for by a large error in the total longitude covered by the Weddell Sea. On the modern map this amounts to 40°; on the Piri Re'is Map only to about 10°. We therefore now subtract 10° from the west longitude readings.

80. Mt. Ropke, Queen Maud Land	72.5 S	42.5 S	+25	5.0 N
	4.0 W	15.0 W	−10	1.0 W
81. The Regula Range	72.5 S	42.5 S	+25	5.0 N
	2.5 W	12.5 W	−10	0.0
82. Muhlig–Hofmann Mountains	71–73 S	41–43 S	+25	4.0 N
	1–6 E	7–10 W	−10	0.0
83. Penck Trough	73.0 S	44.0 S	+25	4.0 N
	2.5 W	12.0 W	−10	0.0
84. Neumeyer Escarpment	73.5 S	45.0 S	+25	3.5 N
	2.0 W	12.0 W	−10	0.0
85. Drygalski Mountains	71–73 S	40.0 S	+25	7.0 N
	8–14 E	2.0 E	+10	0.0
86. Vorposten Peak	71.5 S	42.5 S	+25	3.0 N
	16.0 E	6.0 E	+10	0.0
87. Boreas, Passat Nunataks	71.5 S	37–38 S	+25	4.0 N
	4.0 W	11–14 W	−10	0.0

Locality	True Position	Piri Re'is Map	Errors
(k) SOUTH ATLANTIC ISLANDS (on the main grid of the map)			
88. Tristan d'Acunha	37.0 S	33.0 S	4.0 N
	12.5 W	12.5 W	0.0
89. Gough Island	40.3 S	35.5 S	4.8 N
	10.0 W	9.0 W	1.0 E
77. South Georgia	54.5 S	36.0 S	18.5 N
	37.0 W	37–38 W	0.0
95. Fernando da Naronha	4.0 S	10.0 S	6.0 S
	31.0 W	30.0 W	1.0 E

The latitude error suggests that the island was placed on the map with reference to Amazon No. 1, but on the scale of the main grid.

(l) PACIFIC COAST OF SOUTH AMERICA

90a, 90b. Coastal ranges of the Andes

91. Peninsula of Paracas

92. Valparaiso

(m) NON-EXISTENT ISLANDS

93. Island over Mid-Atlantic Ridge

94. Island labeled "Antillia" by Piri Re'is

Table 2: The Oronteus Finaeus World Map of 1531

	Antarctica		
Geographical Localities	True Position	Oronteus Finaeus Position	Errors
(a) QUEEN MAUD LAND			
1. Cape Norvegia	71.5 S	66.5 S	5.0 N
	12.0 W	6.0 W	6.0 E
2. Regula Range	72–73 S	68–69 S	4.0 N
	2–5 W	0–3 E	c. 5.0 E
3. Penck Trough	71–74 S	66.5–69 S	c. 3.0 N
	3.0 W	4.0 E	7.0 E
4. Neumeyer Escarpment	73.0 E	68–69 S	4.0 N
	0–4 W	0.5 E	c. 4.0 E

Geographical Localities	True Position	Oronteus Finaeus Position	Errors
5. Muhlig–Hofmann and Wohlthat Mountains	71–73 S 0–15 E	70–71 S 10–15 E	0.0 0.0
6. Sor Rondanne and Belgica Mts.	72.0 S 22–33 E	72–73 S 20–30 E	0.0 0.0
7. Prince Harald Coast, Lützow-Holm Bay, Shirase Glacier	69–70 S 35--40 E	70.0 S 35–37 E	0.0 0.0
8. Queen Fabiola Mountains	76–77 S 35–36 E	73.0 S 30–40 E	c. 3.0 N 0.0

(b) ENDERBY LAND

9. Casey Bay (Lena Bay) or Amundsen Bay	67.0 S 48–50 E	70.0 S 48.0 E	3.0 N 0.0
10. Nye Mountains, Sandercook Nunataks	73.0 S 49.0 E	72–73 S 50.0 E	0.0 1.0 E
11. Edward VIII Bay (Kemp Coast)	66–67 S 58–60 E	69–70 S 55.0 E	3.0 S 4.0 W
12. Schwartz Range, Rayner Peak, Dismal Mountains, Leckie Range, Knuckley Peaks, etc.	71–74 S 54–57 E	71–73 S 60–75 E	0.0 12.0 E
13. Amery Ice Shelf, MacKenzie-Prydz Bays	73–78 S 70–75 E	67.0 S 73.0 E	c. 8.5 N 0.0
14. Prince Charles Mountains and adjacent peaks	72–74 S 60–69 E	68–70 S 70–75 E	c. 4.0 N c. 5.0 E

(c) WILKES LAND

15. Philippi Glacier, Posadowsky Bay	67.0 S 88–89 E	66.0 S 77.0 E	1.0 N 11.5 W
16. Denman-Scott Glaciers (Shakleton Ice Shelf)	66–67 S 99–101 E	66.0 S 85.0 E	0.0 15.0 W
18. Vincennes Bay	66–67 S 109 E	65.0 S 105.0 E	1.5 N 4.0 W
19. Totten Glacier	67.0 S 115–117 E	66.0 S 112.0 E	1.0 N 4.0 W
20. Porpoise Bay	67.0 S 128–130 E	67.0 S 122.0 E	0.0 7.0 W

Geographical Localities	True Position	Oronteus Finaeus Position	Errors
21. Merz Glacier	68.0 S 144–145 E	70.0 S 140.0 E	2.0 S 4.5 W
22. McLean, Carroll Nunataks, Aurora Peak, Medigan Nunatak, etc.	67–68 S 143–145 E	70–71 S 132–134 E	3.0 S 11.0 W
23. Pennell Glacier, Lauritzen Bay	69.0 S 157–158 E	70.0 S 150.0 E	1.0 S 7.5 W
(d) VICTORIA LAND and THE ROSS SEA			
24. Rennick Bay	70.5 S 162.0 E	72.5 S 155.0 E	2.0 S 7.0 W
25. Arctic Institute Range	70–73 S 161–162 E	74–75 S 150–155 E	3.5 S 7.0 W
26. Newnes Iceshelf and Glacier	73.5 S 167.0 E	74.5 S 170.0 E	1.0 S 3.0 E
27. Ross I. (Mount Erebus)	77.5 S 168.0 E	76.5 S 172.0 E	1.0 S 4.0 E
28. Ferrer Taylor Glacier	77–78 S 159–163 E	77.0 S 160–170 E	0.0 0.0
29. Boomerang Range and adjacent peaks (Escalade Peak, Portal Mt., Mt. Harmsworth, etc.)	77–78 S 159–163 E	77–79 S 142–152 E	0.0 14.0 E
30. Mountain group: Mt. Christmas, Mt. Nares, Mt. Albert Markham, Pyramid Mountains, Mt. Wharton, Mt. Field, Mt. Hamilton	80–82 S 157–160 E	79–80 S 140–150 E	1.5 N c. 13.5 W
31. Queen Alexandra Range	84–85 S 160–165 E	84–87 S 145–155 E	0.0 c. 10.0 W
32. Queen Maud Range	87.0 S 140 W–180 E/W	81–82 S 140–160 E	5.5 N c. 20–40 E
33. Nimrod Glacier	82.5 S 157–163 E	81.0 S 160–175 E	1.5 S c. 0.0
34. Beardmore Glacier	84–85 S 170.0 E	82–84 S 170 E–170 W	0.0 c. 0.0
36. Leverett Glacier	85.5 S 150.0 W	81–82 S 140–160 W	4.0 N c. 0.0

Geographical Localities	True Position	Oronteus Finaeus Position	Errors
37. Supporting Party Mountain, or Mt. Gould	85.3 S 150.0 W	80.0 S 160.0 W	5.3 N 10.0 W
38. Thiel and Horlich Mts.	86.0 S 90–130 W	83–85 S 130–150 W	3.0 N c. 30.0 W
39. Bay, unnamed, west coast of Ross Sea	80–81 S 150.0 W	77.5 S 150–160 W	3.0 N c. 0.0
40. Prestrude Inlet, Kiel Glacier	78.0 S 157–159 W	74.0 S 160.0 W	4.0 N 2.0 W
42. Edward VII Peninsula	77–78 S 155–158 W	72–73 S 158–165 W	5.0 N c. 4.0 W
43. Sulzberger Bay	77.0 S 146–154 W	73.5 S 150–155 W	3.5 N c. 0.0
44. Land now submerged?			

(e) MARIE BYRD LAND

45. Edsel Ford Range	76–78 S 142.0 W	73–74 S 135–143 W	3.5 N 0.0
46. Executive Committee Range	77.0 S 125–130 W	73.0 S 130–135 W	4.0 N c. 5.0 W
47. Cape Dart, Wrigley Gulf, and Getz Ice Shelf	75.0 S 130.0 W	70.5 S 130.0 W	4.5 N 0.0
48. Cape Herlacher, Martin Peninsula	74.0 S 114.0 W	72.0 S 108.0 W	2.0 N 6.0 E
49. Kohler Range and Crary Mountains	76–77 S 111–118 W	74–75 S 108–115 W	2.5 N 0.0
50. Canisteo Peninsula	74.0 S 102.0 W	74.0 S 98.0 W	0.0 4.0 E
51. Hudson Mountains	74–75 S 99.0 W	72–74 S 105–110 W	1.0 N c. 8.0 W

(f) ELLSWORTH LAND

52. Jones Mountains	73.5 S 94.0 W·	74.0 S 100.0 W	0.5 S 6.0 W
53. Inlet in Ellsworth Land, indicated to exist under the present ice cap	73–80 ?S 80–95 W	72–74 S 85–95 W	0.0 0.0

Geographical Localities	True Position	Oronteus Finaeus Position	Errors

54. Base of the Antarctic (Palmer) Peninsula?

55. The Weddell Sea, shown almost connected with the Ross Sea?

56. Duplicated coastline of Ellsworth Land?

57. Duplicated base of Antarctic Peninsula?

58. Berkner Island in the Weddell Sea shown extending over the continental shelf to the north?

Table 3: The Hadji Ahmed Map of 1559, Exclusive of the Antarctic

Locality	True Position	Turkish Map	Errors
(a) POSITIONS CLOSE TO THE PRIME MERIDIAN			
Magellan Strait	54 S	49 S	5.0 S
	68–73 W	68–73 W	0.0
Cape Horn	55 S	51 S	4.0 S
	65 W	70 W	5.0 W
Amazon River	1 S	1 S	0.0
	50 W	50 W	0.0
Rio de la Plata	35 S	35 S	0.0
	55 W	51 W	4.0 E
Cape Frio	23.5 S	23 S	5.0 N
	43 W	43 W	0.0
Peninsula of Paria	10.5 N	10 N	0.5 S
	62 W	70 W	8.0 W
Hudson River	41 N	51 N	10.0 N
	72 W	72 W	0.0
Gibraltar	35 N	31 N	4.0 S
	7 W	10 W	3.0 W
(b) POSITIONS DISTANT FROM THE PRIME MERIDIAN			
Chile Coast	25–50 S	40–49 S	c. 7.5 S
	72 W	80 W	8.0 W
Haiti	19–20 N	19–20 N	0.0
	70 W	80 W	10.0 W

Locality	True Position	Turkish Map	Errors
Coast of Texas	29 N	29 N	0.0
	99 W	110 W	11.0 W
Ceylon	6 N	9 N	3.0 N
	80 E	110 E	30.0 E
Aden	12 N	12 N	0.0
	60 E	45 E	15.0 W
Gulf of California	23.5 N	30 N	6.5 N
	130–140 W	110 W	25.0 E

Table 4: Mercator's World Map of 1569 (Antarctic coast on a polar or circular projection)

Locality	True Position	Mercator	Errors
1. Cape Dart (Mt. Siple) (Marie Byrd Land)	73.5 S	63.0 S	10.0 N
	121.0 W	95.0 W	26.0 E
2. Cape Herlacher (Marie Byrd Land)	74.0 S	65.0 S	11.0 N
	114.0 W	92.0 W	22.0 E
3. Amundsen Sea	72.0 S	69–71 S	1.0 N
	90–104 W	90–100 W	0.0
4. Thurston Island (Eights Coast) (Ellsworth Land)	72.0 S	70.0 S	2.0 N
	88–93 W	81–84 W	c. 7.0 E
5. Fletcher Islands (Bellingshausen Sea)	71–82 S	72–75 S	0.0
	71–82 W	70–80 W	0.0
6. Alexander I Island	72.0 S	72.5 S	0.5 S
	67.0 W	58.0 W	9.0 E
7. Palmer (Antarctic) Peninsula*	70.0 S	72.0 S	1.5 S
	60.0 W	47.0 W	13.0 E
8. Weddell Sea†	72–73 S	72–75 S	0.0
	35–50 W	35–40 W	0.0
9. Cape Norvegia (Queen Maud Land)	71.0 S	75.0 S	4.0 S
	28.0 W	20.0 W	8.0 E
10. Regula Range (Queen Maud Land)	72.0 S	77.0 S	5.0 S
	4.0 W	5.0 W	1.0 W

* Truncated.

† Longitudes of places west of the Weddell Sea may be off 10° because of an error in the width of the Weddell Sea; the same error is found on the Piri Re'is Map.

Locality	True Position	Mercator	Errors
11. Muhlig–Hofmann Mts. (Queen Maud Land)	72.0 S 3–8 E	77.0 S 2–7 E	5.0 S 0.0
12. Prince Harald Coast	70.0 S 20–31 E	74.0 S 30–40 E	4.0 S 10.0 E
13. Shirase Glacier (Prince Harald Coast)	70.0 S 38–40 E	73.5 S 45.0 E	3.5 S 6.0 E
14. Padda Island (Lutzow-Holm Bay)	69.5 S 35.0 E	71.0 S 42.0 E	1.5 S 7.0 E
15. Prince Olaf Coast (Enderby Land)	68.5 S 33–40 E	73.0 S 45–50 E	4.5 S 11.0 E

Table 5: Mercator's World Map of 1538

Identified Localities	True Longitudes	Mercator's Longitudes
(a) ANTARCTICA		
1. Antarctic (Palmer) Peninsula, truncated at 70° S.	60–70 W	74.0 W
2. Weddell Sea	25–55 W	26–42 W
3. Caird Coast	20–30 W	10–18 W
4, 5. Princess Martha and Princess Astrid Coasts, Queen Maud Land	20 W–20 E	0.0
6. Prince Harald Coast	30–35 E	2 W–6 E
7. MacKenzie-Prydz Bays	70–80 E	38–46 E
8. Denman Glacier and Ice Tongue	100.0 E	68.0 E
9. Vincennes Bay	110.0 E	78.0 E
10. Beardmore Glacier, Ross Sea	170.0 E	142.0 E
11. Robert Scott Glacier, Ross Sea	150.0 W	150.0 E
12. Amundsen Sea	110–120 W	162–170 W
13. Alexander Island	70–75 W	82.0 W
(b) SOUTH AMERICA		
14. The Chile Coast	70–75 W	74.0 W
15. Falklands? (as one large island)	60.0 W	58.66 W
16. Gulf of San Gorge, Argentina	65–67 W	50.0 W
17. Gulf of San Mathias, Argentina	64–65 W	42.0 W
18. Arica (Peru-Bolivia)	70.0 W	66.0 W
19. Pt. Aguia	81.0 W	74.0 W
20. Gulf of Guayaquil (Ecuador)	80.0 W	74.0 W
21. Ensenada di Tumaco (Colombia)	79.0 W	71.0 W
(c) NEW ZEALAND		
22. New Zealand: "Los roccos Insula"	165–180 E	102–110 E

Identified Localities	True Longitudes	Mercator's Longitudes
(d) AFRICA		
23. Cape of Good Hope	18.5 E	22.0 E
24. St. Helena Bay	18.0 E	21.0 E
25. Walvis Bay	15.0 E	12.0 E
26. Laurence Marques Bay	33.0 E	36.0 E
27. Cape San Sebastian	35.0 E	42.0 E
28. Madagascar	45.0 E	46–50 E

Table 6a: Dulcert Portolano of 1339

Locality	True Position	Dulcert Portolano	Errors
1. Malin Head, Ireland	55.2 N	56.5 N	1.3 N
	7.0 W	7.0 W	0.0
2. Galway, Ireland	53.0 N	53.0 N	0.0
	9.0 W	8.0 W	1.0 E
3. Cape Clear, Ireland	51.5 N	50.5 N	1.0 S
	9.0 W	8.0 W	1.0 E
4. The Hebrides	57–58.5 N	57–58 N	0.0
	6–7 W	4–5 W	2.0 E
5. Moray Firth, Scotland	57.7 N	57.0 N	0.7 S
	3.5–4 W	1.5 W	c. 2.0 E
6. Solway Firth	55.0 N	55.0 N	0.0
	3–4 W	3.5 W	c. 0.0
7. Lands End, England	50.0 N	50.0 N	0.0
	6.0 W	4.5 W	1.5 E
8. Scilly Islands	50.0 N	50.0 N	0.0
	6.5 W	5–6 W	1.0 E
9. The Wash	53.0 N	53.5 N	0.5 N
	0.3 E	1.0 W	1.3 W
10. Thames River	51.3 N	51.5 N	0.2 N
	0.5 E	0.5 W	1.0 W
11. Isle of Wight	50.6 N	50.4 N	0.2 S
	1.3 W	2.0 W	0.7 W
12. Calais, France	50.7 N	50.4 N	0.3 S
	2.0 E	0.0	2.0 W
13. Brest	48.5 N	47.5 N	1.0 S
	4.5 W	4.0 W	0.5 E

Locality	True Position	Dulcert Portolano	Errors
14. Belle Isle	47.2 N 3.0 W	46.2 N 3.0 W	1.0 S 0.0
15. The Loire River	47.2 N 2.0 W	46.2 N 2.5 W	1.0 S 0.5 W
16. La Gironde River	45.5 N 1.0 W	44.0 N 2.0 W	1.5 S 1.0 W
17. Cape Finisterre, Spain	43.0 N 9.0 W	42.5 N 8.5 W	0.5 S 0.5 W
18. Tagus River, Portugal	39.0 N 9.0 W	37.0 N 7.0 W	2.0 S 2.0 E
19. Guadalquivir River	37.0 N 6.3 W	35.5 N 7.0 W	1.5 S 0.7 W
20. Gibraltar	36.0 N 5.5 W	34.9 N 5.0 W	1.1 S 0.5 E
21. Sebu River	35.0 N 6.0 W	32.0 N 7.5 W	3.0 S 1.5 W
22. Madeira Islands	32.5 N 16–17 W	30.0 N 13.0 W	2.5 S 3.5 E
23. Cape Juby	27.0 N 13.0 W	27.0 N 11.5 W	0.0 2.0 E
24. Fuerteventura (Canary Islands)	28.0 N 14.0 W	30.0 N 13.0 W	2.0 N 1.0 E
25. Cartagena, Spain	37.5 N 1.0 W	36.5 N 0.5 W	1.0 S 0.5 E
26. Majorca (Mallorca)	39.9 N 3.0 E	39.0 N 3.0 E	0.9 S 0.0
27. Marseilles	42.0 N 5.0 E	44.0 N 6.0 E	2.0 N 1.0 E
28. Cape Corse, Corsica	43.5 N 9.3 E	43.5 N 9.5 E	0.0 0.2 E
29. Cagliari, Sardinia	39.0 N 9.0 E	37.0 N 9.5 E	2.0 S 0.5 E
30. Cape Bon, Tunisia	37.0 N 11.0 E	36.0 N 11.5 E	1.0 S 0.5 E
31. Cape Passero, Sicily	36.5 N 15.0 E	35.5 N 15.0 E	1.0 S 0.0

Locality	True Position	Dulcert Portolano	Errors
32. Bengazi, Libya	32.0 N	30.0 N	2.0 S
	20.0 E	20.0 E	0.0
33. Trieste, Italy	46.0 N	47.5 N	1.5 N
	14.0 E	14.5 E	0.5 E
34. Corfu	40.2 N	40.0 N	0.2 S
	20.0 E	20.0 E	0.0
35. Kalamai, Peloponnesus	37.0 N	37.0 N	0.0
	22.0 E	22.5 E	0.5 E
36. The Bosphorus	41.0 N	44.0 N	3.0 N
	29.0 E	28.5 E	0.5 W
37. Danube River	45.0 N	50.0 N	5.0 N
	34.5 E	29.5 E	5.0 W
38. Sevastopol	44.5 N	49.3 N	4.8 N
	33.5 E	33.5 E	0.0
39. Rostov	47.3 N	53.0 N	5.7 N
	39.5 E	38.0 E	1.5 W
40. Don River bend	49.0 N	52.5 N	3.5 N
	44.0 E	44.5 E	0.5 E
41. Stalingrad (Volga River)	49.0 N	55.0 N	6.0 N
	44.5 E	47.0 E	2.5 E
42. Batum	41.5 N	45.0 N	3.5 N
	41.5 E	41.5 E	0.0
43. Cape Andreas (Cyprus)	36.0 N	37.0 N	1.0 N
	34.5 E	35.5 E	1.0 E
44. Crete, South Coast	37.5 N	35.0 N	2.5 S
	24.5–26 E	24–26 E	0.0
45. Rhodes	36.0 N	37.0 N	1.0 N
	28.0 E	28.0 E	0.0
46. Antalya, Turkey	37.0 N	37.5 N	0.5 N
	30.7 E	31.5 E	0.8 E
47. Alexandria	31.0 N	Correct by assumption	
	30.0 E	32.0 E	2.0 E
48. Aswan	24.0 N	24.0 N	0.0
	33.0 E	34.0 E	1.0 E
49. Ras Muhammad (Sinai)	27.5 N	27.5 N	0.0
	34.0 E	35.5 E	1.5 E
50. Coast of India?*			

---

* Appears to resemble the coast of India from about 25° North Latitude, 62° East Longitude, to about 22° North Latitude, 70° East Longitude.

Analysis of the Dulcert Portolano

If we compare a group of the northernmost Atlantic points on this chart, with respect to latitude, we find errors as follows:

Table 6b

	Geographical Localities	Lat. Error
1.	Malin Head, Ireland	1.3 N
2.	Galway, Ireland	0.0
3.	Cape Clear, Ireland	1.0 S
4.	The Hebrides	0.0
5.	Moray Firth, Scotland	0.0
6.	Solway Firth	0.0
7.	Lands End	0.0
8.	Scilly Islands	0.0
9.	The Wash	0.5 N
10.	The Thames	0.2 N
11.	Isle of Wight	0.2 S

We note that Ireland seems to be represented on too large a scale, but that otherwise latitudes are essentially correct. They are based on the parallel of Alexandria, our base line of latitude, and are therefore correct with respect to the latitude of that city. Latitude errors of the more southern points on the map do not indicate any considerable error in the length of the degree of latitude.

With respect to longitude, the location of a group of western localities distributed from northern Ireland to Cape Juby shows evidence of a remarkable knowledge of their comparative longitudes:

Table 6c

	Geographical Localities	Long. Errors
1.	Malin Head, Ireland	0.0
2.	Galway, Ireland	1.0 E
3.	Cape Clear, Ireland	1.0 E
4.	The Hebrides	2.0 E
7.	Lands End	1.5 E
8, 9.	The Wash	1.3 W
10.	Thames	1.0 W
11.	Isle of Wight	0.7 W
12.	Calais	2.0 W
13.	Brest	0.5 E
14.	Belle Isle	0.0
15.	Loire River	0.5 W
16.	La Gironde	1.0 W
17.	Cape Finisterre	0.5 W
18.	Tagus River	2.0 E
19.	Guadalquivir River	0.7 W
20.	Gibraltar	0.5 E
21.	Sebu River	1.5 W
22.	Madeira Islands	3.5 E
23.	Cape Juby	2.0 E

There is an indication here of an error in the scale of England, points on the west being too far east, and points on the east being too far west.

A comparison of the longitudes found for the localities on the eastern part of the map reveals the same order of accuracy:

Table 6d

	Geographical Localities	Long. Errors
34.	Corfu	0.0
35.	Kalamai	0.5 E
36.	The Bosphorus	0.5 W
37.	Danube River	5.0 W
38.	Sevastopol	0.0
39.	Rostov	1.5 W
40.	Don River Bend	0.5 E
41.	Stalingrad	2.5 E
42.	Batum	0.0
45.	Rhodes	0.0
46.	Antalya	0.8 E
48.	Aswan	1.0 E

It seems that these places are in remarkably correct longitudinal accuracy with respect to each other. But, it may be noted, they are also in equally correct relationship to the points on the Atlantic coast; in other words, the mapmaker achieved high accuracy in finding the longitudes of places distributed over an east-west distance of some 3,000 miles.

Table 7: De Canerio Map of 1502 (grid drawn by spherical trigonometry)

Geographical Localities	True Position	De Canerio Map	Errors
1. Cape of Good Hope	35.5 S	37.0 S	1.5 S
	18.5 E	14.0 E	4.5 W
2. Gt. Paternoster Pt.	33.0 S	35.5 S	2.5 S
	18.0 E	14.0 E	4.0 W
3. Walvis Bay, Pelican Point	23.0 S	22.5 S	0.5 N
	15.0 E	10.0 E	5.0 W
4. Congo River	6.0 S	13.0 S	7.0 S
	12.2 E	14.0 E	1.8 E
5. Cap Lopez	1.0 S	5.0 S	4.0 S
	9.0 E	10.0 E	1.0 E
6. São Tomé (island)	0.0	4.0 S	4.0 S
	6.0 E	8.0 E	2.0 E
7. Niger Delta	4.0 N	1.0 N	3.0 S
	6.0 E	8.0 E	2.0 E

Geographical Localities	True Position	De Canerio Map	Errors
8. Cape Three Points	5.0 N 2.0 W	3.5 N 5.0 W	1.5 S 3.0 W
9. Cap Palmas	4.5 N 8.0 W	4.0 N 9.0 W	0.5 S 1.0 W
10. Freetown	8.0 N 13.5 W	10.0 N 17.5 W	2.0 N 4.0 W
11. Dakar	15.0 N 17.0 W	17.0 N 20.0 W	2.0 N 3.0 W
12. Cap Blanc	21.0 N 17.0 W	24.0 N 20.0 W	3.0 N 3.0 W
13. Cap Yubi	28.0 N 13.0 W	28.0 N 15.5 W	0.0 2.5 W
14. Gibraltar	36.0 N 5.5 W	34.5 N 5.5 W	1.5 S 0.0
15. Cap Bon	37.0 N 11.0 E	34.0 N 11.0 E	3.0 S 0.0
16. Bengazi	32.0 N 20.0 E	28.0 N 19.5 E	4.0 S 0.5 W
17. Alexandria	31.0 N 30.0 E	26.5 N 30.0 E	4.5 S (assumption)
18. Cyprus	35.0 N 32–34 E	32.0 N 32–34 E	3.0 S 0.0
19. Crete	35.0 N 24–26 E	32.0 N 25.0 E	3.0 S 0.0
20. Lesbos (Aegean Sea)	39.0 N 26.0 E	36.0 N 26.5 E	3.0 S 0.5 E
21. Bosphorus	41.0 N 29.0 E	39.0 N 29.5 E	2.0 S 0.5 E
22. Sevastopol (Crimea)	44.5 N 34.5 E	41.0 N 35.0 E	3.5 S 0.5 E
23. Batum (Caucasus)	42.0 N 42.0 E	38.0 N 43.0 E	4.0 S 1.0 E
24. Sicily	36–37 N 13–15 E	34.0 N 15.0 E	2.5 S 0.0
25. Sardinia	39–41 N 8–10 E	35–38 N 9–10 E	3.0 S 0.0

Geographical Localities	True Position	De Canerio Map	Errors
26. Cape St. Vincent	37.0 N	36.0 N	1.0 S
	11.0 W	10.0 W	1.0 E
27. Cap Finisterre	43.0 N	42.0 N	1.0 S
	9.0 W	9.0 W	0.0
28. Brest	48.0 N	47.0 N	1.0 S
	5.0 W	4.0 W	1.0 E
29. Cape Clear (Ireland)	52.0 N	49.5 N	2.5 S
	10.0 W	8.0 W	2.0 E
30. Londonderry	55.0 N	54.5 N	0.5 S
	7.5 W	6.5 W	1.0 E
31. Denmark (northern coast)	50–53 N	56.0 N	4.5 N
	8–10 E	7.5 E	2.5 W
32. Gulf of Riga	57.5 N	52.5 N	5.0 S
	23–24 E	25.0 E	1.5 E
33. Saarma Island (Gulf of Riga)	57.5 N	55.0 N	2.5 S
	22.5 E	24.0 E	1.5 E

AFRICAN EAST COAST

Geographical Localities	True Position	De Canerio Map	Errors
34. Laurenço Marques	25.0 S	27.0 S	2.0 S
	33.0 E	35.0 E	2.0 E
35. Beira*	20.0 S	15.0 S	5.0 N
	35.0 E	45.0 E	10.0 E
36. Capo Guardafui* (Ras Assir)	12.0 N	1.0 N	11.0 S
	51.0 E	69.0 E	18.0 E
37. Al Hadd, Oman*	22.5 N	19.0 N	3.5 S
	60.0 E	79.5 E	19.5 E

*Analysis of De Canerio Map, with Grid Based on
Spherical Trigonometry*

The map is composed of three main sections. Its most accurate part, based on the ancient portolan tradition, consists of the Atlantic coasts of Africa and Europe, and the coasts of the Mediterranean and Black Seas. An inaccurate eastern section, based on Ptolemy, consists of the east coast of Africa (north of Laurenço Marques), the Red Sea, and Arabia. The third section is an inaccurate map of part of the Baltic. Disregarding the unscientific parts of the map, and considering only the parts apparently related to the ancient portolan tradition, the errors are:

---

* Ptolemaic part of the map.

		Latitude	Longitude
a. West Coast of Africa			
1.	Cape of Good Hope	1.5 S	4.5 W
2.	Gt. Paternoster Pt.	2.5 S	4.0 W
3.	Walvis Bay, Pelican Pt.	0.5 N	5.0 W
4.	Congo River	7.0 S	1.8 E
5.	Cap Lopez	4.0 S	1.0 E
6.	São Tomé	4.0 S	2.0 E
7.	Niger Delta	3.0 S	2.0 E
8.	Cape Three Points	1.5 S	3.0 W
9.	Cap Palmas	0.5 S	1.0 W
10.	Freetown	2.0 N	4.0 W
11.	Dakar	2.0 N	3.0 W
12.	Cap Blanc	3.0 N	3.0 W
13.	Cap Yubi	0.0	2.5 W
b. Atlantic Coast of Europe			
14.	Gibraltar	1.5 S	0.0
26.	Cap St. Vincent	1.0 S	1.0 E
27.	Cap Finisterre	1.0 S	0.0
28.	Brest	1.0 S	0.0
29.	Cape Clear	2.5 S	2.0 E
30.	Londonderry	0.5 S	1.0 E
c. Mediterranean and Black Seas, west to east			
14.	Gibraltar	1.5 S	0.0
15.	Cap Bon	3.0 S	0.0
16.	Bengazi	4.0 S	0.5 W
17.	Alexandria	4.5 S	0.0 (assumption)
18.	Cyprus	3.0 S	0.0
19.	Crete	3.0 S	0.0
20.	Lesbos	3.0 S	0.5 E
21.	Bosphorus	2.0 S	0.5 E
22.	Sevastopol	3.5 S	0.5 E
23.	Batum	4.0 S	1.0 E
24.	Sicily	2.5 S	0.0
25.	Sardinia	3.0 S	0.0

Table 8: The Venetian Map of 1484

Shown are the positions of identified geographical points, as found on a grid worked out empirically. The length of the degree of latitude was found between Villa Cisneros, at 24° North, and the equator as found by comparison with the geography. The length of the degree of longitude was found from the longitude distance between Dakar and Cape Lopez. These degrees turned out to be practically the same. For the grid, longitude was set with reference to the longitude of Dakar, and latitude with reference to the equator as found. The map was reoriented to True North from its former magnetic orientation.

Localities	True Position	Venetian Map	Errors
1. Villa Cisneros	24.0 N 16.0 W	24.0 N 16.0 W	by assumption 0.0
2. Cape Blanc	21.0 N 17.0 W	21.5 N 17.0 W	0.5 N 0.0
3. Cape Minik	19.3 N 16.0 W	20.0 N 16.0 W	0.7 N 0.0
4. Senegal River	16.0 N 17.0 W	16.0 N 17.0 W	0.0 0.0
5. Dakar	14.0 N 17.5 W	15.5 N 17.5 W	1.5 N by assumption
6. Gambia River	12.5 N 17.0 W	14.5 N 17.0 W	2.0 N 0.0
7. São Nicolau, Cape Verde Islands	16.5 N 24.0 W	18.5 N 24.0 W	2.0 N 0.0
8. São Tiago, Cape Verde Islands	16.0 N 24.0 W	17.0 N 23.0 W	1.0 N 1.0 E
9. Cape Roxe	12.5 N 17.0 W	13.0 N 17.0 W	0.5 N 0.0
10. Bijagos Islands	11.0 N 16.0 W	12.0 N 16.0 W	1.0 N 0.0
11. Freetown	8.5 N 13.0 W	8.5 N 13.5 W	0.0 0.5 W
12. Cape Palmas	4.5 N 7.5 W	5.0 N 7.5 W	0.5 N 0.0
13. Cape Three Points	5.0 N 2.0 W	5.0 N 2.5 W	0.0 0.5 W
14. Volta River Estuary	6.0 N 0.5 E	6.0 N 0.5 E	0.0 0.0
15. Lagos	6.5 N 3.5 E	7.0 N 4.0 E	0.5 N 0.5 E
16. Niger River Delta	4.5 N 6–7 E	5.0 N 6–7 E	0.5 N 0.0
17. Fernando Po Island	3.5 N 9.0 E	3.5 N 9.5 E	0.0 0.5 E

Localities	True Position	Venetian Map	Errors
18. São Tomé	0.0	0.0	by assumption
	7.0 E	6.0 E	1.0 W

19. A non-existent island, possibly a duplicate of São Tomé.

Localities	True Position	Venetian Map	Errors
20. Cape Lopez	1.0 S	1.0 S	0.0
	9.0 E	9.0 E	by assumption
21. Congo Estuary	6.0 S	9.0 S	3.0 S
	12.0 E	12.0 E	0.0
22. Benguela	12.5 S	16.0 S	3.5 S
	13.0 E	11.0 E	2.0 W

Analysis of Errors in Venetian Map

This table suggests that the mapmaker achieved astonishing accuracy in relative latitudes and longitudes for most of the coast covered by the map. If the anomalous southern extension of the coast be disregarded, the only serious distortion is in the latitudes of Dakar, the Gambia River, and the Cape Verde Islands.

Localities	Latitude	Longitude
1. Villa Cisneros	——*	0.0
2. C. Blanc	0.5 N	0.0
3. Cape Minik	0.7 N	0.0
4. Senegal River	0.0	0.0
5. Dakar	1.5 N	——*
6. Gambia River	2.0 N	0.0
7. São Nicolau	2.0 N	0.0
8. São Tiago	1.0 N	1.0 E
9. Cape Roxe	0.5 N	0.0
10. Bijagos Isl.	1.0 N	0.0
11. Freetown	0.0	0.5 W
12. Cape Palmas	0.5 N	0.0
13. Cape Three Points	0.0	0.5 W
14. Volta River	0.0	0.0
15. Lagos	0.5 N	0.0
16. Niger River	0.5 N	0.0
17. Fernando Po I.	0.0	0.5 E
18. São Tomé	——*	1.0 W
21. Congo Estuary	3.0 S†	0.0 E†
22. Benguela	3.5 S†	2.0 W†

* By assumption.
† Anomalous. Recent additions to map?

Table 9: Reinel Map of the Indian Ocean

Localities	True Location	Portuguese Map	Errors
1. Rocks of St. Paul	0.0 29.0 W	0.0 33.0 W	0.0 4.0 W
2. Ascension Island	8.0 S 14.5 W	16.0 S 15.0 W	8.0 S 0.5 W
3. São Tomé	0.0 0.0	0.0 7.0 W	0.0 7.0 W
4. Tristan d'Achuna	37.0 S 12.5 W	40.0 S 8–13 W	3.0 S 0.0
5. The Congo River	6.0 S 12.0 E	6.0 S 8.0 E	0.0 4.0 W
6. Cape Town	34.5 S 18.5 E	35.0 S 15.0 E	0.5 S 3.5 W
7. Cape St. Marie, Madagascar	25.5 S 45.0 E	26.0 S 45.0 E	0.5 S 0.0
8. Dar es Salaam (Tanganyika)	7.0 S 39.5 E	9.0 S 40.0 E	2.0 S 0.5 E
9. Zanzibar			
10. Comore Islands	11–13 S 43–45 E	10–16 S 41–46 E	0.0 0.0
11. Cape St. André (Madagascar)	16.0 S 44.5 E	17.0 S 44.5 E	1.0 S 0.0
12. Cape d'Ambre	12.0 S 49.5 E	12.0 S 53.0 E	0.0 3.5 E
13. Cape Guardafui (Ras Assir)	12.0 N 51.0 E	12.0 N 57.0 E	0.0 6.0 E
14. Seychelles and Amirante Islands	4–7 S 53–56 E	3–5 S 55–59 E	0.0 2.0 E
15. Al Hadd (Arabia)	27.5 N 60.0 E	22.0 N 70.0 E	5.5 S 10.0 E
16. Reunion Island	21.0 S 55.5 E	21–25 S 72–74 E	0.0 17.0 E
17. Mauritius	20.0 S 57.0 E	23–24 S 78.0 E	3.5 S 21.0 E

Localities	True Location	Portuguese Map	Errors
18. Mouth of Indus R. (no delta shown)	24.0 N 68.0 E	23.0 N 81.0 E	1.0 S 13.0 E
19. Laccadive Islands	10–13 N 72–74 E	8–13 N 78–82 E	0.0 6.0 E
20. Maldive Islands	1–6 N 73–74 E	4 N–5 S 80–85 E	0.0 10.0 E
21. Ceylon	6–8 N 80–82 E	7–10 N 89–90 E	2.0 N 9.0 E
23. Cape Leeuwin, Australia	34.5 S 115.0 E	17.0 S 104.0 E	17.5 N 11.0 W
24. Northwest Cape, Australia	22.0 S 114.0 E	1.0 S 106.0 E	21.0 N 8.0 W
25. Caroline Islands (Central: Ulul, Truk)	7–10 N 147–150 E	5 N–5 S 133–140 E	c. 8.0 N c. 10.0 W

Table 10a: The Chinese Map of 1137 A.D.

Locality	True Position	1137 Map	Errors
a. THE NORTHWEST QUADRANT			
18. Junction of the Tatung and Sining Rivers	36.4 N 103.0 E	37.0 N 103.3 E	0.6 N 0.3 E
15. Bend of the R. Hwang near Ningsia	38.5 N 107.2 E	37.0 N 107.5 E	1.5 S 0.3 E
16. Junction of R. Hwang and R. Tsingshui	38.0 N 106.0 E	38.0 N 105.5 E	0.0 0.5 W
17. Junction of R. Hwang and R. Fen	35.5 N 110.5 E	35.3 N 110.5 E	0.2 S 0.0
19. Eastward turn of R. Hwang at Tali	34.5 N 110.0 E	34.5 N 110.5 E	0.0 0.5 E
20. Junction of R. King and R. Wei (Siking Province)	34.4 N 109.0 E	35.0 N 109.0 E	0.6 N 0.0
28. Bend in the R. Tao	34.5 N 104.0 E	35.0 N 103.0 E	0.5 N 1.0 W
47. Junction of the R. Hwang and the R. Tao	36.0 N 103.0 E	36.0 N 103.0 E	0.0 0.0

Locality	True Position	1137 Map	Errors
b. THE NORTHEAST QUADRANT			
1. Penglai	37.7 N	37.7 N	0.0
	120.6 E	119.5 E	1.1 W
2. Chenshan Tow (tip of Shantung Peninsula)	37.4 N	37.7 N	0.3 N
	122.5 E	122.0 E	0.5 W
11. Lake (Tunga) on former course of the R. Hwang	36.0 N	35–36 N	0.0
	116.0 E	116.0 E	0.0
13. Southward turn of the Hwang at Tokoto (Suiyuan P.)	40.0 N	40.0 N	0.0
	111.2 E	111.6 E	0.4 E
22. Junction of R. Hwang and R. Chin	35.0 N	35.2 N	0.2 N
	113.3 E	113.3 E	0.0
26. Source of the R. Tzeya	37.0 N	36.6 N	0.4 S
	113.2 E	113.3 E	0.1 E
29. Island in Lake Tai Hu	30.7 N	31.0 N	0.3 N
	120.5 E	120.3 E	0.2 W
42. Mouth of the R. Yangtze	31.5 N	32.0 N	0.5 N
	122.0 E	121.0 E	1.0 W
43. Lake Hungtze Hu	33.2 N	32.0 N	1.2 S
	118.5 E	118.0 E	0.5 W
44. Taku, former mouth of the R. Hwang (1852–1938)	37.7 N	37.8 N	0.1 N
	118.8 E	117.5 E	1.3 W
c. SOUTHWEST QUADRANT			
8. Junction of R. Kwei and R. Yu at Kwei Ping	23.5 N	23.3 N	0.2 S
	110.0 E	113.0 E	3.0 E
21. Chungking, at junction of R. Yangtze and R. (Fow)	29.3 N	29.3 N	0.0
	106.0 E	106.5 E	0.5 E
23. Junction of the R. Yangtze and the R. Wu (Kweichow Province)	30.0 N	29.5 N	0.5 S
	107.5 E	106.5 E	1.0 W
24. Westward bend of the R. Wu	28.0 N	26.7 N	1.3 S
	106.2 E	107.5 E	1.3 E
27. Junction of the R. Changti and the R. Chu	27.9 N	27.1 N	0.8 S
	110.1 E	111.9 E	1.8 E
35. Junction of the R. Yangtze and R. Min near Ipin	29.7 N	27.5 N	2.2 S
	104.5 E	104.5 E	0.0
36. Chengtu on the R. Min	30.5 N	28.5 N	2.0 S
	104.0 E	104.5 E	0.5 E
25. Junction of the R. Yangtze and the R. Yalung	26.5 N	27.7 N	1.2 N
	101.7 E	103.2 E	1.5 E
50. Source of R. Tzu	26.5 N	25.3 N	1.2 S
	110.5 E	112.2 E	1.7 E

Locality	True Position	1137 Map	Errors

d. SOUTHEAST QUADRANT

Locality	True Position	1137 Map	Errors
9. Lake Tung Ting Hu	29.0 N	28–29 N	0.0
	112–113 E	113.5–114.5 E	1.5 E
10. Junction of R. Yangtze and	30.5 N	30.7 N	0.2 N
R. Han at Hankow	114.0 E	115.0 E	1.0 E
4. Mouth of R. Fushun	30.4 N	29.7 N	0.7 S
	121.0 E	121.0 E	0.0
6. Mouth of the R. Kwei	22.0 N	22.6 N	0.6 N
	113.0 E	115.5 E	2.5 E
39. Mouth of the R. Wu (Chekiang	23.0 N	25.7 N	2.7 N
Province)	121.0 E	122.5 E	1.5 E
46. Lake Pohang Hu	29.0 N	30.0 N	1.0 N
	116.5 E	117.5 E	1.0 E
49. Kanhsien, Kiang Prov.	25.8 N	26.7 N	0.9 N
	115.0 E	116.5 E	1.5 E

Table 10b: Northernmost and Southernmost Latitude Errors in Chinese Map of 1137 A.D.

The averages of these errors of the northernmost and southernmost positions, about 0.4° of latitude on the north (some northernly and some southernly), and a little more than one degree on the south, about equally distributed northward and southward, suggest that we have approximately the right length for the degree of latitude.

	Errors
NORTHERNMOST POSITIONS: NORTH OF 35TH PARALLEL	
18. Junction of the R. Tatung and the R. Sining	0.6 N
15. Bend of the R. Hwang near Ningsia	1.5 S
16. Junction of R. Hwang and R. Tsingshui	0.0
17. Junction of R. Hwang and R. Fen	0.2 S
47. Junction of R. Hwang and R. Tao	0.0
1. Penglai	0.0
2. Chenchan Tow	0.3 N
11. Lake Tunga	0.0
13. Southward turn of R. Hwang	0.0
22. Junction of R. Hwang and R. Chin	0.2 N
26. Source of R. Tzeya	0.4 S
44. Taku, former mouth of the Hwang	0.1 N
SOUTHERNMOST POSITIONS: SOUTH OF 28TH PARALLEL	
8. Junction of R. Kwei and R. Yu	0.2 S
25. Junction of the R. Yangtze and R. Yalung	1.0 N
50. Source of the R. Tzu	1.2 S
24. Westward bend of R. Wu	1.3 S
27. Junction of R. Yangtze and the R. Chu	0.8 S
6. Mouth of River Si	0.6 N
39. Mouth of R. Wu (Chekiang Prov.)	2.5 N
49. Kanhsien, Kiang Prov.	0.9 N

Table 10c: Longitude Errors on Easternmost and Westernmost Positions on Chinese Map of 1137 A.D.

These longitude positions suggest that we have approximately a correct length for the degree of longitude and, therefore, that the lengths of the degrees of latitude and longitude differ, according to the Mercator Projection, although we have not computed the projection mathematically.

Errors

POSITIONS IN SOUTH, 104TH MERIDIAN OR FARTHER WEST

18. Junction of R. Tatung and R. Sining	0.3 E
28. Bend in R. Tao	1.0 W
47. Junction of R. Hwang and R. Tao	0.0
35. Junction of the Yangzte and R. Min	0.0
36. Chengtu	0.5 E
25. Junction of the Yangtze and R. Yalung	1.5 E

POSITIONS ON 116TH MERIDIAN OR FARTHER EAST

1. Penglai	1.1 W
2. Chenshan Tow	0.5 W
11. Lake Tunga	0.0
29. Island in Lake Tai Hu	0.2 W
4. Mouth of R. Fushun	0.0
39. Mouth of R. Wu (Chekiang)	1.5 E
46. Lake Pohang Hu	1.0 E

Longitude errors on west:	on east:
0.3 E	1.1 W
1.0 W	0.5 W
0.0	0.0
0.0	0.2 W
0.5 E	0.0
1.5 E	1.5 E
	1.0 E

Table 11a: The Zeno Map of the North (polar projection)*

Locality	Latitude	Longitude
WEST COAST OF GREENLAND	*Errors*	
1. Cape Farewell	(Ref. Point)	(Ref. Point)
	Assumed correct for Latitude and Longitude	
2. Nanortalik	0.1 S	3.0 E
3. Julianehaab	1.0 N	1.5 W
4. Kebberline Bugt		
5. Godthaab	0.0	4.0 E
6. Kangamiut	0.0	3.0 E
7. Disco Island	1.0 S	8.0 E
8. Karrats Fjord	0.5 N	13.0 E
9. Cape Atholl	3.5 S	25.0 E

* Errors according to polar projection with straight meridians, constructed from the geography.

Locality	Latitude	Longitude
EAST COAST OF GREENLAND	*Errors*	
10. Tingmiamiut	0.0	0.5 W
11. Dannebrogs	1.5 S	1.0 W
12. Cape Dan		
13. Cape Gustav Holm	0.5 S	2.5 W
14. Gunnbiorn's Fjord	0.5 N	2.5 E
15. Cape Brewster	0.0	3.0 W
16. Ymirs, Geographical Society and Traill Islands		
17. Kong Oscar's Fjord	0.5 N	0.0
18. Hold with Hope Pen.	1.0 S	9.0 E
19. Germania Land	3.0 S	20.0 E
ICELAND		
20. Keflavik (as an island)	0.5 S	7.0 W
21. Vik	0.5 N	8.0 W
22. Seydisfjordur	0.5 N	7.0 W
23. Raufarhofn	3.0 N	4.0 W
NORWAY, DENMARK, GERMANY, SHETLANDS, FAROES, SCOTLAND		
31. Cape Lindesnes	(correct by assumption)	
32. Oslo Fjord	2.0 N	1.5 W
33. Copenhagen	0.5 S	0.5 E
34. Elbe River	2.0 S	0.0
35. Weser River	0.5 S	1.0 W
36. Shetland Islands	0.0	c. 6.0 W
37. The Faroes	0.0	c. 6.0 W
38. Scotland (north tip)	3.5 S	c. 10.0 W

Table 11b: The Zeno Map of the North (portolan projection)*

Locality	Latitude	Longitude
WEST COAST OF GREENLAND	*Errors*	
1. Cape Farewell	0.0	0.0 (assumption: base line of latitude)
2. Julianehaab	0.0	2.0 E
3. Sarmiligarssak Fiord	1.0 N	2.0 E
4. Godthaab	1.5 S	3.0 E
5. Disko Bay	2.0 S	3.0 E
6. Melville Bay	2.0 S	14.0 E
7. Cape Atholl	1.5 S	22.0 E

---

* Errors according to the 3rd Grid, on the portolan projection, with two norths, based on the meridian of Alexandria, as found by the 2nd Grid.

Locality	Latitude	Longitude

EAST COAST OF GREENLAND

8. Sermilik Fjord	3.0 N	5.0 W
9. Scoresby Sound	5.0 N	11.0 W

NORWAY, DENMARK, GERMANY, ETC.

10. North Cape?	5.5 S?	15.0 W?
11. Serja	2.5 S	9.0 W
12. Trondheim Fjord	1.0 N	14.0 W
13. Cape Lindesnes	0.0	10.0 W
14. Alborg (Denmark)	0.5 S	10.5 W
15. Elbe River (Germany)	4.0 S	16.0 W

Table 12: Andrea Benincasa Portolano of 1508

Localities	True Position	Benincasa Map	Errors
1. Cape Yubi	28.0 N 13.0 W	27.0 N 15.0 W	1.0 S 2.0 W
2. Cape Guir	30.5 N 10.0 W	30.5 N 10.0 W	0.0 0.0
3. Mojador	31.5 N 10.0 W	32.0 N 10.0 W	0.5 N 0.0
4. Mazagan (Oum er Rbia River)	33.5 N 8.5 W	34.0 N 6.5 W	0.5 N 2.0 W
5. Gibraltar	36.5 N 5.5 W	36.0 N 6.0 W	0.5 S 0.5 W
6. Guadalquivir	37.0 N 6.5 W	37.0 N 6.5 W	0.0 0.0
7. Cap St. Vincent	37.5 N 9.0 W	37.5 N 9.0 W	0.0 0.0
8. Lisbon	38.5 N 9.5 W	39.5 N 8.5 W	1.0 N 1.0 E
9. Cap Finisterre	43.0 N 9.5 W	44.5 N 9.0 W	1.5 N 0.5 E
10. Cape de Penas	43.5 N 6.0 W	45.0 N 6.0 W	1.5 N 0.0

Localities	True Position	Benincasa Map	Errors
11. Brest	48.5 N 5.0 W	51.0 N 5.0 W	2.5 N 0.0
12. Cherbourg	49.5 N 1.5 W	52.0 N 2.5 W	2.5 N 1.0 W
13. Calais	51.0 N 2.0 E	53.5 N 2.0 E	2.5 N 0.0
14. Lands End	50.0 N 5.5 W	53.0 N 5.5 W	3.0 N 0.0
15. Cape Clear, Ireland	51.5 N 9.5 W	54.0 N 9.0 W	2.5 N 0.5 E
16. Malin Head, Ireland	55.0 N 7.5 W	59.0 N 7.5 W	4.0 N 0.0
17. Firth of Forth	56.0 N 3.0 W	57.5 N 1.0 W	1.5 N 2.0 E
18. Kinnard's Head	57.5 N 2.0 W	61.0 N 1.5 W	3.5 N 0.5 E
19. The Rhine	52.0 N 4.0 E	54.0 N 2.5 E	2.0 N 1.5 W
20. The Elbe	54.0 N 9.0 E	56.0 N 2.0 E	2.0 N 7.0 W
21. Erz Gebirge (Mts.)	50–51 N 12–16 E	56–58 N 13–15 E	5.0 N 0.0
22. Swedish Coast (south coast)	c. 58.0 N 5–10 E	63.0 N 2 W–3 E	5.0 N 6.0 W
24. The Wash (England)	53.0 N 0.0	56.0 N 1.0 W	3.0 N 1.0 W

25. The Danube runs an entirely fictitious course from the Alps to the Black Sea, only the terminal points being correct; it may be a medieval addition to the source map.

26. The Alps are drawn in the medieval tradition, and probably were added by the modern geographer (as also the mountains in Farica).

THE MEDITERRANEAN AND BLACK SEA

Localities	True Position	Benincasa Map	Errors
27. River Moulouya	35.0 N 2.5 W	35.0 N 2.5 W	0.0 0.0
28. Cape Bon	37.0 N 11.0 E	37.5 N 10.0 E	0.5 N 1.0 W

Localities	True Position	Benincasa Map	Errors
29. Ile Djerba	34.0 N 11.0 E	35.0 N 11.0 E	1.0 N 0.0
30. Cape Misurata	32.5 N 15.3 E	32.0 N 15.0 E	0.5 S 0.3 W
31. Bengazi	32.0 N 20.0 E	31.5 N 19.0 E	0.5 S 1.0 W
32. Alexandria	31.0 N 30.0 E	32.0 N 31.0 E	1.0 N 1.0 E
33. Cyprus	34.5–35.5 N 32.5–34.5 E	37–38 N 32–34.5 E	2.5 N 0.0
34. Crete (southern coast)	35.0 N 23–27 E	36.0 N 23–26.5 E	1.0 N 1.0 W
35. The Bosphorus	41.0 N 29.0 E	45.0 N 28.5 E	4.0 N 0.5 W
36. Yalta	44.5 N 34.0 E	50.0 N 33.0 E	5.5 N 1.0 W
37. Batum	41.5 N 41.5 E	46.5 N 42.0 E	5.0 N 0.5 E

CENTRAL MEDITERRANEAN (NORTH COAST)

	True Position	Benincasa Map	Errors
38. Trieste	46.0 N 14.0 E	49.0 N 13.0 E	3.0 N 1.0 W
39. Marseilles	43.0 N 5.0 E	45.0 N 5.0 E	3.0 N 0.0

40. The Red Sea is shown without the Gulf of Aqaba or the Gulf of Suez. Latitude 28° N (northern end) and Longitude 35–40° E. Correct.

Table 13: The Map of Iehudi Ibn Ben Zara of Alexandria

Locality	True Position	Zara Map	Errors
1. Cape Yubi	28.0 N 13.0 W	c. 28.0 N c. 14.0 W	0.0 1.0 W
2. Cape Guir	30.5 N 10.0 W	31.5 N 10.0 W	1.0 N 0.0
3. Majador	31.5 N 10.0 W	32.5 N 9.0 W	1.0 N 1.0 E

Locality	True Position	Zara Map	Errors
4. Mazagan (Oum er Rbia River)	33.5 N 8.5 W	34.0 N 7.0 W	0.5 N 1.5 E
5. Gibraltar	36.5 N 5.5 W	35.5 N 4.5 W	1.0 S 1.0 E
6. Guadalquivir	37.0 N 6.5 W	37.0 N 5.0 W	0.0 1.5 E
7. Cape St. Vincent	37.5 N 9.0 W	37.5 N 7.5 W	0.0 1.5 E
8. Tagus River (Lisbon)	38.5 N 9.5 W	39.5 N 7.0 W	1.0 N 2.5 E
9. Oporto (Douro River)	41.0 N 8.5 W	41.5 N 6.0 W	0.5 N 2.5 E
10. Cap Finisterre	43.0 N 9.5 W	44.0 N 7.0 W	1.0 N 2.5 E
11. Cape de Penas	43.5 N 6.0 W	45.0 N 4.0 W	1.5 N 2.0 E
12. Cape Machichaco	43.5 N 3.0 W	44.0 N 1.5 W	0.5 N 1.5 E
13. Arcachon, France	44.5 N 1.0 W	45.0 N 0.0	0.5 N 1.0 E
14. Gironde River	45.5 N 1.0 W	46.0 N 0.0	0.5 N 1.0 E
15. Loire River	47.0 N 3.0 W	48.5 N 0.5 W	1.5 N 2.5 E
16. Isle d'Ouessant (off Pointe de St. Mathieu) (Brest)	48.5 N 5.0 W	50.0 N 3.0 W	1.5 N 2.0 E
17. Cherbourg	49.5 N 1.5 W	51.5 N 0.0	2.0 N 1.5 E
18. Calais	51.0 N 2.0 E	52.0 N 2.5 E	1.0 N 0.5 E
19. Isle of Wight, England	50.5 N 1.0 W	52.5 N 0.5 E	2.0 N 1.5 E
20. Lands End	50.0 N 5.5 W	52.5 N 2.5 W	2.5 N 3.0 E
21. Scilly Islands			

Locality	True Position	Zara Map	Errors
22. St. Bride's Bay	52.0 N 5.0 W	54.5 N 2.0 W	2.5 N 3.0 E
23. Cape Clear, Ireland	51.5 N 9.5 W	53.0 N 7.0 W	1.5 N 2.5 E
24. Carnsore Point	52.3 N 6.5 W	54.0 N 4.0 W	1.7 N 2.5 E
25. The Rhine	52.0 N 4.0 E	52.5 N 4.0 E	0.5 N 0.0
26. The Elbe River	54.0 N 9.0 E	55.0 N 5.0 E	1.0 N 4.0 W
27. Erz Gebirge (Mts.)	50–51 N 12–16 E	53–56 N 13–16 E	c. 2.0 N 0.0
28. The Alps*	44–48 N 6–16 E	46–50 N 6–16 E	2.0 N 0.0

THE MEDITERRANEAN

Locality	True Position	Zara Map	Errors
29. Cape Tres Forcas, Morocco	35.5 N 3.0 W	35.0 N 2.0 W	0.5 S 1.0 E
30. River Moulouya			
31. Cape Bon	37.0 N 11.0 E	36.0 N 10.5 E	1.0 S 0.5 W
32. Iles Kerkennah			
33. Ile Djerba	34.0 N 11.0 E	33.0 N 10.5 E	1.0 S 0.5 W
34. Cape Misurata	32.5 N 15.3 E	31.0 N 15.0 E	1.5 S 0.3 W
35. Bengazi	32.0 N 20.0 E	30.0 N 19.0 E	2.0 S 1.0 W
36. Ras et Tin	32.5 N 23.0 E	31.0 N 22.5 E	1.5 S 0.5 W
37. Alexandria	31.0 N 30.0 E	29.0 N correct by assumption	2.0 S
38. Cyprus	34.5–35.5 N 32.5–34.5 E	33–34 N 32–34 E	1.5 S 0.0

Locality	True Position	Zara Map	Errors
39. The Bosphorus	41.0 N 29.0 E	41.0 N 28.5 E	0.0 0.5 W
40. Yalta	44.5 N 34.0 E	46.0 N 34.5 E	1.5 N 0.5 E
41. Dolzhanskaya (Sea of Azov)	46.0 N 38.0 E	48.0 N 38.0 E	2.0 N 0.0

THE CENTRAL MEDITERRANEAN

Locality	True Position	Zara Map	Errors
42. Kythera (Cythera) Island (s. of Peloponnesus)	36.3 N 23.0 E	35.0 N 22.5 E	1.3 S 0.5 W
43. Trieste	46.0 N 14.0 E	47.5 N 15.0 E	1.5 N 1.0 E
44. Genoa	44.3 N 9.0 E	44.5 N 9.0 E	0.2 N 0.0
45. Sicily, no thern coast	38.0 N 12.5–15.5 E	37.0 N 12.5–15 E	1.0 S 0.0
46. The Rhone River	43.5 N 5.0 E	44.5 N 5.0 E	1.0 N 0.0
47. Mallorca	39.5–40 N 3.0 E	39.5 N 3.0 E	0.0 0.0

Table 14: The Map of Ibn Ben Zara (Spanish section)

Locality	True Position	Zara Map	Errors
1. Tarifa	36.0 N 5.5 W	35.8 N 6.4 W	0.2 S 0.9 W
2. La Línea	36.2 N 5.3 W	35.8 N 6.0 W	0.4 S 0.7 W
3. Málaga	36.7 N 4.3 W	36.2 N 5.0 W	0.5 S 0.7 W
4. Motril	36.7 N 3.5 W	36.1 N 3.7 W	0.6 S 0.3 E
5. Almería	36.8 N 2.5 W	36.3 N 2.2 W	0.5 S 0.3 E

Locality	True Position	Zara Map	Errors
6. Cape of Gata	36.7 N 2.2 W	36.1 N 2.0 W	0.6 S 0.2 E
7. Cape of Palos	37.7 N 0.7 W	36.9 N 0.0	0.8 S 0.7 E
8. Alicante	38.2 N 0.5 W	37.5 N 0.3 E	0.7 S 0.8 E
9. Cape of Nao?	38.7 N 0.2 E	38.2 N 1.2 E	0.5 S 1.0 E
10. Valencia	39.5 N 0.2 W	39.0 N 0.8 E	0.5 S 1.0 E
12. Cape of Tortosa	40.7 N 0.2 E	40.3 N 2.3 E	0.4 S 2.1 E
13. Cape Trafalgar	36.2 N 6.0 W	36.0 N 6.7 W	0.2 S 0.7 W
14. Cádiz	36.5 N 6.2 W	36.2 N 6.7 W	0.3 S 0.5 W
15. Guadalquiver Delta	36.7 N 6.2 W	36.7 N 6.7 W	0.0 0.5 W
16. Huelva	37.2 N 7.0 W	37.2 N 7.5 W	0.0 0.5 W
17. Faro	37.0 N 8.0 W	36.8 N 8.7 W	0.2 S 0.7 W
18. Cape of São Vicente	37.0 N 9.0 W	37.0 N 9.7 W	0.0 0.7 W
19. Cape Sines	38.0 N 8.7 W	38.0 N 9.0 W	0.0 0.3 W
20. Cape Espichel	38.5 N 9.2 W	38.6 N 9.2 W	0.1 N 0.0
21. Lisbon	38.7 N 9.2 W	38.8 N 9.2 W	0.1 N 0.0
22. Cape Carvoeiro	39.5 N 9.5 W	39.7 N 9.3 W	0.2 N 0.2 E
23. Farilhões Is.	39.5 N 9.7 W	39.7 N 9.6 W	0.2 N 0.1 E
24. Cape Mondego	40.2 N 9.0 W	40.2 N 8.9 W	0.0 0.1 E

Locality	True Position	Zara Map	Errors
25. Oporto	41.2 N 8.7 W	41.0 N 8.2 W	0.2 S 0.5 E
26. Vigo	42.2 N 8.7 W	42.0 N 8.6 W	0.2 S 0.1 E
27. Cap Finisterre	42.7 N 9.2 W	43.0 N 8.9 W	0.3 N 0.3 E
28. La Coruña	43.5 N 8.3 W	43.2 N 7.8 W	0.3 S 0.5 E
29. Cape Ortegal	43.7 N 7.7 W	43.7 N 7.4 W	0.0 0.3 E
30. Cape of Peñas	43.7 N 6.0 W	43.5 N 5.3 W	0.2 S 0.7 E
31. Santander	43.5 N 3.7 W	42.9 N 3.7 W	0.6 S 0.0
32. Cape Machichaco	43.5 N 2.7 W	43.0 N 2.3 W	0.5 S 0.4 E
33. San Sebastian	43.2 N 2.7 W	42.8 N 1.3 W	0.4 S 1.4 E
34. Biarritz	43.5 N 1.5 W	42.9 N 1.0 W	0.6 S 0.5 E

Analysis of Errors in the Ben Zara Map of Spain

This suggests that a slight error in the length of the degree of longitude may have produced errors in the longitudes of places across Spain amounting to about 20 miles, while a smaller error in the length of the degree of latitude may have accounted for latitude errors averaging about six miles.

	Latitude	Longitude
(a) East Coast	0.5 S*	0.2 E
(b) West Coast	0.0	0.1 E
(c) Northern Coast	0.4 S	0.4 E
(d) Southern Coast	0.3 S†	0.0

Error in Long. Distance, East and West coasts: 0.3°
Error in Lat. Distance, North and South coasts: 0.1°

---

* Two localities anomalous, not averaged.
† One locality anomalous, not averaged.

Table 15: Alternative Grid for the East Coast of South America (Piri Re'is Map)

Locality	True Position	Alternative Grid	Errors
1. The Amazon (Para River)	00.0 48.0 W	00.0 48.0 W	Correct by Assumption
2. Bahia de São Marcos	2.0 S 44.0 W	3.0 S 44.0 W	1.0 S 0.0
3. Parnaiba	3.0 S 42.0 W	3.0 S 4.0 W	0.0 1.0 E
4. Fortaleza	3.5 S 38.5 W	3.0 S 37.5 S	0.5 N 1.0 E
5. C. de São Roque	5.0 S 35.5 W	2.5 S 36.0 W	2.5 N 0.5 W
6. Recife	8.0 S 35.0 W	4.5 S 34.5 W	3.5 N 0.5 E
7. Rio São Francisco	10.5 S 36.5 W	7.5 S 36.5 W	3.0 N 0.0
8. Salvador	13.0 S 38.5 W	10.0 S 38.0 W	3.0 N 0.5 E
9–11. Ponta de Baleia	17.6 S 39.0 W	16.0 S 39.0 W	1.6 N 0.0
12. C. de São Tomé (and Rio Paraiba)	22.0 S 41.0 W	19.0 S 40.0 W	3.0 N 1.0 E
13. C. Frio	23.0 S 42.0 W	22.0 S 41.0 W	1.0 N 1.0 E
14a. Rio de Janeiro	23.0 S 43.0 W	23.0 S 44.0 W	Correct by Assumption 1.0 W
14b. Bahia de Ilha Grande	23.0 S 44.5 W	24.0 S 45.0 W	1.0 S 0.5 W

Bibliography

This list includes not only works referred to in the text, but a selection of the relevant literature (with no attempt at comprehensiveness). It is hoped that it may be helpful at least as a starting point for those wishing to pursue further studies.

1. "A Columbus Controversy" (Piri Re'is Map), *Illustrated London News*, February 27, 1932.

2. Afet Inan, Dr. *The Oldest Map of America, Drawn by Piri Re'is.* Translated by Dr. Leman Yolac. Ankara: The National Library, 1954.

3. *Afrique* [Physical map of Africa]. (Sheets 1 and 3: Northwest Africa and West Africa to the Gold Coast.) Paris: Institut Géographique Nationale, 1940–1941.

4. Ainalov, D. V. *The Hellenistic Origins of Byzantine Art.* Translated from the Russian by E. and S. Sobolovitch. New Brunswick, N.J.: Rutgers University Press, 1961.

5. Akçura, Yusuf. *Piri Re'is Haritasi hakkinda izahname.* "Turk Tarihi Arastirma Kurumu yayinlarindan," No. 1. Istanbul, 1935.

6. ———. "Turkish Interest in America in 1513: Piri Re'is' Chart of the Atlantic," *Illustrated London News,* July 23, 1932.

7. Amalgia, Roberto, ed. See *Monumenta Cartographica Vaticana.*

8. *American Encyclopedia of History,* ed. Ridpath, J. C., et al. Vol. I. Philadelphia: Encyclopedia Publishing Company, 1919.

9. *American Oxford Atlas.* New York: Oxford University Press, 1951.

10. Andrews, E. Wylls. "Chronology and Astronomy in the Maya Area," *The Maya and Their Neighbors.* Memorial volume to Prof. Alfred M. Tozzer. New York and London: D. Appleton Century, n.d.

11. Andrews, M. C. "The British Isles in the Nautical Charts of the 14th and 15th Centuries," *Geographical Journal,* LXVIII (1926), 474–481.

12. ———. "The Study and Classification of Medieval Mappae Mundi," *Archeologia,* LXXXV (1926), 61–76.

13. ———. "Scotland in the Portolan Charts," *Scottish Geographical Magazine,* XLII (1926), 129–153, 193–213.

14. Anthiaume, l'Abbé A. *Cartes marines, Constructions navales, Voyages de découvertes chez les Normandes, 1500–1650.* Paris: Dumont, 1916.

15. Ashe, Geoffrey. *Land to the West.* New York: The Viking Press, 1962.

16. Asimov, Isaac. *Asimov's Biographical Encyclopedia of Science and Technology.* Garden City, N.Y.: Doubleday, 1964.

17. *Atlas of Mountain Glaciers of the Northern Hemisphere.* (Technical Report EP-92.) Headquarters, Quartermaster Research and Engineering Command, U.S. Army, Natick, Mass., n.d.

18. *Atlas Över Sverige.* (Utgiven av Svenska Sallskapet For Anthropologi och Geografi.) Loose sheets.

19. d'Avezac-Macaya, Marie Armand Pascal. *Note sur un Mappemonde Turke du XVIᵉ Siècle, conservée à la bibliothèque de Saint-Marc à Venise.* Paris: Imprimerie de E. Martinet, 1866.

20. Ayusawa, Shintaro. "The Types of World Maps Made in Japan's Age of National Isolation," *Imago Mundi*, X (1953), 125–128.

21. Babcock, William H. *Legendary Islands of the Atlantic.* ("American Geographical Society Research Series," No. 8.) New York: American Geographical Society, 1922.

22. Bagrow, Leo. *The History of Cartography*, ed. R. A. Skelton. Cambridge, Mass.: The Harvard University Press, 1964.

23. ———. "A Tale from the Bosphorus. Some Impressions of My Work at the Topkapu Saray Library, Summer of 1954," *Imago Mundi*, XII (1955).

24. Barnett, Lincoln. *Treasure of Our Tongue.* New York: Knopf, 1964.

25. Beazley, Charles Raymond. *The Dawn of Modern Geography.* 3 vols. London: J. Murray, 1897–1906.

26. ———. *Prince Henry the Navigator.* New York and London, 1895.

27. Belov, M. "Mistake or Intention?" (translated for E. A. Kendall), *Priroda*, No. 1 (November, 1960). (Moscow: Soviet Academy of Sciences.)

28. ———. "A Windfall Among Manuscripts" (translated for E. A. Kendall), *Vednyi Transport* (August 17, 1961).

29. Bernardini-Sjoestedt, Armand. *Christophe Colomb.* Paris: Les Sept Couleurs, 1961.

30. Bibby, Geoffrey. *The Testimony of the Spade.* New York: Knopf; London: Collins, 1956.

31. Bigelow, John. "The So-Called Bartholomew Columbus Map of 1506," *Geographical Review*, October, 1935, pp. 643–656.

32. Blacket, W. S. *Lost Histories of America* [etc.]. London: Trübner and Co., 1883. Now obtainable from University Microfilms, Ann Arbor, Mich. (Xerox).

33. Blundeville [Blundivile], M. *Blundeville His Exercises . . . A New and Necessarie Treatise of Navigation . . .* London, 1594.

34. Boland, Charles Michael. *They All Discovered America.* New York: Doubleday, 1961; Permabook Edition, 1963.

35. Boyd, William C. *Genetics and the Races of Man.* Boston: Little, Brown, 1952.

36. Braunlich, Erick. *Zwei turkischen Weltkarten aus dem Zeitalter der grossen Entdeckungen.* Leipsic: S. Hirzel, 1937.

37. Brooks, C. E. P. *Climate through the Ages.* New York: McGraw-Hill, 1949.

38. Brosses, C. de. *Histoire des Navigations aus Terres Australes.* 2 vols. Paris: Durand, 1756.

39. Brown, Lloyd Arnold. *The Story of Maps.* Boston: Little, Brown, 1950.

40. Bunski, Hans-Albrecht von. "Kemal Re'is, ein Betrag zur Geschichte der turkischen Flotte." Bonn, 1928. (Ph.D. thesis.)

41. Calahan, Harold Augustin. *The Sky and the Sailor.* New York: Harper, 1952.

42. Callis, Helmut G. *China, Confucian and Communist.* New York: Henry Holt, 1959.

43. Caras, Roger A. *Antarctica—Land of Frozen Time.* Philadelphia: Chilton Books, 1962.

44. Carli, Carlo Giovanni Dinaede. *Lettres Americaines.* (French translation of his Italian work.) 2 vols. Boston, 1788.

45. *Cartografia de Ultramar.* Vol. I, 1949; Vol. II, 1953; Vol. III, 1955; Vol. IV, 1957. Madrid: Ejercito Servicio Geographico.

46. Cary, M., and E. H. Warmington. *The Ancient Explorers.* London: Methuen, 1929; New York: Pelican Books, 1963.

47. Chamberlin, Wellman. *Round Earth on Flat Paper.* Washington, D.C.: National Geographic Society, 1950.

48. Charlesworth, J. K. *The Quaternary Era with Special Reference to Its Glaciation.* 2 vols. London: Edward Arnold, Ltd., 1957. Vol. II.

49. Chatterton, E. Keble. *Sailing Ships and Their Story.* New ed. London: Sidgewick and Jackson, Ltd., 1923.

50. Chevallier, M. "L'Aimantation des lavas de l'Etna et l'orientation du champ terrestre en Sicile du XIIe au XVIIe Siècle," *Annales de Physique*, IV, 1925, pp. 5–162.

51. *China, Map of, Compiled and Drawn in the Cartographic Section of the National Geographic Society, for the National Geographic Magazine*. Washington, D.C.: National Geographic Society, 1945.

52. Claggett, Marshall. *Science in Antiquity*. New York: Collier Books, 1963.

53. Commission on the Bibliography of Ancient Maps, International Geophysical Union. 2 vols. Paris, 1952. Pp. 63, 93.

54. Cortesao, Armand. *The Nautical Chart of 1424 and the Early Discovery and Cartographical Representation of America*. Coimbra, Portugal: University of Coimbra, 1954.

55. Crombie, A. C., ed. *Scientific Change*. London: Heinemann, 1963.

56. Cummings, Byron S. "Cuicuilco and the Archaic Culture of Mexico," *Bulletin*, University of Arizona, IV, No. 8, Nov. 15, 1933.

57. ———. "Obituary," *Science*, CXX, No. 3115, Sept. 10, 1954, pp. 407–408.

58. Dampier, Sir William. *A History of Science* (3rd ed. rev.). New York: The Macmillan Company, 1944.

59. Davis, H. T. *Alexandria, the Golden City*. 2 vols. Evanston, Ill.: The Principia Press of Illinois.

60. Deetz, Charles H., and Oscar Adams. *Elements of Map Projection with Applications to Map and Chart Construction*. (U.S. Coast and Geodetic Survey "Special Publication," No. 68, Serial No. 146.) Washington, D.C.: Government Printing Office, 1921.

61. Deissmann, A. *Forschungen und Funde im Serai*. Berlin-Leipsic: 1933.

62. Denuce, Jean. *Les Origines de la Cartographie portugaise et les Cartes de Reinel*. Ghent: E. van Goethem, 1908.

63. Diller, Aubrey. "The Ancient Measurements of the Earth," *Isis*, XL (1948), pp. 6–9.

64. *Directory of Special Libraries and Research Centers*, ed. Anthony T. Kruzas. Detroit, Mich.: Gale Research Company, 1963.

65. Durand, Dana B. "The Origins of German Cartography in the 15th Century." 3 vols. and portfolio. (Thesis, Harvard University.)

66. ———. *The Vienna-Klosterneuburg Map Corpus of the 15th Century*. Leiden: E. J. Brill, 1952.

67. Eames, Wilberforce. *A List of the Editions of Ptolemy's Geography, 1475–1730*. New York, 1886.

68. *Encyclopaedia Britannica* (11th ed.). VII, p. 869.

69. *Encyclopedia of Islam*, ed. M. Th. Houtsma, et al. Leiden: E. J. Brill, 1936.

70. *Expéditions Polaires Françaises, Missions Paul-Emile Victor*. "Campagne au Groenland, 1950, Rapports Préliminaires." "Campagne au Groenland, 1951, Rapports Préliminaires." Paris, 1953.

71. *Facsimiles of Portolan Charts*. Publication No. 114 of the Hispanic Society of America (Introduction by Edward Luther Stevenson). New York: The Hispanic Society, 1916.

72. Fairbridge, Rhodes E. "Dating the Latest Movements of the Quaternary Sea Level." *Transactions*, New York Acad. Sci., Ser. II, XX, No. 6, pp. 471–482, April, 1958.

73. Finé, Oronce (Oronteus Finaeus). *Sphaera mundi . . . Lvtetiae Parisorum apud Michaëlem Vascasanum*, 1551.

74. Fischer, Teobald. (Atlas-Series.) Venice: Ferd. Ongania, editore, 1881–1886.

75. ———. *Sammlung Mittelalterlicher Welt und Seekarten italienischen Ursprungs und aus italienischen Bibliotheken und Archiven*. Venice: F. Ongania, 1886.

76. Gallois, Lucien. *De Orontio Finaeo Gallico Geographo*. Paris: E. Lerous, 1890.

77. Glanville, S. R. K., ed. *The Legacy of Egypt.* Oxford: The Clarendon Press, 1957 (reprint).

78. Goodwin, William A. Letter to James A. Robertson, May 26, 1933 (analysis of the Piri Re'is Map). (Unpublished: Library of the American Geographical Society.)

79. Gordon, Cyrus H. "The Decipherment of Minoan," *Natural History,* LXXII, No. 9, November, 1963.

80. Gruneisen, W. de. *Les caractéristiques de l'art Copte.* Florence, Italy: Instituto di Edizione Artistiche Fratelli Alinari, 1922.

81. Guest, Edwin. *Origines Celticae and other contributions to the History of Britain.* 2 vols. London: The MacMillan Co., 1883. (Caesar's destruction of the Druid library.)

82. *A Guide to Historical Cartography.* Washington: Map Division, Library of Congress, 1954. Revised, 1960.

83. Hakki, Ibrahim. *Topkapi Sarayinde deri uzerine yapilmis eski Haritalar.* Istanbul, 1936.

84. Hamy, Jules Theodore Ernest. *La Mappemonde d'Angelino Dulcert de Majorque, 1339* (2nd ed.). Paris: H. Champion, 1903.

85. Hapgood, Charles H. *Earth's Shifting Crust.* New York: Pantheon, 1958.

86. Hawkes, Jacquetta. *The World of the Past.* New York: Knopf, 1963.

87. Hawkins, Gerald S. "Stonehenge: A Neolithic Computer," *Nature,* CCII, No. 4939 (June 27, 1964), 1258–1261. Reprinted by the Smithsonian Astrophysical Observatory, Cambridge, Mass.

88. ———. "Stonehenge Decoded," *Nature,* CC, No. 4909 (October 26, 1963), 306–308. Reprinted by the Smithsonian Astrophysical Observatory, Cambridge, Mass.

89. Heathecote, N. H. de V. "Early Nautical Charts," *Annals of Science,* I (1963), 13–28.

90. ———. "Christopher Columbus and the Discovery of Magnetic Variation," *Science Progress,* XXVII (1932), 82–103.

91. Heidel, William. *The Frame of the Ancient Greek Maps.* American Geographical Society Research Series No. 20. New York: The Society, 1937.

92. Heizer, R. F., and J. A. Bennyhoff. "Archaeological Investigation of Cuicuilco, Valley of Mexico, 1957," *Science,* v. 127, No. 3292, pp. 232–233.

93. Hobbs, William H. "The Fourteenth Century Discoveries by Antonio Zeno," *The Scientific Monthly,* XXII, Jan., 1951, pp. 24–31.

94. ———. "Zeno and the Cartography of Greenland," *Imago Mundi,* VI.

95. Holtzscherer, J. J., and G. de Q. Robin. "Depth of Polar Ice Caps," *Geographical Journal,* CXX, Part 2, June, 1954, pp. 193–202.

96. Hough, Jack. "Pleistocene Lithology of Antarctic Ocean Bottom Sediments," *Journal of Geology,* LVIII, 257–259.

97. Humboldt, Alexander von. *Cosmos: A Sketch of a Physical Description of the Universe* (trans. E. C. Otte). New York: Harpers, 1852.

98. *Idrisi, World Map.* (1150 A.D. in Arabic.) Map Division, Library of Congress, 1951 (six sheets).

99. Irwin, Constance. *Fair Gods and Stone Faces: Ancient Seafarers and the World's Most Intriguing Riddle.* New York: St. Martin's Press, 1963.

100. Jeans, Sir James. *The Growth of Physical Science.* A Premier Book: Fawcett World Library, 1958.

101. Jervis, W. W. *The World in Maps.* London: George Philip and Son, 1936.

102. Johnston, Thomas Crawford. *Did the Phoenicians Discover America? Embracing the origin of the Aztecs, with some further light on Phoenician civilization and colonization. The Story of the Mariner's Compass.* San Francisco, 1892. 2nd ed. London: James Nisbet, 1913.

103. Jomard, Edme François. *Les Monuments de la Géographie.* Paris: Du Prat, 1862.

104. Kahle, Paul. "La Carte Mondiale de Piri Re'is." *Actes du XVIII Congress international des orientalistes.* Leiden, 1932.

105. ———. "Die Verschollene Columbus-Karte von Amerika von Jahre 1498 in einer turkischen Weltkarten von 1513," *Forschungen und Fortschritte,* 8 Jahrg. No. 19 (1 Juli 1932), pp. 248–249.

106. ———. "The Lost Map of Columbus," *The Geographical Review,* XXIII, No. 4 (1933).

107. Kamal, Prince Youssouf. *Hallucinations Scientifiques (Les Portulans).* Leiden: E. J. Brill, 1937.

108. ———. *Monumenta Cartographica Africae et Aegypti.* 16 vols. in 5. Cairo, 1926–1951. (Vol. IV: Epoque des Portulans.)

109. Ketman, Georges. "The Disturbing Maps of Piri Re'is," *Science et Vie* (Paris), Sept., 1960.

110. Keuning, Johannes. "The History of Geographical Map Projection until 1600," *Imago Mundi,* XII (1955).

111. Kimble, George Herbert Tinsley. *Geography in the Middle Ages.* London: Methuen, 1938.

112. Kitson, Arthur. *Capt. James Cook, the Circumnavigator.* London: L. Murray, 1907.

113. Koeman, Cornelis. *Collections of Maps and Atlases in the Netherlands.* Leiden: E. J. Brill, 1961.

114. Konjnenburg, E. van. *Shipbuilding from Its Beginnings.* 3 vols. Brussels: Permanent International Association of Congresses of Navigation, Executive Committee, 1913.

115. Krachkovsky, J. "The Columbus Map of America, Worked in Turkish," *Isvestia* (Moscow), VGO, XVI (1954), pp. 184–186.

116. Kretschmer, Konrad. *Die italienischen Portolane des Mittelalters; ein Beitrag zur Geschichte der Kartographie und Nautik.* Berlin: E. S. Mittler, 1909.

117. ———. "Die verschollene Kolumbuskarte von 1498 in einer turkischen Weltkarte von 1513" (A. Petermans), *Mitteilungen,* LXXX (1934).

119. La Roncière, Charles de. *La Découverte de l'Afrique au Moyen Age: Cartographie et Explorateurs.* 3 vols. (in Memoires of the Royal Society of the Geography of Egypt, Vols. V, VI, VII).

120. Leithauser, Joachim G. von. *Mappae Mundi: die geistige Eroberung der Welt.* Berlin: Safari-Verlag, 1958.

121. Lelewel, Joachim. *Géographie du Moyen Age accompagné d'Atlas et de cartes dans chaque Volume.* Brussels: V. et J. Pilliet, 1850–52.

122. ———. *Pytheas de Marseille et la géographie de son temps.* Paris: Straszéwicz, 1836.

123. Lewis, W. N. *The Splendid Century. Life in the France of Louis XIV.* Garden City, N.Y.: Doubleday Anchor Books, 1957.

124. Libby, Willard F. *Radiocarbon Dating.* Chicago: The Chicago University Press, 1952.

125. Lucas, Fred W. *Annals of the Brothers Niccolo and Antonio Zeno.* London: H. Stevens, Son and Stiles, 1898.

126. Lynam, Edward. *The Map-Maker's Art.* London: The Batchworth Press, 1953.

127. McElroy, John W. "The Ocean Navigation of Columbus on His First Voyage," *The American Neptune,* Vol. I, No. 3 (July, 1941).

128. Maggiolo, Visconte de. *Atlas of Portolan Charts.* New York: The Hispanic Society of America, 1911.

129. Major, R. H. *Voyages of the Zeno Brothers.* London: The Hakluyt Society, 1873; Boston: The Massachusetts Historical Society, 1875.

130. Mallery, Arlington H. *Lost America.* Washington, D.C.: The Overlook Printing Co., 1951.

131. Mallery, Arlington H., et al. "New and Old Discoveries in Antarctica," *Georgetown University Forum of the Air,* Aug. 26, 1956 (mimeographed verbatim).

132. *Mapas Espanoles de America, Siglos XV–XVII.* Vorwick Jacobo Maria del Pilar Carlos Manuel Stuart, Duque de FitzJames, ed., 1878. Madrid, 1941.

133. Marshak, Alexander. "Lunar Notation on Upper Paleolithic Remains," *Science*, CXLVI (Nov. 6, 1964).

134. Means, Philip Ainsworth. "Pre-Spanish Navigation off the Andean Coast," *American Neptune*, II, No. 2 (April, 1942).

135. Mercator, Gerardus. *Correspondence Mercatorienne*, ed. M. van Durme. Anvers: De Nederlandsche Boekhandel, 1959.

136. Miller, Konrad, ed. *Mappae Arabicae*. Stuttgart (privately printed), 1926–1929.

137. ———. *Mappaemundi: Die altesten Weltkarten*. 6 vols. Stuttgart: Roth, 1895–1898.

138. *Monumenta Cartographica Africae et Aegypti*, see Kamal.

139. *Monumenta Cartographica Vaticana*, ed. Roberto Amalgia. 5 vols. Vatican City, 1944.

140. Morison, Samuel E. *Admiral of the Ocean Sea*. Boston: Little, Brown, 1942.

141. ———. *Christopher Columbus, Mariner*. Mentor Books, 1956.

143. Motzo, Bacchisio R., ed. *Il compasso da navigare; opera italiana della metà del secolo XIII* [etc.]. Cagliari: Universita, 1947.

144. Needham, Joseph. "Poverties and Triumphs of Chinese Scientific Tradition." (In Crombie.)

145. ———. *Science and Civilization in China*. 3 vols. Cambridge University Press, 1959.

146. Nordenskiöld, A. E. *Facsimile-Atlas to the Early History of Cartography, with Reproductions of the Most Important Maps printed in the XVth and XVIth Centuries*. Translated from the Swedish original by J. A. Ekelbf and C. R. Markham. Stockholm, 1889.

147. ———. *Periplus: An Essay in the Early History of Charts and Sailing Directions*. Translated from the Swedish original by F. A. Bathev. Stockholm: Norstedt, 1897.

148. Norlund, Niels Erick. *Islands Kortlaegning en Historisk Fremstilling*. (Geodaetsk Instituts Publication VII.) Copenhagen: Ejnar Munksgaard, 1944.

149. "Notice of a British Discovery of Antarctica in 1819," *Blackwood's Magazine*, VII (August, 1920), p. 566.

150. *Nouvelle Biographie Universelle*. Paris: Firmin Didot Frères, 1858.

151. Nowell, Charles E. *The Great Discoveries and the First Colonial Empires*. Ithaca, N.Y.: Cornell Univ. Press, 4th Printing, 1964.

152. Nunez, Pedro. *Tratado da Sphera*. Lisbon, 1537.

153. Nunn, George E. *The Geographical Conceptions of Columbus*. New York: American Geographical Society, 1924.

154. Oberhummer, A. K. "Eine turkische Karte zur Endeckung Amerikas," *Anzeiger der Akademie der Wissenschaften* (Vienna), Philos-histor-Kl 68, 1931, pp. 99–112.

155. *Oceanographic Atlas of the Polar Seas*. (U.S. Hydrographic Office Publication No. 705.) Part I, "Antarctica," 1957.

156. Oronteus Finaeus. Biography. See Finé, Oronce, in *Nouvelle Biographie Universelle*, Paris: Firmin Didot Frères, 1858.

157. ———. *Cartes Géographiques désignées par O. Finé: Galliae totius nova descriptio*. Paris, 1525, 1557; Venice, 1561, 1566; Cosmographia universalis: Paris, 1536, 1566 (lost?)

158. Parry, J. H. *The Age of Reconnaissance*. Mentor Books (No. 597), 1963.

159. Parsons, Edward Alexander. *The Alexandrian Library*. Amsterdam, London, New York: The Elsevier Press, 1952.

160. Pauvels, Louis, and Jacques Bergier. *The Dawn of Magic*. Trans. Rollo Myers. London: Anthony Gibbs and Phillips, 1963.

161. Pears, Edwin. *The Destruction of the Greek Empire and the Story of the Capture of Constantinople by the Turks*. New York: Longmans, Green and Co., 1903.

162. Penrose, Boies. *Travel and Discovery in the Renaissance, 1420–1620*. Cambridge, Mass.: Harvard University Press, 1955.

163. Petermans, A. *Mitteilungen aus Justis Perthes' Geographischer Anstalt.* XXXIX, No. 182 (Mercator). Gotha: Justus Perthes, 1915.

164. Piri Re'is (Muhiddin ibn Mahmud or Ahmet Muhiddin). *Kitabe Bahriye.* 2 vols. Berlin: De Gruyter, 1926.

165. Pohl, Frederick J. *Atlantic Crossings Before Columbus.* New York: W. W. Norton, 1961.

166. *Portugaliae Monumenta Cartographica,* ed. Armando Cortesao and Avelino Teixeira da Mota. 4 vols. (Portuguese and English) Lisbon, 1960.

167. Price, Derek de Solla. "An Ancient Greek Computer," *The Scientific American.* CC, No. 6 (June, 1959).

168. Ptolemy, Claudius. *The Geography,* translated and edited by Edward Luther Stevenson, with reproductions of the maps. New York: The New York Public Library, 1932.

169. Pullè, Francesco Lorenzo. *La Cartographia antica dell'India.* (Vol. I: Byzantine and Arab; Vol. II: Medieval and Early Renaissance; Vol. II, Supplement: The Orient; Vol. III: The Century of the Discovery) 1901–1932.

170. "Radiocarbon": *Supplement of the American Journal of Science,* I–VII (1959–1965), New Haven: The Yale University Press.

171. Rainaud, Armand. *Le Continent Austral, hypothèses et découvertes.* Paris: Armand Colin, 1893.

172. Ravenstein, E. G. *Martin Behaim, His Life and Works.* London: George Philip and Son, 1908.

173. Rawlinson, George. *The Five Great Monarchies of the Ancient Eastern World.* 4th ed., 3 vols. New York: Scribner and Welford, 1880.

175. Reymond, Arnold. *History of the Sciences in Graeco-Roman Antiquity.* Trans. Ruth de Bray. New York: E. P. Dutton, 1932.

176. Ristow, W. W., and C. E. Legear. *A Guide to Historical Cartography* (2nd rev. ed.). Washington, D.C.: The Library of Congress, 1960.

177. Robbins, Roland W., and Evan Jones. *Hidden America.* New York: Knopf, 1959.

178. Roncière. See La Roncière (119).

179. Rotz, John. *John Rotz: His Books of Hydrography* (1542). (Brit. Mus. Ms. 20. E.IX).

180. Rouse, Irving. "Prehistory of the West Indies," *Science,* CXLIV, no. 3618, May 11, 1964.

181. Salinari, Marina Emiliani. "An Atlas of the 15th Century Preserved in the Library of the former Serail in Constantinople," *Imago Mundi,* Vol. III, 1951, pp. 101–102.

182. Santarem, Vicomte de. *Atlas Composé de Mappemondes, de portulans et de cartes hydrographiques et historiques depuis le VI° jusqu'au XVII° Siècle [etc.].*

183. Sarton, George. *A Guide to the History of Science.* New York: The Ronald Press Co., 1952.

184. ———. *Hellenistic Science and Culture in the Last Three Centuries B.C.* Cambridge, Mass.: Harvard University Press, 1952.

185. Schlumberger, Gustave. *Le siège, la prise et le sac de Constantinople par les turcs en 1453.* Paris: Librarie Plon, 1913 (reprinted with slight changes—no date—before 1935).

186. Schütt, Gudmund. *Ptolemy's Maps of Northern Europe: A Reconstruction of the Prototypes.* (Royal Danish Geographical Society) Copenhagen: H. Hagerup, 1917.

187. Selen, H. Sadi. *Piri Re'is in Simali Amerika.* Haritasi Bulletin, No. II'den ayri basim Istanbul: Devlet Basimevi, 1937.

189. Sharpe, Samuel. *Alexandrian Chronology from the Building of the City until Its Conquest by the Arabs (A.D. 640).* London, 1857 (no publisher).

190. Silverberg, Robert. *Lost Cities and Vanished Civilizations.* Philadelphia: Chilton Books, 1962.

191. Stahl, William Harris. *Ptolemy's Geography: A select bibliography.* New York: The New York Public Library, 1953.

192. Stefansson, Vilhjalmur. *Greenland.* New York: Doubleday, Doran and Company, 1942.

193. Steger, Ernst. *Untersuchungen über italienische seekarten des mittelalters auf grund der kartometrischen methode*. Gottingen: W. F. Kaestner, 1896.

194. Stevens, Henry. *Johann Schöner: a reproduction of his globe of 1523 long lost*. London: H. Stevens & Son, 1888.

195. Stevenson, Edward Luther. *Marine World Chart of Nicolo de Canerio Januensis (1502)*. New York: The De Vinne Press, 1908.

196. ———. *Terrestrial and Celestial Globes, Their History and Construction* [etc.]. 2 vols. New Haven: published for the Hispanic Society of America by the Yale University Press, 1921.

197. Strabo. *The Geography of Strabo*. (Trans. by Horace Leonard Jones) The Loeb Classical Library. London: Heinemann; New York: Putnam, 1923–28.

199. Taylor, E. G. R. *The Haven-Finding Art: A History of Navigation from Odysseus to Captain Cook*. New York: Abelard-Schuman Ltd., 1957.

200. ———. "Jean Rotz and the Marine Chart," *Journal of the Institute of Navigation*, VII, No. 2 (April, 1954), pp. 136–143.

201. ———. "The Navigating Manual of Columbus," *Journal of the Institute of Navigation*, V, No. 1 (January, 1952), pp. 42–54.

202. ———. "The Oldest Mediterranean Pilot (Carta Pisana)," *Journal of the Institute of Navigation*, IV, 81 (1951).

203. Teleki, Paul. *Atlas zur Geschichte der Kartographes der japonischen Inseln*. Budapest, 1909; Leipsic: K. W. Hiersemann, 1909.

204. Thalamas, A. *La Géographie* [de *Eratosthenes*]. (Ph.D. Thesis, Harvard University, 1921.)

205. Thompson, James W. *Ancient Libraries*. University of California Press, 1940.

206. Tooley, R. V. *Maps of Antarctica: A list of early maps of the South Polar Regions*. Map Collectors' Circle, Durrant House, Chiswell Street, London E.C. 1, England.

207. ———. *Maps and Map Makers*. London: B. T. Botsford, Ltd., 1949, 1952.

208. True, David O. "Correspondence relating to the Piri Re'is Map." (In files of the American Geographic Society, New York City.)

209. Tseukernik, David. (An article entitled "Did Columbus Discover America?" summarizing an article written by this author and previously published in the Soviet publication *Novy Mir* was published in the English language paper, USSR, October, 1963.)

210. Uhden, Richard. "The Oldest Original Portuguese Chart of the Indian Ocean, A.D. 1509," *Imago Mundi*, III, 1939.

211. Verrill, A. Hyatt. *America's Ancient Civilizations*. New York: Putnam, 1953.

212. Villiers, Alan John. *Men, Ships and the Sea*. Washington, D.C.: National Geographic Society, 1962.

213. ———. *Wild Ocean*. New York: McGraw-Hill, 1957.

214. Wadler, Arnold D. *The Origin of Language*. New York: The American Press for Art and Science, 1948.

215. Waltari, Mika. *The Wanderer*. New York: Putnam, 1951; Pocket Books, 1964. (Chapter on Piri Re'is)

216. Waerden, B. L. van der. *Basic Ideas and Methods in Babylonian and Greek Astronomy* (in Crombie).

217. Wieder, F. C., ed. *Monumenta Cartographica: Reproductions of Unique and Rare Maps*. 5 vols. The Hague: Nijhoff, 1925–1933.

218. Wilson, A. Tuzo. "Continental Drift," *Scientific American*, CCVIII, no. 4, April, 1963, pp. 2–16.

219. Winsor, Justin. *A Bibliography of Ptolemy's Geography*. Cambridge, Mass.: Harvard University Press, 1884.

220. ———. *Narrative and Critical History of America*. Boston and New York: Houghton Mifflin & Co., 1889.

221. Winter, Heinrich. "Catalan Portolan Maps and Their Place in the Total View of Cartographic Development," *Imago Mundi*, XI (1954).

222. ———. "A Late Portolan Chart at Madrid, and Late Portolan Charts in General," *Imago Mundi*, VII (1950).

223. ———. "The Origin of the Sea-Chart," *Imago Mundi*, XIII (1951), pp. 39 ff. (See also *Imago Mundi*, Vol. V, for his reply to Richard Uhden [Vol. I] on question of compass.)

224. ———. "The True Position of H. Wagner in the Controversy of the Compass Chart," *Imago Mundi*, V (1948).

225. Worcester, Donald E., and Wendell C. Schaeffer. *The Growth and Culture of Latin America*. New York: Oxford University Press, 1956.

226. *World Map, Miller Cylindrical Projection, 1 Degree Grid Prepared by the American Geographical Society for the U.S. Department of State*. "Outline Series No. 7." Drawn by William Briesemeister.

227. Wright, Helen S. *The Seventh Continent*. Boston: Richard G. Badger; New York: The Gorham Press, 1918.

228. Wright, John Kirtland. *The Geographical Lore of the Time of the Crusades*. New York: The American Geographical Society, 1925.

229. ———. *The Leardo Map of the World, 1452–1453*. "American Geographical Society Library Series," No. 4, 1928.

230. Wright, John Kirtland, and Elizabeth T. Platt. *Aids to Geographical Research*. New York: American Geographical Society Research Series, No. 22, 1947.

231. Wytfleet, Corneille. *Descriptionis PröIemaicae Augmentum*. Louvain: Ionannis Bogardi, 1598 (Antarctic map, pp. 100–101). Republished 1964 in Amsterdam and Israel with introduction by R. A. Skelton.

Index

About the Author

Charles H. Hapgood, a professor at Keene State College in Keene, New Hampshire, is a lifelong student of the history of science. In a previous book, *Earth's Shifting Crust*, Hapgood offered answers to long-standing enigmas of geology. In a foreword to that book, Albert Einstein said Hapgood's idea could be "of great importance to everything that is related to the history of the earth's surface." In *Maps of the Ancient Sea Kings*, Hapgood offers theories of importance to the history of mankind itself.

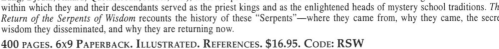

ATLANTIS STUDIES

MAPS OF THE ANCIENT SEA KINGS
Evidence of Advanced Civilization in the Ice Age
by Charles H. Hapgood

Charles Hapgood's classic 1966 book on ancient maps produces concrete evidence of an advanced world-wide civilization existing many thousands of years before ancient Egypt. He has found the evidence in the Piri Reis Map that shows Antarctica, the Hadji Ahmed map, the Oronteus Finaeus and other amazing maps. Hapgood concluded that these maps were made from more ancient maps from the various ancient archives around the world, now lost. Not only were these unknown people more advanced in mapmaking than any people prior to the 18th century, it appears they mapped all the continents. The Americas were mapped thousands of years before Columbus. Antarctica was mapped when its coasts were free of ice.

316 PAGES. 7X10 PAPERBACK. ILLUSTRATED. BIBLIOGRAPHY & INDEX. $19.95. CODE: MASK

PATH OF THE POLE
Cataclysmic Pole Shift Geology
by Charles Hapgood

Maps of the Ancient Sea Kings author Hapgood's classic book *Path of the Pole* is back in print! Hapgood researched Antarctica, ancient maps and the geological record to conclude that the Earth's crust has slipped in the inner core many times in the past, changing the position of the pole. *Path of the Pole* discusses the various "pole shifts" in Earth's past, giving evidence for each one, and moves on to possible future pole shifts. Packed with illustrations, this is the sourcebook for many other books on cataclysms and pole shifts.

356 PAGES. 6X9 PAPERBACK. ILLUSTRATED. $16.95. CODE: POP.

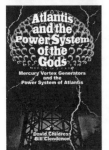

ATLANTIS & THE POWER SYSTEM OF THE GODS
Mercury Vortex Generators & the Power System of Atlantis
by David Hatcher Childress and Bill Clendenon

Atlantis and the Power System of the Gods starts with a reprinting of the rare 1990 book *Mercury: UFO Messenger of the Gods* by Bill Clendenon. Clendenon takes on an unusual voyage into the world of ancient flying vehicles, strange personal UFO sightings, a meeting with a "Man In Black" and then to a centuries-old library in India where he got his ideas for the diagrams of mercury vortex engines. The second part of the book is Childress' fascinating analysis of Nikola Tesla's broadcast system in light of Edgar Cayce's "Terrible Crystal" and the obelisks of ancient Egypt and Ethiopia. Includes: Atlantis and its crystal power towers that broadcast energy; how these incredible power stations may still exist today; inventor Nikola Tesla's nearly identical system of power transmission; Mercury Proton Gyros and mercury vortex propulsion; more. Richly illustrated, and packed with evidence that Atlantis not only existed—it had a world-wide energy system more sophisticated than ours today.

246 PAGES. 6X9 PAPERBACK. ILLUSTRATED. $15.95. CODE: APSG

ATLANTIS IN AMERICA
Navigators of the Ancient World
by Ivar Zapp and George Erikson

This book is an intensive examination of the archeological sites of the Americas, an examination that reveals civilization has existed here for tens of thousands of years. Zapp is an expert on the enigmatic giant stone spheres of Costa Rica, and maintains that they were sighting stones similar to those found throughout the Pacific as well as in Egypt and the Middle East. They were used to teach star-paths and sea navigation to the world-wide navigators of the ancient world. While the Mediterranean and European regions "forgot" world-wide navigation and fought wars, the Mesoamericans of diverse races were building vast interconnected cities without walls. This Golden Age of ancient America was merely a myth of suppressed history—until now. Profusely illustrated, chapters are on Navigators of the Ancient World; Pyramids & Megaliths: Older Than You Think; Ancient Ports and Colonies; Cataclysms of the Past; Atlantis: From Myth to Reality; The Serpent and the Cross: The Loss of the City States; Calendars and Star Temples; and more.

360 PAGES. 6X9 PAPERBACK. ILLUSTRATED. BIBLIOGRAPHY & INDEX. $17.95. CODE: AIA

FAR-OUT ADVENTURES   *REVISED EDITION*
The Best of World Explorer Magazine

This is a compilation of the first nine issues of *World Explorer* in a large-format paperback. Authors include: David Hatcher Childress, Joseph Jochmans, John Major Jenkins, Deanna Emerson, Katherine Routledge, Alexander Horvat, Greg Deyermenjian, Dr. Marc Miller, and others. Articles in this book include Smithsonian Gate, Dinosaur Hunting in the Congo, Secret Writings of the Incas, On the Trail of the Yeti, Secrets of the Sphinx, Living Pterodactyls, Quest for Atlantis, What Happened to the Great Library of Alexandria?, In Search of Seamonsters, Egyptians in the Pacific, Lost Megaliths of Guatemala, the Mystery of Easter Island, Comacalco: Mayan City of Mystery, Professor Wexler and plenty more.

580 PAGES. 8X11 PAPERBACK. ILLUSTRATED. REVISED EDITION. $25.00. CODE: FOA

RETURN OF THE SERPENTS OF WISDOM
by Mark Amaru Pinkham

According to ancient records, the patriarchs and founders of the early civilizations in Egypt, India, China, Peru, Mesopotamia, Britain, and the Americas were the Serpents of Wisdom—spiritual masters associated with the serpent—who arrived in these lands after abandoning their beloved homelands and crossing great seas. While bearing names denoting snake or dragon (such as Naga, Lung, Djedhi, Amaru, Quetzalcoatl, Adder, etc.), these Serpents of Wisdom oversaw the construction of magnificent civilizations within which they and their descendants served as the priest kings and as the enlightened heads of mystery school traditions. *The Return of the Serpents of Wisdom* recounts the history of these "Serpents"—where they came from, why they came, the secret wisdom they disseminated, and why they are returning now.

400 PAGES. 6X9 PAPERBACK. ILLUSTRATED. REFERENCES. $16.95. CODE: RSW

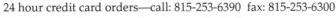

ATLANTIS REPRINT SERIES

ATLANTIS: MOTHER OF EMPIRES
Atlantis Reprint Series
by Robert Stacy-Judd

Robert Stacy-Judd's classic 1939 book on Atlantis is back in print in this large-format paperback edition. Stacy-Judd was a California architect and an expert on the Mayas and their relationship to Atlantis. He was an excellent artist and his work is lavishly illustrated. The eighteen comprehensive chapters in the book are: The Mayas and the Lost Atlantis; Conjectures and Opinions; The Atlantean Theory; Cro-Magnon Man; East is West; And West is East; The Mormons and the Mayas; Astrology in Two Hemispheres; The Language of Architecture; The American Indian; Pre-Panamanians and Pre-Incas; Columns and City Planning; Comparisons and Mayan Art; The Iberian Link; The Maya Tongue; Quetzalcoatl; Summing Up the Evidence; The Mayas in Yucatan.
340 PAGES. 8X11 PAPERBACK. ILLUSTRATED. INDEX. $19.95. CODE: AMOE

MYSTERIES OF ANCIENT SOUTH AMERICA
Atlantis Reprint Series
by Harold T. Wilkins

The reprint of Wilkins' classic book on the megaliths and mysteries of South America. This book predates Wilkin's book *Secret Cities of Old South America* published in 1952. *Mysteries of Ancient South America* was first published in 1947 and is considered a classic book of its kind. With diagrams, photos and maps, Wilkins digs into old manuscripts and books to bring us some truly amazing stories of South America: a bizarre subterranean tunnel system; lost cities in the remote border jungles of Brazil; legends of Atlantis in South America; cataclysmic changes that shaped South America; and other strange stories from one of the world's great researchers. Chapters include: Our Earth's Greatest Disaster, Dead Cities of Ancient Brazil, The Jungle Light that Shines by Itself, The Missionary Men in Black: Forerunners of the Great Catastrophe, The Sign of the Sun: The World's Oldest Alphabet, Sign-Posts to the Shadow of Atlantis, The Atlanean "Subterraneans" of the Incas, Tiahuanacu and the Giants, more.
236 PAGES. 6X9 PAPERBACK. ILLUSTRATED. INDEX. $14.95. CODE: MASA

SECRET CITIES OF OLD SOUTH AMERICA
Atlantis Reprint Series
by Harold T. Wilkins

The reprint of Wilkins' classic book, first published in 1952, claiming that South America was Atlantis. Chapters include Mysteries of a Lost World; Atlantis Unveiled; Red Riddles on the Rocks; South America's Amazons Existed!; The Mystery of El Dorado and Gran Payatiti—the Final Refuge of the Incas; Monstrous Beasts of the Unexplored Swamps & Wilds; Weird Denizens of Antediluvian Forests; New Light on Atlantis from the World's Oldest Book; The Mystery of Old Man Noah and the Arks; and more.
438 PAGES. 6X9 PAPERBACK. ILLUSTRATED. BIBLIOGRAPHY & INDEX. $16.95. CODE: SCOS

THE SHADOW OF ATLANTIS
The Echoes of Atlantean Civilization Tracked through Space & Time
by Colonel Alexander Braghine

First published in 1940, *The Shadow of Atlantis* is one of the great classics of Atlantis research. The book amasses a great deal of archaeological, anthropological, historical and scientific evidence in support of a lost continent in the Atlantic Ocean. Braghine covers such diverse topics as Egyptians in Central America, the myth of Quetzalcoatl, the Basque language and its connection with Atlantis, the connections with the ancient pyramids of Mexico, Egypt and Atlantis, the sudden demise of mammoths, legends of giants and much more. Braghine was a linguist and spends part of the book tracing ancient languages to Atlantis and studying little-known inscriptions in Brazil, deluge myths and the connections between ancient languages. Braghine takes us on a fascinating journey through space and time in search of the lost continent.
288 PAGES. 6X9 PAPERBACK. ILLUSTRATED. $16.95. CODE: SOA

RIDDLE OF THE PACIFIC
by John Macmillan Brown

Oxford scholar Brown's classic work on lost civilizations of the Pacific is now back in print! John Macmillan Brown was an historian and New Zealand's premier scientist when he wrote about the origins of the Maoris. After many years of travel throughout the Pacific studying the people and customs of the south seas islands, he wrote *Riddle of the Pacific* in 1924. The book is packed with rare turn-of-the-century illustrations. Don't miss Brown's classic study of Easter Island, ancient scripts, megalithic roads and cities, more. Brown was an early believer in a lost continent in the Pacific.
460 PAGES. 6X9 PAPERBACK. ILLUSTRATED. $16.95. CODE: ROP

THE HISTORY OF ATLANTIS
by Lewis Spence

Lewis Spence's classic book on Atlantis is now back in print! Spence was a Scottish historian (1874-1955) who is best known for his volumes on world mythology and his five Atlantis books. *The History of Atlantis* (1926) is considered his finest. Spence does his scholarly best in chapters on the Sources of Atlantean History, the Geography of Atlantis, the Races of Atlantis, the Kings of Atlantis, the Religion of Atlantis, the Colonies of Atlantis, more. Sixteen chapters in all.
240 PAGES. 6X9 PAPERBACK. ILLUSTRATED WITH MAPS, PHOTOS & DIAGRAMS. $16.95. CODE: HOA

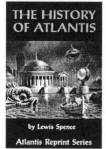

ATLANTIS IN SPAIN
A Study of the Ancient Sun Kingdoms of Spain
by E.M. Whishaw

First published by Rider & Co. of London in 1928, this classic book is a study of the megaliths of Spain, ancient writing, cyclopean walls, sun worshiping empires, hydraulic engineering, and sunken cities. An extremely rare book, it was out of print for 60 years. Learn about the Biblical Tartessus; an Atlantean city at Niebla; the Temple of Hercules and the Sun Temple of Seville; Libyans and the Copper Age; more. Profusely illustrated with photos, maps and drawings.
284 PAGES. 6X9 PAPERBACK. ILLUSTRATED. TABLES OF ANCIENT SCRIPTS. $15.95. CODE: AIS

24 hour credit card orders—call: 815-253-6390 fax: 815-253-6300

email: auphq@frontiernet.net   www.adventuresunlimitedpress.com   www.wexclub.com

ANCIENT SCIENCE

THE GIZA DEATH STAR
The Paleophysics of the Great Pyramid & the Military Complex at Giza
by Joseph P. Farrell

Physicist Joseph Farrell's amazing book on the secrets of Great Pyramid of Giza. *The Giza Death Star* starts where British engineer Christopher Dunn leaves off in his 1998 book, *The Giza Power Plant*. Was the Giza complex part of a military installation over 10,000 years ago? Chapters include: An Archaeology of Mass Destruction, Thoth and Theories; The Machine Hypothesis; Pythagoras, Plato, Planck, and the Pyramid; The Weapon Hypothesis; Encoded Harmonics of the Planck Units in the Great Pyramid; High Freqguency Direct Current "Impulse" Technology; The Grand Gallery and its Crystals: Gravito-acoustic Resonators; The Other Two Large Pyramids; the "Causeways," and the "Temples"; A Phase Conjugate Howitzer; Evidence of the Use of Weapons of Mass Destruction in Ancient Times; more.

290 PAGES. 6X9 PAPERBACK. ILLUSTRATED. $16.95. CODE: GDS

THE GIZA DEATH STAR DEPLOYED
The Physics & Engineering of the Great Pyramid
by Joseph P. Farrell

Physicist Joseph Farrell's amazing sequel to *The Giza Death Star* which takes us from the Great Pyramid to the asteroid belt and the so-called Pyramids of Mars. Farrell expands on his thesis that the Great Pyramid was a chemical maser, designed as a weapon and eventually deployed—with disastrous results to the solar system. Includes: Exploding Planets: The Movie, the Mirror, and the Model; Dating the Catastrophe and the Compound; A Brief History of the Exoteric and Esoteric Investigations of the Great Pyramid; No Machines, Please!; The Stargate Conspiracy; The Scalar Weapons; Message or Machine?; A Tesla Analysis of the Putative Physics and Engineering of the Giza Death Star; Cohering the Zero Point, Vacuum Energy, Flux: Synopsis of Scalar Physics and Paleophysics; Configuring the Scalar Pulse Wave; Inferred Applications in the Great Pyramid; Quantum Numerology, Feedback Loops and Tetrahedral Physics; and more.

290 PAGES. 6X9 PAPERBACK. ILLUSTRATED. BIBLIOGRAPHY. INDEX. $16.95. CODE: GDSD

PIRATES & THE LOST TEMPLAR FLEET
The Secret Naval War Between the Templars & the Vatican
by David Hatcher Childress

The lost Templar fleet was originally based at La Rochelle in southern France, but fled to the deep fiords of Scotland upon the dissolution of the Order by King Phillip. This banned fleet of ships was later commanded by the St. Clair family of Rosslyn Chapel (birthplace of Free Masonry). St. Clair and his Templars made a voyage to Canada in the year 1398 AD, nearly 100 years before Columbus! Chapters include: 10,000 Years of Seafaring; The Knights Templar & the Crusades; The Templars and the Assassins; The Lost Templar Fleet and the Jolly Roger; Maps of the Ancient Sea Kings; Pirates, Templars and the New World; Christopher Columbus—Secret Templar Pirate?; Later Day Pirates and the War with the Vatican; Pirate Utopias and the New Jerusalem; more.

320 PAGES. 6X9 PAPERBACK. ILLUSTRATED. BIBLIOGRAPHY. $16.95. CODE: PLTF

CLOAK OF THE ILLUMINATI
Secrets, Transformations, Crossing the Star Gate
by William Henry

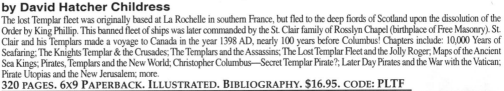

Thousands of years ago the stargate technology of the gods was lost. Mayan Prophecy says it will return by 2012, along with our alignment with the center of our galaxy. In this book: Find examples of stargates and wormholes in the ancient world; Examine myths and scripture with hidden references to a stargate cloak worn by the Illuminati, including Mari, Nimrod, Elijah, and Jesus; See rare images of gods and goddesses wearing the Cloak of the illuminati; Learn about Saddam Hussein and the secret missing library of Jesus; Uncover the secret Roman-era eugenics experiments at the Temple of Hathor in Denderah, Egypt; Explore the duplicate of the Stargate Pillar of the Gods in the Illuminists' secret garden in Nashville, TN; Discover the secrets of manna, the food of the angels; Share the lost Peace Prayer posture of Osiris, Jesus and the Illuminati; more. Chapters include: Seven Stars Under Three Stars; The Long Walk; Squaring the Circle; The Mill of the Host; The Miracle Garment; The Fig; Nimrod: The Mighty Man; Nebuchadnezzar's Gate; The New Mighty Man; more.

238 PAGES. 6X9 PAPERBACK. ILLUSTRATED. BIBLIOGRAPHY. INDEX. $16.95. CODE: COIL

THE CHRONOLOGY OF GENESIS
A Complete History of the Nefilim
by Neil Zimmerer

Follow the Nefilim through the Ages! This is a complete history of Genesis, the gods and the history of Earth — before the gods were destroyed by their own creations more than 2500 years ago! Zimmerer presents the most complete history of the Nefilim ever developed — from the Sumerian Nefilim kings through the Nefilim today. He provides evidence of extraterrestrial Nefilim monuments, and includes fascinating information on pre-Nefilim man-apes and man-apes of the world in the present age. Includes the following subjects and chapters: Creation of the Universe; Evolution: The Greatest Mystery; Who Were the Nefilim?; Pre-Nefilim Man-Apes; Man-Apes of the World—Present Age; Extraterrestrial Nefilim Monuments; The Nefilim Today; All the Sumerian Nefilim Kings listed in chronological order, more. A book not to be missed by researchers into the mysterious origins of mankind.

244 PAGES. 6X9 PAPERBACK. ILLUSTRATED. REFERENCES. $16.95. CODE: CGEN

LEY LINE & EARTH ENERGIES
An Extraordinary Journey into the Earth's Natural Energy System
by David Cowan & Anne Silk

The mysterious standing stones, burial grounds and stone circles that lace Europe, the British Isles and other areas have intrigued scientists, writers, artists and travellers through the centuries. They pose so many questions: Why do some places feel special? How do ley lines work? How did our ancestors use Earth energy to map their sacred sites and burial grounds? How do ghosts and poltergeists interact with Earth energy? How can Earth spirals and black spots affect our health? This exploration shows how natural forces affect our behavior, how they can be used to enhance our health and well being, and ultimately, how they bring us closer to penetrating one of the deepest mysteries being explored. A fascinating and visual book about subtle Earth energies and how they affect us and the world around them.

368 PAGES. 6X9 PAPERBACK. ILLUSTRATED. BIBLIOGRAPHY. INDEX. $18.95. CODE: LLEE

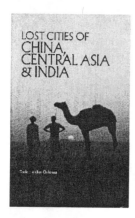

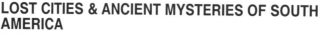
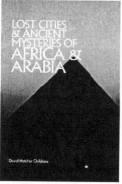
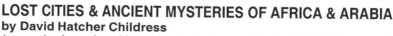

TECHNOLOGY OF THE GODS
The Incredible Sciences of the Ancients
by David Hatcher Childress

Popular *Lost Cities* author David Hatcher Childress takes us into the amazing world of ancient technology, from computers in antiquity to the "flying machines of the gods." Childress looks at the technology that was allegedly used in Atlantis and the theory that the Great Pyramid of Egypt was originally a gigantic power station. He examines tales of ancient flight and the technology that it involved; how the ancients used electricity; megalithic building techniques; the use of crystal lenses and the fire from the gods; evidence of various high tech weapons in the past, including atomic weapons; ancient metallurgy and heavy machinery; the role of modern inventors such as Nikola Tesla in bringing ancient technology back into modern use; impossible artifacts; and more.

356 PAGES. 6X9 PAPERBACK. ILLUSTRATED. BIBLIOGRAPHY. $16.95. CODE: TGOD

VIMANA AIRCRAFT OF ANCIENT INDIA & ATLANTIS
by David Hatcher Childress, introduction by Ivan T. Sanderson

Did the ancients have the technology of flight? In this incredible volume on ancient India, authentic Indian texts such as the *Ramayana* and the *Mahabharata* are used to prove that ancient aircraft were in use more than four thousand years ago. Included in this book is the entire Fourth Century BC manuscript *Vimaanika Shastra* by the ancient author Maharishi Bharadwaaja, translated into English by the Mysore Sanskrit professor G.R. Josyer. Also included are chapters on Atlantean technology, the incredible Rama Empire of India and the devastating wars that destroyed it. Also an entire chapter on mercury vortex propulsion and mercury gyros, the power source described in the ancient Indian texts. Not to be missed by those interested in ancient civilizations or the UFO enigma.

334 PAGES. 6X9 PAPERBACK. RARE PHOTOGRAPHS, MAPS AND DRAWINGS. $15.95. CODE: VAA

LOST CONTINENTS & THE HOLLOW EARTH
I Remember Lemuria and the Shaver Mystery
by David Hatcher Childress & Richard Shaver

Lost Continents & the Hollow Earth is Childress' thorough examination of the early hollow earth stories of Richard Shaver and the fascination that fringe fantasy subjects such as lost continents and the hollow earth have had for the American public. Shaver's rare 1948 book *I Remember Lemuria* is reprinted in its entirety, and the book is packed with illustrations from Ray Palmer's *Amazing Stories* magazine of the 1940s. Palmer and Shaver told of tunnels running through the earth—tunnels inhabited by the Deros and Teros, humanoids from an ancient spacefaring race that had inhabited the earth, eventually going underground, hundreds of thousands of years ago. Childress discusses the famous hollow earth books and delves deep into whatever reality may be behind the stories of tunnels in the earth. Operation High Jump to Antarctica in 1947 and Admiral Byrd's bizarre statements, tunnel systems in South America and Tibet, the underground world of Agartha, the belief of UFOs coming from the South Pole, more.

344 PAGES. 6X9 PAPERBACK. ILLUSTRATED. $16.95. CODE: LCHE

A HITCHHIKER'S GUIDE TO ARMAGEDDON
by David Hatcher Childress

With wit and humor, popular Lost Cities author David Hatcher Childress takes us around the world and back in his trippy finalé to the Lost Cities series. He's off on an adventure in search of the apocalypse and end times. Childress hits the road from the fortress of Megiddo, the legendary citadel in northern Israel where Armageddon is prophesied to start. Hitchhiking around the world, Childress takes us from one adventure to another, to ancient cities in the deserts and the legends of worlds before our own. Childress muses on the rise and fall of civilizations, and the forces that have shaped mankind over the millennia, including wars, invasions and cataclysms. He discusses the ancient Armageddons of the past, and chronicles recent Middle East developments and their ominous undertones. In the meantime, he becomes a cargo cult god on a remote island off New Guinea, gets dragged into the Kennedy Assassination by one of the "conspirators," investigates a strange power operating out of the Altai Mountains of Mongolia, and discovers how the Knights Templar and their off-shoots have driven the world toward an epic battle centered around Jerusalem and the Middle East.

320 PAGES. 6X9 PAPERBACK. ILLUSTRATED. BIBLIOGRAPHY. INDEX. $16.95. CODE: HGA

THE LAND OF OSIRIS
An Introduction to Khemitology
by Stephen S. Mehler

Was there an advanced prehistoric civilization in ancient Egypt? Were they the people who built the great pyramids and carved the Great Sphinx? Did the pyramids serve as energy devices and not as tombs for kings? Mehler has uncovered an indigenous oral tradition that still exists in Egypt, and has been fortunate to have studied with a living master of this tradition, Abd'El Hakim Awyan. Mehler has also been given permission to present these teachings to the Western world, teachings that unfold a whole new understanding of ancient Egypt and have only been presented heretofore in fragments by other researchers. Chapters include: Egyptology and Its Paradigms; Khemitology—New Paradigms; Asgat Nefer—The Harmony of Water; Khemit and the Myth of Atlantis; The Extraterrestrial Question; more.

272 PAGES. 6X9 PAPERBACK. ILLUSTRATED. COLOR SECTION. BIBLIOGRAPHY. $18.95. CODE: LOOS

IN QUEST OF LOST WORLDS
Journey to Mysterious Algeria, Ethiopia & the Yucatan
by Count Byron Khun de Prorok

Finally, a reprint of Count Byron de Prorok's classic archeology/adventure book first published in 1936 by E.P. Dutton & Co. in New York. In this exciting and well illustrated book, de Prorok takes us into the deep Sahara of forbidden Algeria, to unknown Ethiopia, and to the many prehistoric ruins of the Yucatan. Includes: Tin Hinan, Legendary Queen of the Tuaregs; The mysterious A'Haggar Range of southern Algeria; Jupiter, Ammon and Tripolitania; The "Talking Dune"; The Land of the Garamantes; Mexico and the Poison Trail; Seeking Atlantis—Chichen Itza; Shadowed by the "Little People"—the Lacandon Pygmie Maya; Ancient Pyramids of the Usamasinta and Piedras Negras in Guatemala; In Search of King Solomon's Mines & the Land of Ophir; Ancient Emerald Mines of Ethiopia. Also included in this book are 24 pages of special illustrations of the famous—and strange—wall paintings of the Ahaggar from the rare book *The Search for the Tassili Frescoes* by Henri Lhote (1959). A visual treat of a remote area of the world that is even today forbidden to outsiders!

324 PAGES. 6X9 PAPERBACK. ILLUSTRATED. $16.95. CODE: IQLW

TESLA TECHNOLOGY

THE FANTASTIC INVENTIONS OF NIKOLA TESLA
Nikola Tesla with additional material by David Hatcher Childress

This book is a readable compendium of patents, diagrams, photos and explanations of the many incredible inventions of the originator of the modern era of electrification. In Tesla's own words are such topics as wireless transmission of power, death rays, and radio-controlled airships. In addition, rare material on German bases in Antarctica and South America, and a secret city built at a remote jungle site in South America by one of Tesla's students, Guglielmo Marconi. Marconi's secret group claims to have built flying saucers in the 1940s and to have gone to Mars in the early 1950s! Incredible photos of these Tesla craft are included. The Ancient Atlantean system of broadcasting energy through a grid system of obelisks and pyramids is discussed, and a fascinating concept comes out of one chapter: that Egyptian engineers had to wear protective metal head-shields while in these power plants, hence the Egyptian Pharoah's head covering as well as the Face on Mars!
•His plan to transmit free electricity into the atmosphere. •How electrical devices would work using only small antennas mounted on them.
•Why unlimited power could be utilized anywhere on earth. •How radio and radar technology can be used as death-ray weapons in Star Wars.
•Includes an appendix of Supreme Court documents on dismantling his free energy towers. •Tesla's Death Rays, Ozone generators, and more…
342 PAGES. 6X9 PAPERBACK. ILLUSTRATED. BIBLIOGRAPHY AND APPENDIX. $16.95. CODE: FINT

THE TESLA PAPERS
Nikola Tesla on Free Energy & Wireless Transmission of Power
by Nikola Tesla, edited by David Hatcher Childress

In the tradition of *The Fantastic Inventions of Nikola Tesla, The Anti-Gravity Handbook* and *The Free-Energy Device Handbook*, science and UFO author David Hatcher Childress takes us into the incredible world of Nikola Tesla and his amazing inventions. Tesla's rare article "The Problem of Increasing Human Energy with Special Reference to the Harnessing of the Sun's Energy" is included. This lengthy article was originally published in the June 1900 issue of *The Century Illustrated Monthly Magazine* and it was the outline for Tesla's master blueprint for the world. Tesla's fantastic vision of the future, including wireless power, anti-gravity, free energy and highly advanced solar power.
Also included are some of the papers, patents and material collected on Tesla at the Colorado Springs Tesla Symposiums, including papers on:
•The Secret History of Wireless Transmission •Tesla and the Magnifying Transmitter
•Design and Construction of a half-wave Tesla Coil •Electrostatics: A Key to Free Energy
•Progress in Zero-Point Energy Research •Electromagnetic Energy from Antennas to Atoms
•Tesla's Particle Beam Technology •Fundamental Excitatory Modes of the Earth-Ionosphere Cavity
325 PAGES. 8X10 PAPERBACK. ILLUSTRATED. $16.95. CODE: TTP

LOST SCIENCE
by Gerry Vassilatos

Secrets of Cold War Technology author Vassilatos on the remarkable lives, astounding discoveries, and incredible inventions of such famous people as Nikola Tesla, Dr. Royal Rife, T.T. Brown, and T. Henry Moray. Read about the aura research of Baron Karl von Reichenbach, the wireless technology of Antonio Meucci, the controlled fusion devices of Philo Farnsworth, the earth battery of Nathan Stubblefield, and more. What were the twisted intrigues which surrounded the often deliberate attempts to stop this technology? Vassilatos claims that we are living hundreds of years behind our intended level of technology and we must recapture this "lost science."
304 PAGES. 6X9 PAPERBACK. ILLUSTRATED. BIBLIOGRAPHY. $16.95. CODE: LOS

SECRETS OF COLD WAR TECHNOLOGY
Project HAARP and Beyond
by Gerry Vassilatos

Vassilatos reveals that "Death Ray" technology has been secretly researched and developed since the turn of the century. Included are chapters on such inventors and their devices as H.C. Vion, the developer of auroral energy receivers; Dr. Selim Lemstrom's pre-Tesla experiments; the early beam weapons of Grindell-Mathews, Ulivi, Turpain and others; John Hettenger and his early beam power systems. Learn about Project Argus, Project Teak and Project Orange; EMP experiments in the 60s; why the Air Force directed the construction of a huge Ionospheric "backscatter" telemetry system across the Pacific just after WWII; why Raytheon has collected every patent relevant to HAARP over the past few years; more.
250 PAGES. 6X9 PAPERBACK. ILLUSTRATED. $15.95. CODE: SCWT

HAARP
The Ultimate Weapon of the Conspiracy
by Jerry Smith

The HAARP project in Alaska is one of the most controversial projects ever undertaken by the U.S. Government. Jerry Smith gives us the history of the HAARP project and explains how works, in technically correct yet easy to understand language. At best, HAARP is science out-of-control; at worst, HAARP could be the most dangerous device ever created, a futuristic technology that is everything from super-beam weapon to world-wide mind control device. Topics include Over-the-Horizon Radar and HAARP, Mind Control, ELF and HAARP, The Telsa Connection, The Russian Woodpecker, GWEN & HAARP, Earth Penetrating Tomography, Weather Modification, Secret Science of the Conspiracy, more. Includes the complete 1987 Eastlund patent for his pulsed super-weapon that he claims was stolen by the HAARP Project.
256 PAGES. 6X9 PAPERBACK. ILLUSTRATED. $14.95. CODE: HARP

HARNESSING THE WHEELWORK OF NATURE
Tesla's Science of Energy
by Thomas Valone, Ph.D., P.E.

Chapters include: Tesla: Scientific Superman who Launched the Westinghouse Industrial Firm by John Shatlan; Nikola Tesla—Electricity's Hidden Genius, excerpt from The Search for Free Energy; Tesla's History at Niagara Falls; Non-Hertzian Waves: True Meaning of the Wireless Transmission of Power by Toby Grotz; On the Transmission of Electricity Without Wires by Nikola Tesla; Tesla's Magnifying Transmitter by Andrija Puharich; Tesla's Self-Sustaining Electrical Generator and the Ether by Oliver Nichelson; Self-Sustaining Non-Hertzian Longitudinal Waves by Dr. Robert Bass; Modification of Maxwell's Equations in Free Space; Scalar Electromagnetic Waves; Disclosures Concerning Tesla's Operation of an ELF Oscillator; A Study of Tesla's Advanced Concepts & Glossary of Tesla Technology Terms; Electric Weather Forces: Tesla's Vision by Charles Yost; The New Art of Projecting Concentrated Non-Dispersive Energy Through Natural Media; The Homopolar Generator: Tesla's Contribution by Thomas Valone; Tesla's Ionizer and Ozonator: Implications for Indoor Air Pollution by Thomas Valone; How Cosmic Forces Shape Our Destiny by Nikola Tesla; Tesla's Death Ray plus Selected Tesla Patents; more.
288 PAGES. 6X9 PAPERBACK. ILLUSTRATED. $16.95. CODE: HWWN

24 hour credit card orders—call: 815-253-6390 fax: 815-253-6300
email: auphq@frontiernet.net   www.adventuresunlimitedpress.com   www.wexclub.com

STRANGE SCIENCE

UNDERGROUND BASES & TUNNELS
What is the Government Trying to Hide?
by Richard Sauder, Ph.D.

Working from government documents and corporate records, Sauder has compiled an impressive book that digs below the surface of the military's super-secret underground! Go behind the scenes into little-known corners of the public record and discover how corporate America has worked hand-in-glove with the Pentagon for decades, dreaming about, planning, and actually constructing, secret underground bases. This book includes chapters on the locations of the bases, the tunneling technology, various military designs for underground bases, nuclear testing & underground bases, abductions, needles & implants, military involvement in "alien" cattle mutilations, more. 50 page photo & map insert.

201 PAGES. 6x9 PAPERBACK. ILLUSTRATED. $15.95. CODE: UGB

UNDERWATER & UNDERGROUND BASES
Surprising Facts the Government Does Not Want You to Know
by Richard Sauder

Dr. Richard Sauder's brand new book *Underwater and Underground Bases* is an explosive, eye-opening sequel to his best-selling, *Underground Bases and Tunnels: What is the Government Trying to Hide?* Dr. Sauder lays out the amazing evidence and government paper trail for the construction of huge, manned bases offshore, in mid-ocean, and deep beneath the sea floor! Bases big enough to secretly dock submarines! Official United States Navy documents, and other hard evidence, raise many questions about what really lies 20,000 leagues beneath the sea. Many UFOs have been seen coming and going from the world's oceans, seas and lakes, implying the existence of secret underwater bases. Hold on to your hats: Jules Verne may not have been so far from the truth, after all! Dr. Sauder also adds to his incredible database of underground bases onshore. New, breakthrough material reveals the existence of additional clandestine underground facilities as well as the surprising location of one of the CIA's own underground bases. Plus, new information on tunneling and cutting-edge, high speed rail magnetic-levitation (MagLev) technology. There are many rumors of secret, underground tunnels with MagLev trains hurtling through them. Is there truth behind the rumors? *Underwater and Underground Bases* carefully examines the evidence and comes to a thought provoking conclusion!

264 PAGES. 6x9 PAPERBACK. ILLUSTRATED. BIBLIOGRAPHY. INDEX. $16.95. CODE: UUB

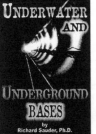

KUNDALINI TALES
by Richard Sauder, Ph.D.

Underground Bases and Tunnels author Richard Sauder's second book on his personal experiences and provocative research into spontaneous spiritual awakening, out-of-body journeys, encounters with secretive governmental powers, daylight sightings of UFOs, and more. Sauder continues his studies of underground bases with new information on the occult underpinnings of the U.S. space program. The book also contains a breakthrough section that examines actual U.S. patents for devices that manipulate minds and thoughts from a remote distance. Included are chapters on the secret space program and a 130-page appendix of patents and schematic diagrams of secret technology and mind control devices.

296 PAGES. 7x10 PAPERBACK. ILLUSTRATED. BIBLIOGRAPHY. $14.95. CODE: KTAL

QUEST FOR ZERO-POINT ENERGY
Engineering Principles for "Free Energy"
by Moray B. King

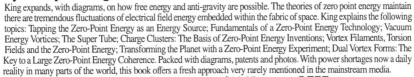

King expands, with diagrams, on how free energy and anti-gravity are possible. The theories of zero point energy maintain there are tremendous fluctuations of electrical field energy embedded within the fabric of space. King explains the following topics: Tapping the Zero-Point Energy as an Energy Source; Fundamentals of a Zero-Point Energy Technology; Vacuum Energy Vortices; The Super Tube; Charge Clusters: The Basis of Zero-Point Energy Inventions; Vortex Filaments, Torsion Fields and the Zero-Point Energy; Transforming the Planet with a Zero-Point Energy Experiment; Dual Vortex Forms: The Key to a Large Zero-Point Energy Coherence. Packed with diagrams, patents and photos. With power shortages now a daily reality in many parts of the world, this book offers a fresh approach very rarely mentioned in the mainstream media.

224 PAGES. 6x9 PAPERBACK. ILLUSTRATED. $14.95. CODE: QZPE

HITLER'S FLYING SAUCERS
A Guide to German Flying Discs of the Second World War
by Henry Stevens

Learn why the Schriever-Habermohl project was actually two projects and read the written statement of a German test pilot who actually flew one of these saucers; about the Leduc engine, the key to Dr. Miethe's saucer designs; how U.S. government officials kept the truth about foo fighters hidden for almost sixty years and how they were finally forced to "come clean" about the foo fighter's German origin. Learn of the Peenemuende saucer project and how it was slated to "go atomic." Read the testimony of a German eyewitness who saw "magnetic discs." Read the U.S. government's own reports on German field propulsion saucers. Read how the post-war German KM-2 field propulsion 'rocket' worked. Learn details of the work of Karl Schappeller and Viktor Schauberger. Learn how their ideas figure in the quest to build field propulsion flying discs. Find out what happened to this technology after the war. Find out how the Canadians got saucer technology directly from the SS. Find out about the surviving "Third Power" of former Nazis. Learn of the U.S. government's methods of UFO deception and how they used the German "Sonderbueroll" as the model for Project Blue Book.

388 PAGES. 6x9 PAPERBACK. ILLUSTRATED. INDEX. $18.95. CODE: HFS

THE TIME TRAVEL HANDBOOK
A Manual of Practical Teleportation & Time Travel
edited by David Hatcher Childress

In the tradition of *The Anti-Gravity Handbook* and *The Free-Energy Device Handbook*, science and UFO author David Hatcher Childress takes us into the weird world of time travel and teleportation. Not just a whacked-out look at science fiction, this book is an authoritative chronicling of real-life time travel experiments, teleportation devices and more. *The Time Travel Handbook* takes the reader beyond the government experiments and deep into the uncharted territory of early time travellers such as Nikola Tesla and Guglielmo Marconi and their alleged time travel experiments, as well as the Wilson Brothers of EMI and their connection to the Philadelphia Experiment—the U.S. Navy's forays into invisibility, time travel, and teleportation. Childress looks into the claims of time travelling individuals, and investigates the unusual claim that the pyramids on Mars were built in the future and sent back in time. A highly visual, large format book, with patents, photos and schematics. Be the first on your block to build your own time travel device!

316 PAGES. 7x10 PAPERBACK. ILLUSTRATED. $16.95. CODE: TTH

EXTRATERRESTRIALS & UFOS

EXTRATERRESTRIAL ARCHAEOLOGY NEW EDITION!
by David Hatcher Childress

Using official NASA and Soviet photos, as well as other photos taken via telescope, this book seeks to prove that many of the planets (and moons) of our solar system are in some way inhabited by intelligent life. The book includes many blow-ups of NASA photos and detailed diagrams of structures—particularly on the Moon.
•NASA PHOTOS OF PYRAMIDS AND DOMED CITIES ON THE MOON. •PYRAMIDS AND GIANT STATUES ON MARS. •HOLLOW MOONS OF MARS AND OTHER PLANETS. •ROBOT MINING VEHICLES THAT MOVE ABOUT THE MOON PROCESSING VALUABLE METALS. •NASA & RUSSIAN PHOTOS OF SPACE-BASES ON MARS AND ITS MOONS. •A BRITISH SCIENTIST WHO DISCOVERED A TUNNEL ON THE MOON, AND OTHER "BOTTOMLESS CRATERS." •EARLY CLAIMS OF TRIPS TO THE MOON AND MARS. •STRUCTURAL ANOMALIES ON VENUS, SATURN, JUPITER, MERCURY, URANUS & NEPTUNE. •NASA, THE MOON AND ANTI-GRAVITY. PLUS MORE. HIGHLY ILLUSTRATED WITH PHOTOS, DIAGRAMS AND MAPS!
320 PAGES. 8X11 PAPERBACK. BIBLIOGRAPHY & APPENDIX. $19.95. CODE: ETA

THE CASE FOR THE FACE
Scientists Examine the Evidence for Alien Artifacts on Mars
edited by Stanley McDaniel and Monica Rix Paxson. Mars Imagery by Mark Carlotto

The ultimate compendium on artificial structures in the Cydonia region of Mars. *The Case for the Face* unifies the research and opinions of a remarkably accomplished group of scientists, including a former NASA astronaut, a quantum physicist who is the chair of a space science program, leading meteor researchers, nine Ph.D.'s, the best-selling science author in Germany and more. The book includes: NASA research proving we're not the first intelligent race in this solar system; 120 amazing high resolution images never seen before by the general public; three separate doctoral statistical studies demonstrating the likelihood of artificial objects at the Cydonian site to be over 99%; and other definitive proof of life on Mars.
320 PAGES. 6X9 PAPERBACK. ILLUSTRATED. INDEX & BIBLIOGRAPHY. $17.95. CODE: CFF

DARK MOON
Apollo and the Whistleblowers
by Mary Bennett and David Percy

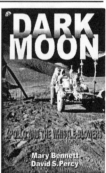

•Was Neil Armstrong really the first man on the Moon?
•Did you know a second craft was going to the Moon at the same time as Apollo 11?
•Do you know that potentially lethal radiation is prevalent throughout deep space?
•Did you know that 'live' color TV from the Moon was not actually live at all?
•Do you know that lighting was used in the Apollo photographs—yet no lighting equipment was taken to the Moon?
All these questions, and more, are discussed in great detail by British researchers Bennett and Percy in *Dark Moon*, the definitive book (nearly 600 pages) on the possible faking of the Apollo Moon missions. Bennett and Percy delve into every possible aspect of this beguiling theory, one that rocks the very foundation of our beliefs concerning NASA and the space program. Tons of NASA photos analyzed for possible deceptions.
568 PAGES. 6X9 PAPERBACK. ILLUSTRATED. BIBLIOGRAPHY. INDEX. $25.00. CODE: DARK

THE CHRONOLOGY OF GENESIS
A Complete History of Nefilim
by Neil Zimmerer

Follow the Nefilim through the Ages! This is a complete history of Genesis, the gods and the history of Earth — before the gods were destroyed by their own creations more than 2500 years ago! Zimmerer presents the most complete history of the Nefilim ever developed — from the Sumerian Nefilim kings through the Nefilim today. He provides evidence of extraterrestrial Nefilim monuments, and includes fascinating information on pre-Nefilim man-apes and man-apes of the world in the present age. Includes the following subjects and chapters: Creation of the Universe; Evolution: The Greatest Mystery; Who Were the Nefilim?; Pre-Nefilim Man-Apes; Man-Apes of the World— Present Age; Extraterrestrial Nefilim Monuments; The Nefilim Today; All the Sumerian Nefilim Kings listed in chronological order, more. A book not to be missed by researchers into the mysterious origins of mankind.
244 PAGES. 6X9 PAPERBACK. ILLUSTRATED. REFERENCES. $16.95. CODE: CGEN

FACES OF THE VISITORS
An Illustrated Reference to Alien Contact
by Kevin Randle and Russ Estes

A visual encyclopedia of reports of alien visitors with detailed drawings of each entity. Includes some photos, a unique reliability rating and meticulous documentation of source material. Includes virtually every photo and illustration of extraterrestrial entities.
309 PAGES. 6X9 PAPERBACK. ILLUSTRATED. INDEX. $12.00. CODE: FOVS

THE ROSWELL MESSAGE
50 Years On—The Aliens Speak
by René Coudris

This strange book, imported from Britain, is a compilation of material on the Roswell, New Mexico UFO crash of 1947. With lots of good photos, including the Santilli alien autopsy pics (the strange instrument panel for two six-fingered hands is worth the price of the book alone), plus other material with an impressive reconstruction of the crash and messages from the "aliens" themselves, who say: "We are your future."
223 PAGES. 8X10 PAPERBACK. ILLUSTRATED. INDEX. $19.95. CODE: RMES

WAKE UP DOWN THERE!
The Excluded Middle Anthology
by Greg Bishop

The great American tradition of dropout culture makes it over the millennium mark with a collection of the best from *The Excluded Middle,* the critically acclaimed underground zine of UFOs, the paranormal, conspiracies, psychedelia, and spirit. Contributions from Robert Anton Wilson, Ivan Stang, Martin Kottmeyer, John Shirley, Scott Corrales, Adam Gorightly and Robert Sterling; and interviews with James Moseley, Karla Turner, Bill Moore, Kenn Thomas, Richard Boylan, Dean Radin, Joe McMoneagle, and the mysterious Ira Einhorn (an *Excluded Middle* exclusive). Includes full versions of interviews and extra material not found in the newsstand versions.
420 PAGES. 8X11 PAPERBACK. ILLUSTRATED. $25.00. CODE: WUDT

ANTI-GRAVITY

THE FREE-ENERGY DEVICE HANDBOOK
A Compilation of Patents and Reports
by David Hatcher Childress

A large-format compilation of various patents, papers, descriptions and diagrams concerning free-energy devices and systems. *The Free-Energy Device Handbook* is a visual tool for experimenters and researchers into magnetic motors and other "overunity" devices. With chapters on the Adams Motor, the Hans Coler Generator, cold fusion, superconductors, "N" machines, space-energy generators, Nikola Tesla, T. Townsend Brown, and the latest in free-energy devices. Packed with photos, technical diagrams, patents and fascinating information, this book belongs on every science shelf. With energy and profit being a major political reason for fighting various wars, free-energy devices, if ever allowed to be mass distributed to consumers, could change the world! Get your copy now before the Department of Energy bans this book!

292 PAGES. 8x10 PAPERBACK. ILLUSTRATED. BIBLIOGRAPHY. $16.95. CODE: FEH

THE ANTI-GRAVITY HANDBOOK
edited by David Hatcher Childress, with Nikola Tesla, T.B. Paulicki,
Bruce Cathie, Albert Einstein and others

The new expanded compilation of material on Anti-Gravity, Free Energy, Flying Saucer Propulsion, UFOs, Suppressed Technology, NASA Cover-ups and more. Highly illustrated with patents, technical illustrations and photos. This revised and expanded edition has more material, including photos of Area 51, Nevada, the government's secret testing facility. This classic on weird science is back in a 90s format!
- **How to build a flying saucer.**
- **Arthur C. Clarke on Anti-Gravity.**
- **Crystals and their role in levitation.**
- **Secret government research and development.**
- **Nikola Tesla on how anti-gravity airships could draw power from the atmosphere.**
- **Bruce Cathie's Anti-Gravity Equation.**
- **NASA, the Moon and Anti-Gravity.**

253 PAGES. 7X10 PAPERBACK. BIBLIOGRAPHY/INDEX/APPENDIX. HIGHLY ILLUSTRATED. $16.95. CODE: AGH

ANTI–GRAVITY & THE WORLD GRID

Is the earth surrounded by an intricate electromagnetic grid network offering free energy? This compilation of material on ley lines and world power points contains chapters on the geography, mathematics, and light harmonics of the earth grid. Learn the purpose of ley lines and ancient megalithic structures located on the grid. Discover how the grid made the Philadelphia Experiment possible. Explore the Coral Castle and many other mysteries, including acoustic levitation, Tesla Shields and scalar wave weaponry. Browse through the section on anti-gravity patents, and research resources.

274 PAGES. 7X10 PAPERBACK. ILLUSTRATED. $14.95. CODE: AGW

ANTI–GRAVITY & THE UNIFIED FIELD
edited by David Hatcher Childress

Is Einstein's Unified Field Theory the answer to all of our energy problems? Explored in this compilation of material is how gravity, electricity and magnetism manifest from a unified field around us. Why artificial gravity is possible; secrets of UFO propulsion; free energy; Nikola Tesla and anti-gravity airships of the 20s and 30s; flying saucers as superconducting whirls of plasma; anti-mass generators; vortex propulsion; suppressed technology; government cover-ups; gravitational pulse drive; spacecraft & more.

240 PAGES. 7X10 PAPERBACK. ILLUSTRATED. $14.95. CODE: AGU

ETHER TECHNOLOGY
A Rational Approach to Gravity Control
by Rho Sigma

This classic book on anti-gravity and free energy is back in print and back in stock. Written by a well-known American scientist under the pseudonym of "Rho Sigma," this book delves into international efforts at gravity control and discoid craft propulsion. Before the Quantum Field, there was "Ether." This small, but informative book has chapters on John Searle and "Searle discs;" T. Townsend Brown and his work on anti-gravity and ether-vortex turbines. Includes a forward by former NASA astronaut Edgar Mitchell.

108 PAGES. 6X9 PAPERBACK. ILLUSTRATED. $12.95. CODE: ETT

TAPPING THE ZERO POINT ENERGY
Free Energy & Anti-Gravity in Today's Physics
by Moray B. King

King explains how free energy and anti-gravity are possible. The theories of the zero point energy maintain there are tremendous fluctuations of electrical field energy imbedded within the fabric of space. This book tells how, in the 1930s, inventor T. Henry Moray could produce a fifty kilowatt "free energy" machine; how an electrified plasma vortex creates anti-gravity; how the Pons/Fleischmann "cold fusion" experiment could produce tremendous heat without fusion; and how certain experiments might produce a gravitational anomaly.

190 PAGES. 5x8 PAPERBACK. ILLUSTRATED. $12.95. CODE: TAP

24 hour credit card orders—call: 815-253-6390 fax: 815-253-6300

email: auphq@frontiernet.net   www.adventuresunlimitedpress.com   www.wexclub.com

ANTI-GRAVITY

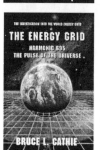

THE A.T. FACTOR
A Scientists Encounter with UFOs: Piece For A Jigsaw Part 3
by Leonard Cramp
British aerospace engineer Cramp began much of the scientific anti-gravity and UFO propulsion analysis back in 1955 with his landmark book *Space, Gravity & the Flying Saucer* (out-of-print and rare). His next books (available from Adventures Unlimited) *UFOs & Anti-Gravity: Piece for a Jig-Saw* and *The Cosmic Matrix: Piece for a Jig-Saw Part 2* began Cramp's in depth look into gravity control, free-energy, and the interlocking web of energy that pervades the universe. In this final book, Cramp brings to a close his detailed and controversial study of UFOs and Anti-Gravity.
324 PAGES. 6X9 PAPERBACK. ILLUSTRATED. BIBLIOGRAPHY. INDEX. $16.95. CODE: ATF

COSMIC MATRIX
Piece for a Jig-Saw, Part Two
by Leonard G. Cramp

Cosmic Matrix is the long-awaited sequel to his 1966 book *UFOs & Anti-Gravity: Piece for a Jig-Saw.* Cramp has had a long history of examining UFO phenomena and has concluded that UFOs use the highest possible aeronautic science to move in the way they do. Cramp examines anti-gravity effects and theorizes that this super-science used by the craft—described in detail in the book—can lift mankind into a new level of technology, transportation and understanding of the universe. The book takes a close look at gravity control, time travel, and the interlocking web of energy between all planets in our solar system with Leonard's unique technical diagrams. A fantastic voyage into the present and future!
364 PAGES. 6X9 PAPERBACK. ILLUSTRATED. BIBLIOGRAPHY. $16.00. CODE: CMX

UFOS AND ANTI-GRAVITY
Piece For A Jig-Saw
by Leonard G. Cramp
Leonard G. Cramp's 1966 classic book on flying saucer propulsion and suppressed technology is a highly technical look at the UFO phenomena by a trained scientist. Cramp first introduces the idea of 'anti-gravity' and introduces us to the various theories of gravitation. He then examines the technology necessary to build a flying saucer and examines in great detail the technical aspects of such a craft. Cramp's book is a wealth of material and diagrams on flying saucers, anti-gravity, suppressed technology, G-fields and UFOs. Chapters include Crossroads of Aerodymanics, Aerodynamic Saucers, Limitations of Rocketry, Gravitation and the Ether, Gravitational Spaceships, G-Field Lift Effects, The Bi-Field Theory, VTOL and Hovercraft, Analysis of UFO photos, more.
388 PAGES. 6X9 PAPERBACK. ILLUSTRATED. $16.95. CODE: UAG

THE ENERGY GRID
Harmonic 695, The Pulse of the Universe
by Captain Bruce Cathie.
This is the breakthrough book that explores the incredible potential of the Energy Grid and the Earth's Unified Field all around us. Cathie's first book, *Harmonic 33*, was published in 1968 when he was a commercial pilot in New Zealand. Since then, Captain Bruce Cathie has been the premier investigator into the amazing potential of the infinite energy that surrounds our planet every microsecond. Cathie investigates the Harmonics of Light and how the Energy Grid is created. In this amazing book are chapters on UFO Propulsion, Nikola Tesla, Unified Equations, the Mysterious Aerials, Pythagoras & the Grid, Nuclear Detonation and the Grid, Maps of the Ancients, an Australian Stonehenge examined, more.
255 PAGES. 6X9 TRADEPAPER. ILLUSTRATED. $15.95. CODE: TEG

THE BRIDGE TO INFINITY
Harmonic 371244
by Captain Bruce Cathie

Cathie has popularized the concept that the earth is crisscrossed by an electromagnetic grid system that can be used for anti-gravity, free energy, levitation and more. The book includes a new analysis of the harmonic nature of reality, acoustic levitation, pyramid power, harmonic receiver towers and UFO propulsion. It concludes that today's scientists have at their command a fantastic store of knowledge with which to advance the welfare of the human race.
204 PAGES. 6X9 TRADEPAPER. ILLUSTRATED. $14.95. CODE: BTF

THE HARMONIC CONQUEST OF SPACE
by Captain Bruce Cathie
Chapters include: Mathematics of the World Grid; the Harmonics of Hiroshima and Nagasaki; Harmonic Transmission and Receiving; the Link Between Human Brain Waves; the Cavity Resonance between the Earth; the Ionosphere and Gravity; Edgar Cayce—the Harmonics of the Subconscious; Stonehenge; the Harmonics of the Moon; the Pyramids of Mars; Nikola Tesla's Electric Car; the Robert Adams Pulsed Electric Motor Generator; Harmonic Clues to the Unified Field; and more. Also included are tables showing the harmonic relations between the earth's magnetic field, the speed of light, and anti-gravity/gravity acceleration at different points on the earth's surface. New chapters in this edition on the giant stone spheres of Costa Rica, Atomic Tests and Volcanic Activity, and a chapter on Ayers Rock analysed with Stone Mountain, Georgia.
248 PAGES. 6X9. PAPERBACK. ILLUSTRATED. BIBLIOGRAPHY. $16.95. CODE: HCS

MAN-MADE UFOS 1944—1994
Fifty Years of Suppression
by Renato Vesco & David Hatcher Childress
A comprehensive look at the early "flying saucer" technology of Nazi Germany and the genesis of man-made UFOs. This book takes us from the work of captured German scientists to escaped battalions of Germans, secret communities in South America and Antarctica to todays state-of-the-art "Dreamland" flying machines. Heavily illustrated, this astonishing book blows the lid off the "government UFO conspiracy" and explains with technical diagrams the technology involved. Examined in detail are secret underground airfields and factories; German secret weapons; "suction" aircraft; the origin of NASA; gyroscopic stabilizers and engines; the secret Marconi aircraft factory in South America; and more. Introduction by W.A. Harbinson, author of the Dell novels *GENESIS* and *REVELATION*.
318 PAGES. 6X9 PAPERBACK. ILLUSTRATED. INDEX & FOOTNOTES. $18.95. CODE: MMU

FREE ENERGY SYSTEMS

LOST SCIENCE
by Gerry Vassilatos
Rediscover the legendary names of suppressed scientific revolution—remarkable lives, astounding discoveries, and incredible inventions which would have produced a world of wonder. How did the aura research of Baron Karl von Reichenbach prove the vitalistic theory and frighten the greatest minds of Germany? How did the physiophone and wireless of Antonio Meucci predate both Bell and Marconi by decades? How does the earth battery technology of Nathan Stubblefield portend an unsuspected energy revolution? How did the geoaetheric engines of Nikola Tesla threaten the establishment of a fuel-dependent America? The microscopes and virus-destroying ray machines of Dr. Royal Rife provided the solution for every world-threatening disease. Why did the FDA and AMA together condemn this great man to Federal Prison? The static crashes on telephone lines enabled Dr. T. Henry Moray to discover the reality of radiant space energy. Was the mysterious "Swedish stone," the powerful mineral which Dr. Moray discovered, the very first historical instance in which stellar power was recognized and secured on earth? Why did the Air Force initially fund the gravitational warp research and warp-cloaking devices of T. Townsend Brown and then reject it? When the controlled fusion devices of Philo Farnsworth achieved the "break-even" point in 1967 the FUSOR project was abruptly cancelled by ITT.
304 PAGES. 6X9 PAPERBACK. ILLUSTRATED. BIBLIOGRAPHY. $16.95. CODE: LOS

SECRETS OF COLD WAR TECHNOLOGY
Project HAARP and Beyond
by Gerry Vassilatos
Vassilatos reveals that "Death Ray" technology has been secretly researched and developed since the turn of the century. Included are chapters on such inventors and their devices as H.C. Vion, the developer of auroral energy receivers; Dr. Selim Lemström's pre-Tesla experiments; the early beam weapons of Grindell-Mathews, Ulivi, Turpain and others; John Hettenger and his early beam power systems. Learn about Project Argus, Project Teak and Project Orange; EMP experiments in the 60s; why the Air Force directed the construction of a huge Ionospheric "backscatter" telemetry system across the Pacific just after WWII; why Raytheon has collected every patent relevant to HAARP over the past few years; more.
250 PAGES. 6X9 PAPERBACK. ILLUSTRATED. $15.95. CODE: SCWT

QUEST FOR ZERO-POINT ENERGY
Engineering Principles for "Free Energy"
by Moray B. King

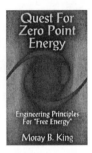

King expands, with diagrams, on how free energy and anti-gravity are possible. The theories of zero point energy maintain there are tremendous fluctuations of electrical field energy embedded within the fabric of space. King explains the following topics: Tapping the Zero-Point Energy as an Energy Source; Fundamentals of a Zero-Point Energy Technology; Vacuum Energy Vortices; The Super Tube; Charge Clusters: The Basis of Zero-Point Energy Inventions; Vortex Filaments, Torsion Fields and the Zero-Point Energy; Transforming the Planet with a Zero-Point Energy Experiment; Dual Vortex Forms: The Key to a Large Zero-Point Energy Coherence. Packed with diagrams, patents and photos. With power shortages now a daily reality in many parts of the world, this book offers a fresh approach very rarely mentioned in the mainstream media.
224 PAGES. 6X9 PAPERBACK. ILLUSTRATED. $14.95. CODE: QZPE

THE TIME TRAVEL HANDBOOK
A Manual of Practical Teleportation & Time Travel
edited by David Hatcher Childress
In the tradition of *The Anti-Gravity Handbook* and *The Free-Energy Device Handbook*, science and UFO author David Hatcher Childress takes us into the weird world of time travel and teleportation. Not just a whacked-out look at science fiction, this book is an authoritative chronicling of real-life time travel experiments, teleportation devices and more. *The Time Travel Handbook* takes the reader beyond the government experiments and deep into the uncharted territory of early time travellers such as Nikola Tesla and Guglielmo Marconi and their alleged time travel experiments, as well as the Wilson Brothers of EMI and their connection to the Philadelphia Experiment—the U.S. Navy's forays into invisibility, time travel, and teleportation. Childress looks into the claims of time travelling individuals, and investigates the unusual claim that the pyramids on Mars were built in the future and sent back in time. A highly visual, large format book, with patents, photos and schematics. Be the first on your block to build your own time travel device!
316 PAGES. 7X10 PAPERBACK. ILLUSTRATED. $16.95. CODE: TTH

THE TESLA PAPERS
Nikola Tesla on Free Energy & Wireless Transmission of Power
by Nikola Tesla, edited by David Hatcher Childress
David Hatcher Childress takes us into the incredible world of Nikola Tesla and his amazing inventions. Tesla's rare article "The Problem of Increasing Human Energy with Special Reference to the Harnessing of the Sun's Energy" is included. This lengthy article was originally published in the June 1900 issue of *The Century Illustrated Monthly Magazine* and it was the outline for Tesla's master blueprint for the world. Tesla's fantastic vision of the future, including wireless power, anti-gravity, free energy and highly advanced solar power. Also included are some of the papers, patents and material collected on Tesla at the Colorado Springs Tesla Symposiums, including papers on: •The Secret History of Wireless Transmission •Tesla and the Magnifying Transmitter •Design and Construction of a Half-Wave Tesla Coil •Electrostatics: A Key to Free Energy •Progress in Zero-Point Energy Research •Electromagnetic Energy from Antennas to Atoms •Tesla's Particle Beam Technology •Fundamental Excitatory Modes of the Earth-Ionosphere Cavity
325 PAGES. 8X10 PAPERBACK. ILLUSTRATED. $16.95. CODE: TTP

THE FANTASTIC INVENTIONS OF NIKOLA TESLA
by Nikola Tesla with additional material by David Hatcher Childress
This book is a readable compendium of patents, diagrams, photos and explanations of the many incredible inventions of the originator of the modern era of electrification. In Tesla's own words are such topics as wireless transmission of power, death rays, and radio-controlled airships. In addition, rare material on German bases in Antarctica and South America, and a secret city built at a remote jungle site in South America by one of Tesla's students, Guglielmo Marconi. Marconi's secret group claims to have built flying saucers in the 1940s and to have gone to Mars in the early 1950s! Incredible photos of these Tesla craft are included. The Ancient Atlantean system of broadcasting energy through a grid system of obelisks and pyramids is discussed, and a fascinating concept comes out of one chapter: that Egyptian engineers had to wear protective metal head-shields while in these power plants, hence the Egyptian Pharoah's head covering as well as the Face on Mars! •His plan to transmit free electricity into the atmosphere. •How electrical devices would work using only small antennas. •Why unlimited power could be utilized anywhere on earth. •How radio and radar technology can be used as death-ray weapons in Star Wars.
342 PAGES. 6X9 PAPERBACK. ILLUSTRATED. $16.95. CODE: FINT

24 hour credit card orders—call: 815-253-6390 fax: 815-253-6300
email: auphq@frontiernet.net   www.adventuresunlimitedpress.com   www.wexclub.com

MYSTIC TRAVELLER SERIES

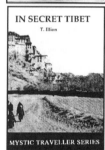

THIS RARE ARCHAEOLOGY BOOK ON
EASTER ISLAND IS BACK IN PRINT!

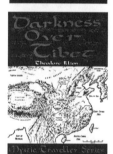

MYSTIC TRAVELLER SERIES

THE MYSTERY OF EASTER ISLAND
by Katherine Routledge

The reprint of Katherine Routledge's classic archaeology book which was first published in London in 1919. The book details her journey by yacht from England to South America, around Patagonia to Chile and on to Easter Island. Routledge explored the amazing island and produced one of the first-ever accounts of the life, history and legends of this strange and remote place. Routledge discusses the statues, pyramid-platforms, Rongo Rongo script, the Bird Cult, the war between the Short Ears and the Long Ears, the secret caves, ancient roads on the island, and more. This rare book serves as a sourcebook on the early discoveries and theories on Easter Island.

432 PAGES. 6X9 PAPERBACK. ILLUSTRATED. $16.95. CODE: MEI

Mystic Travellers Series

MYSTERY CITIES OF THE MAYA
Exploration and Adventure in Lubaantun & Belize
by Thomas Gann

First published in 1925, *Mystery Cities of the Maya* is a classic in Central American archaeology-adventure. Gann was close friends with Mike Mitchell-Hedges, the British adventurer who discovered the famous crystal skull with his adopted daughter Sammy and Lady Richmond Brown, their benefactress. Gann battles pirates along Belize's coast and goes upriver with Mitchell-Hedges to the site of Lubaantun where they excavate a strange lost city where the crystal skull was discovered. Lubaantun is a unique city in the Mayan world as it is built out of precisely carved blocks of stone without the usual plaster-cement facing. Lubaantun contained several large pyramids partially destroyed by earthquakes and a large amount of artifacts. Gann shared Mitchell-Hedges belief in Atlantis and lost civilizations (pre-Mayan) in Central America and the Caribbean. Lots of good photos, maps and diagrams.

252 PAGES. 6X9 PAPERBACK. ILLUSTRATED. $16.95. CODE: MCOM

IN SECRET TIBET
by Theodore Illion

Reprint of a rare 30s adventure travel book. Illion was a German wayfarer who not only spoke fluent Tibetan, but travelled in disguise as a native through forbidden Tibet when it was off-limits to all outsiders. His incredible adventures make this one of the most exciting travel books ever published. Includes illustrations of Tibetan monks levitating stones by acoustics.

210 PAGES. 6X9 PAPERBACK. ILLUSTRATED. $15.95. CODE: IST

DARKNESS OVER TIBET
by Theodore Illion

In this second reprint of Illion's rare books, the German traveller continues his journey through Tibet and is given directions to a strange underground city. As the original publisher's remarks said, "this is a rare account of an underground city in Tibet by the only Westerner ever to enter it and escape alive! "

210 PAGES. 6X9 PAPERBACK. ILLUSTRATED. $15.95. CODE: DOT

DANGER MY ALLY
The Amazing Life Story of the Discoverer of the Crystal Skull
by "Mike" Mitchell-Hedges

The incredible life story of "Mike" Mitchell-Hedges, the British adventurer who discovered the Crystal Skull in the lost Mayan city of Lubaantun in Belize. Mitchell-Hedges has lived an exciting life: gambling everything on a trip to the Americas as a young man, riding with Pancho Villa, questing for Atlantis, fighting bandits in the Caribbean and discovering the famous Crystal Skull.

374 PAGES. 6X9 PAPERBACK. ILLUSTRATED. BIBLIOGRAPHY & INDEX. $16.95. CODE: DMA

IN SECRET MONGOLIA
by Henning Haslund

First published by Kegan Paul of London in 1934, Haslund takes us into the barely known world of Mongolia of 1921, a land of god-kings, bandits, vast mountain wilderness and a Russian army running amok. Starting in Peking, Haslund journeys to Mongolia as part of the Krebs Expedition—a mission to establish a Danish butter farm in a remote corner of northern Mongolia. Along the way, he smuggles guns and nitroglycerin, is thrown into a prison by the new Communist regime, battles the Robber Princess and more. With Haslund we meet the "Mad Baron" Ungern-Sternberg and his renegade Russian army, the many characters of Urga's fledgling foreign community, and the last god-king of Mongolia, Seng Chen Gegen, the fifth reincarnation of the Tiger god and the "ruler of all Torguts." Aside from the esoteric and mystical material, there is plenty of just plain adventure: Haslund encounters a Mongolian werewolf; is ambushed along the trail; escapes from prison and fights terrifying blizzards; more.

374 PAGES. 6X9 PAPERBACK. ILLUSTRATED. BIBLIOGRAPHY & INDEX. $16.95. CODE: ISM

MEN & GODS IN MONGOLIA
by Henning Haslund

First published in 1935 by Kegan Paul of London, Haslund takes us to the lost city of Karakota in the Gobi desert. We meet the Bodgo Gegen, a god-king in Mongolia similar to the Dalai Lama of Tibet. We meet Dambin Jansang, the dreaded warlord of the "Black Gobi." There is even material in this incredible book on the Hi-mori, an "airhorse" that flies through the sky (similar to a Vimana) and carries with it the sacred stone of Chintamani. Aside from the esoteric and mystical material, there is plenty of just plain adventure: Haslund and companions journey across the Gobi desert by camel caravan; are kidnapped and held for ransom; witness initiation into Shamanic societies; meet reincarnated warlords; and experience the violent birth of "modern" Mongolia.

358 PAGES. 6X9 PAPERBACK. ILLUSTRATED. INDEX. $15.95. CODE: MGM